New Methods of Food Preservation

New Methods of Food Preservation

Edited by

G.W. GOULD
Unilever Research Laboratory
Bedford

A Chapman & Hall Food Science Book

An Aspen Publication®
Aspen Publishers, Inc.
Gaithersburg, Maryland
1999

The author has made every effort to ensure the accuracy of the information herein. However, appropriate information sources should be consulted, especially for new or unfamiliar procedures. It is the responsibility of every practitioner to evaluate the appropriateness of a particular opinion in in the context of actual clinical situations and with due considerations to new developments. The author, editors, and the publisher cannot be held responsible for any typographical or other errors found in this book.

Aspen Publishers, Inc., is not affiliated with the American Society of Parenteral and Enteral Nutrition.

Originally published : New York : Chapman & Hall, 1995.
Includes bibliographical references and index.

CIP

For information, address Aspen Publishers, Inc., Permissions Department, 200 Orchard Ridge Drive, Suite 200, Gaithersburg, Maryland 20878.

Orders: (800) 638–8437
Customer Service: (800) 234–1660

Editorial Services: Ruth Bloom

ISBN 978-1-4419-5192-2
Printed in Great Britain at St Edmunsbury Press
1 2 3 4 5

Preface

The major techniques employed for food preservation have a long history of use. They include chilling; freezing; drying; curing; conserving; fermenting or otherwise acidifying; the addition of preservatives; heat-pasteurisation and sterilisation.

Newer techniques more-or-less derived from these traditional procedures include the successful application of combination preservation or 'hurdle' methods, vacuum- and modified atmosphere-packaging, and continuous sterilisation coupled to aseptic packaging. More innovative techniques, such as the use of ionising radiation, are increasingly being employed.

At the same time, there is a reawakening of interest in even more radical approaches. The reasons for this derive principally from consumers' requirements for foods that are higher in quality, so less severely processed; more natural, so less heavily preserved; nutritionally healthier, so containing less salts, sugars and fats; and, with respect to food poisoning, with retained, or preferably improved, assurance of safety.

Some of these more radical approaches are chemically-based, some biological and some physical. A number of them build on current technologies whilst others are completely new. They include, for example, a continual widening of the combination procedures that can be effectively and safely used; new applications of modified atmosphere packaging; use of naturally occurring antimicrobials that are animal-derived (e.g. lysozyme, lactoperoxidase, lactoferrin), plant-derived (e.g. herb, spice and other plant extracts) and microorganism-derived (e.g. bacteriocins); new and improved means for the accurate delivery of heat to foods (e.g. by microwaves, by ohmic heating) so as to achieve the minimal processes necessary to ensure stability and safety; the use of high hydrostatic pressures to inactivate microorganisms in foods without the need for substantial heating, and with consequent minimal damage to product quality; the use of high voltage electric pulses for similar purposes; the direct and synergistic application of ultrasonic radiation to pasteurise and sterilise foods with the minimal application of heat; innovative food surface decontamination procedures aimed at greatly improving the safety of some foods of animal origin; radically new approaches to aseptic processing.

This book covers these major trends in such a way as to summarise progress already made and also to indicate potential for the future. It is directed at food companies involved in production, distribution and sale.

It will be of value to those in government and industry-sponsored food research institutes around the world. It will be useful for teaching courses in food science, home economics, microbiology and process engineering, etc., and for those engaged in food-related research in industry and academia.

I would like to thank the authors of the various chapters for their contributions, and also for their patience and cooperation during the preparation of this volume.

G.W.G.

Contributors

R.G. Board	School of Biological Sciences, University of Bath, Claverton Down, Bath BA2 7AY, UK
J. Burgos	Facultad de Veterinaria, Universidad de Zaragoza, Miguel Servet 177, 50013 Zaragoza, Spain
S. Condón	Facultad de Veterinaria, Universidad de Zaragoza, Miguel Servet 177, 50013 Zaragoza, Spain
A.R. Davies	Department of Food Microbiology, Leatherhead Food Research Association, Randalls Road, Leatherhead, Surrey KT22 7RY, UK
P. Fryer	School of Chemical Engineering, University of Birmingham, Edgbaston, Birmingham B15 2TT, UK
G.W. Gould	Unilever Research Laboratory, Colworth House, Sharnbrook, Bedford MK44 1LQ, UK
C. Hill	National Food Biotechnology Centre, University College, Cork, Ireland
D. Knorr	Department of Food Technology, Berlin University of Technology, Konigin-Luise-Str. 22, D-14195 Berlin, Germany
L. Leistner	An den Weinbergen 20, D-95326 Kulmbach, Germany
P. Loaharanu	Food Preservation Section, International Atomic Energy Agency, Wagamerstrasse 5, PO Box 100, A-1400 Vienna, Austria
P. Lopez	Facultad de Veterinaria, Universidad de Zaragoza, Miguel Servet 177, 50013 Zaragoza, Spain
B. Mertens	Breedstraat 3, RMC Corporation, B-9100 Sirnt-Niklaas, Belgium
J. Mullin	Unilever Research Laboratory, Colworth House, Sharnbrook, Bedford MK44 1LQ, UK
G.J.E. Nychas	National Agricultural Research Foundation, Institute of Technology of Agricultural Products (ITAP), Sof. Venizelou 1str., Lycovrisi 14123, Athens, Greece

J. Raso Facultad de Veterinaria, Universidad de Zaragoza, Miguel Servet 177, 50013 Zaragoza, Spain

D. Rose Department of Produce and Packaging Technology, Campden Food & Drink Association, Chipping Campden GL55 6LD, UK

F.J. Sala Facultad de Veterinaria, Universidad de Zaragoza, Miguel Servet 177, 50013 Zaragoza, Spain

W. Sitzmann Krupp Maschinentechnik GmbH, Division Extraktionstechnik, PO Box 90 05 52, D-21045 Hamburg, Germany

F.J.M. Smulders Department of the Science of Food of Animal Origins, The University of Utrecht, Bilstraat 172, PO Box 80 175, 3508TD Utrecht, The Netherlands

Contents

Overview

G.W. GOULD

Introduction

The major technologies that are employed to preserve the quality and microbiological safety of foods include:

(i) procedures that prevent the access of microorganisms to foods in the first place;
(ii) procedures that inactivate them should they nevertheless have gained access;
(iii) procedures that prevent or slow down their growth should they have gained access and not been inactivated.

Whilst the currently used traditional preservation procedures continue to act in one of these three ways, there has recently been a reawakening of interest in the modification of these technologies, mainly in the direction of reducing the severity of the more extreme techniques. These modifications are being sought primarily to improve the quality of food products, and principally in order to meet the requirements of consumers through the avoidance of the extreme use of any single technique. In addition to the modified techniques, but with the same objective of improving food quality, radically new techniques are also being researched and applied. For both modified and the new techniques it is imperative that they deliver not only the promised improvements in quality but also an equivalent, or preferably an enhanced, level of safety compared with the procedures that they replace.

For these reasons, the summaries of new and improved methods of preservation in the following chapters are opportune.

Consumer requirements

Consumers' requirements constantly change and, with respect to foods in recent years, have encompassed desires for foods that are convenient to store and use and yet have higher quality, are 'fresher', 'more natural' and 'healthier' than hitherto. At the same time, increased awareness of the risks of food poisoning has ensured that a high degree of assurance, and indeed improvement, of safety are perceived as key requirements as well.

Table 1 Consumer requirements impacting on the development of preservation technologies

Major requirements
More convenience
ease of storage
satisfactory shelf-life
Higher quality
better flavour, texture and appearance
Fresher
More natural
Nutritionally healthier
Safer
Means of achievement
Less severe processing
less intensive heating
minimal over-heating
minimal freeze damage
Less use of artificial additives
More use of natural preservation systems
Lower levels of salt, fats and sugars
Elimination of food-poisonoing microorganisms from the most often contaminated foods and raw materials

Table 1 summarises these consumer requirements and indicates some of the likely means for satisfying them. It will be apparent that a number of these means of achievement (e.g. less heat, less salt, less use of preservatives) may actually lead to a *loss* in the intrinsic preservation and safety of a food. It is therefore important that the new and improved technologies effectively build back the preservation that may otherwise be lost.

Existing technologies

The existing technologies for food preservation are summarised in Table 2. Few of these act primarily by restricting the access of microorganisms to foods (item (i) above) except at the terminal phase of production of thermally processed foods, and in the sense that packaging restricts access.

There are more procedures that act via inactivation (item (ii) above) but still, considering the tonnages of foods treated, only heat is used substantially.

Turning to procedures that slow down or prevent the growth of microorganisms in foods (item (iii) above) there are many more procedures available for use, including those that rely on control of the environment (e.g. temperature control), those that result from particular methods of processing (e.g. microstructure control) and those that depend on the intrinsic properties built in to particular formulated foods (e.g. control by the adjustment of water activity or pH value).

Table 2 Major existing and new technologies for food preservation

Restriction of access of microorganisms to products
Aseptic packaging of thermally processed foods
Packaging
Inactivation of microorganisms in products
Heat pasteurisation and sterilisation
Ionising radiation
Addition of enzymes (e.g. lysozyme)
Application of high hydrostatic pressure
Electric shock treatments
Slowing down or prevention of growth of microorganisms in products
Lowered temperature-chilling and freezing
Reduced water activity-curing, conserving, drying
Acidification
Fermentation
Vacuum and modified atmosphere packaging
Addition of preservatives
Microstructure control in water-in-oil emulsions

It is against this background that the new and improved techniques are being developed.

New and improved techniques

With respect to the procedures that *restrict the access* of microorganisms to foods, the employment of aseptic packaging techniques for thermally processed foods has expanded greatly in recent years, both in the numbers of applications and in the numbers of alternative techniques that are commercially available (chapter 13).

With respect to the improvement of techniques for the *inactivation* of microorganisms in foods, most effort and new application has concerned thermal processing. A particular aim has been to minimise damage to product quality. This is being pursued in two, often complementary, ways. Firstly by the wider application of more high temperature–short time processing, with associated aseptic packaging where relevant (chapter 13). Secondly, by delivering heat in new ways, e.g. by microwaves (chapter 6) or by electrical resistance ('ohmic') heating of foods (chapter 10), which allow better control of heat delivery and minimise the over-cooking that commonly occurs in more conventional thermal processes. An important safety consideration that must be borne in mind is the overall reduction in total heat delivery to foods that will result from the wider application of these techniques, as target F_0 values are more and more tightly achieved.

The use of ionising radiation to preserve foods or to eradicate pathogens from them, is already well established. In addition to its value as a preservation technique, it offers a very effective route for the reduction in food poisoning, e.g. via the irradiation of the often *Salmonella*- and *Campylobacter*-contaminated foods such as poultry and other foods of animal origin. Whilst the use of radiation continues to grow worldwide (chapter 5), negative consumer reaction in many countries holds back its wider use.

Radically new procedures for the inactivation of microorganisms in foods include two other physical procedures that offer alternatives to heat: the use of high hydrostatic pressure (chapters 7 and 8) and the use of high voltage electric pulses (chapter 11). Both techniques are highly effective in inactivating vegetative cells of bacteria, yeasts and filamentous fungi, at pressures and at voltage gradients that are compatible with the retention of high quality in some foodstuffs. However, bacterial spores remain more difficult to control by both these procedures, so that their use for the preservation of foods other than relatively short shelf-life or products in which spores are not a problem because they are inhibited by the intrinsic properties of the food (e.g. low pH or low water activity) must await further research.

Finally, concerning novel inactivation procedures, the effectiveness of ultrasonic radiation in inactivating the vegetative forms of microorganisms has been well known for many years. However, the recent research work that has shown that its efficacy can be enhanced by the simultaneous application of (relatively low) hydrostatic pressure is leading to a re-evaluation of its potential as a food preservation aid (chapter 9).

A particularly important new inactivation technique has resulted from the development of surface decontamination procedures that can be applied to meat and poultry carcasses, and to other animal-derived foods which are known to be potentially contaminated with enteric pathogens (chapter 12). In many countries, unacceptably high levels of enteric infection in the human population still occur, and the situation is getting worse rather than getting better. Many food microbiologists have come to realise that although improved hygiene education and the application of Hazard Analysis and Critical Control Point techniques etc. may all help to improve food poisoning statistics, a major reduction will only be achieved if such new elimination techniques are employed. *If the organisms of concern did not enter the home or the catering establishment etc. in the first place, then the momentary lapses of hygiene that will always occur, at some frequency or other, would be of little consequence.*

With respect to procedures that *slow down or prevent* the growth of microorganisms in foods, major successes have been seen, and new applications are steadily being made, in the use of 'combination preservation' techniques or 'hurdle technology' (chapter 1). This has been supported by a greatly improved understanding of the principles underlying the stability and safety of an enormous number of combination-preserved foods that are traditional and indigenous to different parts of the world. It has also been supported by the beginning of an understanding of how many of these combination procedures act at the cellular level, which often seems to involve 'multitarget' interference with the various homeostatic mechanisms that are fundamental to the reaction of microorganisms to the stresses to which the food technologist exposes them in foods (chapter 1).

Though still a relatively new technology, modified atmosphere packaging has grown rapidly in use in some countries, particularly for the extension of the high quality shelf-life of certain chill-stored foods. It remains, however, little used in other countries (chapter 14). Again, considering its wide use, it is surprising that a full understanding of how modified atmospheres (particularly the carbon dioxide component that most of them contain) exert their inhibitory effects at the level of cell biochemistry have not yet been worked out. Elucidation of the mechanisms of action could lead to improved means for effective application.

To some extent, interest in naturally occurring antimicrobial systems has expanded in recent years in response to consumers' requirements for fresher, more natural additive-free foods. With a few notable exceptions, very few of these systems have yet been taken through to application. However, the substantial research efforts underway on animal-derived (chapter 3), plant-derived (chapter 4) and microorganism-derived antimicrobial systems (chapter 2), are demonstrating the efficacy of a wide range of natural mechanisms, many of which have potential for use in food preservation. So far, few such natural systems have been included as components in combination studies, i.e. as additional 'hurdles'. This is a pity, because the food technologist has important opportunities to use these systems in a wide range of combinations with other potential inhibitors. Furthermore, it is arguable that although *in vitro* studies are necessary to investigate mechanisms of action and for intense genetical and biochemical studies, too few studies of natural systems have still been undertaken using actual foodstuffs. Sound and extensive food studies are essential prerequisites before food manufacturers will expend the effort or make the investment necessary to bring new preservation systems into successful application.

Conclusions

This overview serves to highlight the fact that although there is a large and more-or-less traditional and stable range of food preservation techniques available, there is also a surprisingly large and growing number of improved technologies, and also radically new ones, that are being researched or are in the early stages of application.

This is heartening news for food technologists and for research workers, and hopefully for consumers as well, so that their developing requirements can continue to be effectively and safely satisfied in the future.

1 Principles and applications of hurdle technology

L. LEISTNER

1.1 Introduction

The microbial stability and safety of most traditional and novel foods is based on a combination of several factors (hurdles), which should not be overcome by the microorganisms present. This is illustrated by the so-called hurdle effect. The hurdle effect is of fundamental importance for the preservation of foods, since the hurdles in a stable product control microbial spoilage, food-poisoning and, in some instances, the desired fermentation process [1, 2]. Leistner and co-workers acknowledged that the hurdle concept illustrates only the well-known fact that complex interactions of temperature, water activity, pH, redox potential, etc. are significant for the microbial stability of foods.

From an understanding of the hurdle effect, hurdle technology has been derived, which allows improvements in the safety and quality as well as the economic properties (i.e. how much water in a product is compatible with its stability) of foods, by an intelligent combination of hurdles [3–7]. Application of this concept (synonymously called combined methods, combined processes, combination preservation, combination techniques or Hürden-Technologie in German, Technologie des Barrières in French, Tecnologia degli Ostacoli in Italian, and hurdle technology in English) proved very successful, since an intelligent combination of hurdles secures the microbial stability and safety as well as the sensory, nutritive, and economic properties of a food.

1.2 Examples of the hurdle effect

For each stable and safe food a certain set of hurdles is inherent, which differs in quality and intensity depending on the particular product, however, in any case the hurdles must keep the 'normal' population of microorganisms in this food under control. The microorganisms present ('at the start') in a food product should not be able to overcome ('jump over') the hurdles present, otherwise the food will spoil or even cause food-poisoning.

The hurdle effect is illustrated in Figure 1.1, which gives eight examples. Example 1 represents a food which contains six hurdles, and these are:

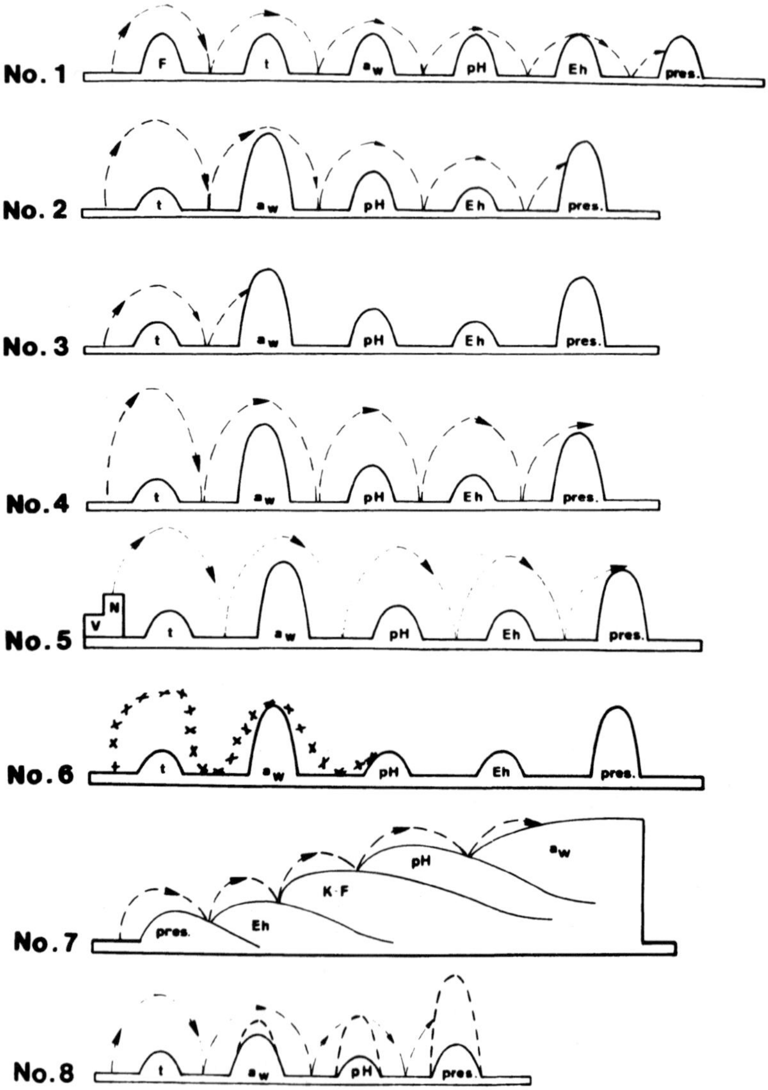

Figure 1.1 Illustration of the hurdle effect, using eight examples. Symbols have the following meaning: F, heating, t, chilling; a_w, water activity; pH, acidification; E_h, redox potential; pres.; preservatives; K–F, competitive flora; V, vitamins; N, nutrients. Figure taken from Leistner [1, 5].

high temperature during processing (F value), low temperature during storage (t value), water activity (a_w), acidity (pH), redox potential (E_h) of the product, as well as preservatives (pres.). The microorganisms present cannot overcome these hurdles, and thus the food is microbiologically stable and safe. However, example 1 is only a theoretical case, because all hurdles are of the same height, i.e. have the same intensity, and this rarely occurs. A more likely situation is presented in example 2, since the microbial stability of this product is based on hurdles of different intensity. In this particular product the main hurdles are a_w and preservatives, whereas other less important hurdles are storage temperature, pH and redox potential. These five hurdles are sufficient to inhibit the usual types and numbers of microorganisms associated with such a product. If there are only a few microorganisms present at the start (example 3), then a few or low hurdles are sufficient for the stability of the product. The superclean or aseptic processing of perishable foods are based on this principle. The same proves true if the initial microbial load of a food (e.g. high moisture fruits) is substantially reduced (e.g. by blanching with steam), because after such a reduction only a few microorganisms are present at the start, which are easily inhibited. On the other hand, as in example 4, if due to bad hygienic conditions too many undesirable microorganisms are initially present, even the usual hurdles inherent in a product may be unable to prevent spoilage or food-poisoning. Example 5 is a food rich in nutrients and vitamins, which will foster the growth of microorganisms (so-called 'trampoline effect'), and thus the hurdles in such a product must be enhanced, otherwise they will be overcome. Example 6 illustrates the behaviour of sublethally damaged organisms in food. If, for instance, bacterial spores in meat products are damaged sublethally by heat (as occurs in F-SSP, discussed later), then the vegetative cells derived from such spores lack 'vitality', and therefore are inhibited by fewer or lower hurdles. In some foods the stability is achieved during processing by a sequence of hurdles, which are important in different stages of the ripening process and lead to a stable final product. Example 7 illustrates the sequence of hurdles in fermented sausages, as will be discussed later. Finally, example 8 illustrates the possible synergistic effect of hurdles, which probably relates to a multi-target disturbance of the homeostasis of microorganisms in foods (see section 1.3).

These examples of the hurdle effect have been given already some time ago [1, 5] but still are useful as an introduction to hurdle technology concept.

1.2.1 *Fermented foods*

In fermented foods – such as fermented sausages, raw hams and ripened cheeses – a sequence of hurdles leads to a stable and safe product. For instance, in fermented sausages (salami) by the sequence of hurdles shown in Figure 1.1 (example 7) the food-poisoning and the spoilage organisms are inhibited, and the desired competitive flora (lactic acid bacteria) is selected. Important hurdles in the early stages of the ripening process of salami are nitrite and salt (pres.), which inhibit many of the bacteria in the batter. However, other bacteria are able to multiply, use up the oxygen and thus cause the redox potential of the product to decrease. This in turn enhances the E_h hurdle, which inhibits aerobic organisms and favours the selection of lactic acid bacteria. They are the competitive flora (K–F) and flourish by metabolizing the added sugars, which causes a decrease in pH value, i.e. an increase of the pH hurdle. In long-ripened salami the nitrite is depleted and the lactic acid bacteria vanish, while the E_h and pH increase again, i.e. all these hurdles become weak during a longer ripening of salami. Only the water activity hurdle (a_w) is strengthened with time, and it is then mainly responsible for the stability of long-ripened raw sausage [4]. Since this sequence of hurdles has been revealed, the production of fermented sausages became less empiric and more advanced [8]. The sequence of hurdles that secures the stability and safety of raw hams is also known [9]. Probably also in other fermented foods, such as ripened cheeses, a sequence of hurdles should be important for the proper ripening process, and it would be challenging to elucidate them.

Another feature peculiar to fermented sausages (and probably for ripened cheeses too) is their microstructure, which influences the desired ripening process as well as the survival of pathogenic bacteria in the product. Therefore, the microstructure is an important hurdle for the stability of salami. Electron microscopy studies [10] have revealed that the natural flora as well as added cultures are not evenly distributed in fermented sausages, but are arrested in little cavities of the product, i.e. the ripening flora only grow in nests (Figure 1.2). These nests are 100–5000 μm apart, and thus large areas of the sausage must be influenced by metabolites (e.g. nitrate reductase, catalase, lactic acid, bacteriocins) accumulated in such nests or cavities. Therefore, the ripening of salami is a 'solid-state-fermentation'. Within these nests the bacteria, either in pure or mixed cultures, are in keen competition for nutrients and impair each other by their metabolic products. In nests of mixed cultures, generally the lactic acid bacteria prevail due to their tolerance of low E_h, pH, and a_w. At the beginning of the sausage fermentation in these nests the lactobacilli appear vigorous and metabolically active, whereas at the end of the ripening process the lactobacilli in their nests have degenerated and many have died. Small and equal distances between nests of desirable bacteria in the sausage matrix should be advantageous, since this would foster the necessary ripening process. The thorough mixing of the meat and fat particles of the sausage batter, before stuffing the sausage mix into casings, would bring about the desired more even distribution of bacteria in the sausage matrix. Moreover, if starter cultures are used, they should

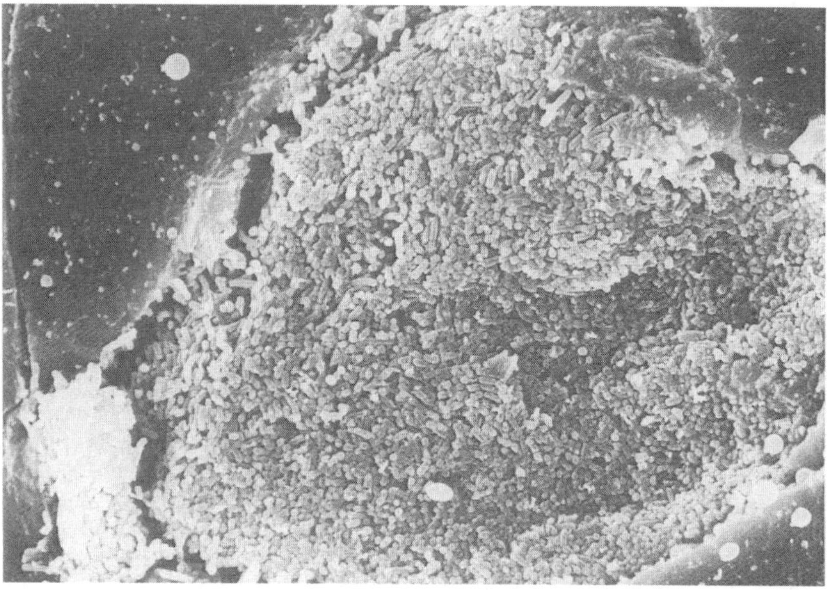

Figure 1.2 Nest of bacteria in a fermented sausage (salami), surrounded by a firm sausage matrix. SEM, ×3600. From these nests desirable bacteria influence by metabolic products (e.g. organic acids, enzymes, bacteriocins) the ripening process of the entire food as well as nests of pathogenic bacteria (e.g. salmonellae or listeria) which are located in distant areas of the food matrix. Figure taken from Katsaras and Leistner [10].

be added in a fashion which favours an even distribution, and this could be achieved by using starter cultures not in a powdery but in a liquid state.

However, the microstructure is not only important for salami (and cheese), but for other foods too. In concentrated oil-in-water emulsions the bacteria form small colonies, and in water-in-oil emulsions the bacterial growth is confined to the water droplets, which might lose their integrity due to coalescence [11]. The impact of microstructure on microbial growth, survival and death in foods has theoretical and practical implications. Certainly, under these circumstances predictive modelling of the behaviour of microorganisms is difficult. On the other hand, it is possible to influence the number, size, and distance of microbial nests in such foods, and thus their safety, stability and quality, by the recipes of the products and the technology applied.

1.2.2 Shelf stable products (SSP)

Heated high moisture foods based on hurdle technology, and thus storable without refrigeration, have been named shelf stable products (SSP), and they offer the following advantages: the mild heat treatment (70°–110°C) improves the sensory and nutritional properties of the food, and the lack of refrigeration simplifies distribution and saves energy during storage. SSP are heated in sealed containers (casings, pouches or cans), which avoid recontamination. However, because of the mild heat treatment these foods still contain viable spores of bacilli and clostridia, which are inhibited by an adjustment of a_w, pH, E_h and, in the case of autoclaved sausages, by sublethal injury of the spores. At present four different types of SSP foods are distinguished, depending on their primary hurdles, though additional hurdles foster the safety and stability of these products too. In one type, the F-SSP [3, 4], the sublethal damage of the spores is the primary hurdle. Examples are autoclaved sausages which have been prevalent in a considerable variety for about 10 years in German supermarkets. They are stored unrefrigerated for several weeks, and have caused no problems with regard to food poisoning or spoilage, because guidelines for their processing have been suggested and followed [12]. The stability of another type, the a_w-SSP [3, 4], is primarily caused by the reduction of the water activity below 0.95, and examples of traditional meats are Italian mortadella and German brühdauerwurst; guidelines for their processing have been suggested [4]. In the third type, the pH-SSP [3, 4], an increased acidity is the primary hurdle. This principle is applied in the Gelderse rookworst, a product which is storable without refrigeration and is exported from the Netherlands in large quantities to the UK. Finally, in the fourth type, the Combi-SSP [5, 13], a combination of equal hurdles is applied, each of which adds a little weight on an imaginary balance [5], which should swing from the unstable to the stable state of the product. In an extensive study conducted for the German army, for military exercises, 75 meat products with fresh-product characteristics, but stable and safe without refrigeration for at least 6 days at 30°C, have been suggested [14], and they include F-SSP, a_w-SSP, pH-SSP, and Combi-SSP [13].

However, not only meat products but other foods too could be stabilized as SSP. For instance, an Italian pasta product (tortellini) was stabilized by using as hurdles a water activity reduction and mild heat processing, as well as modified atmosphere or ethanol vapour during storage, combined with moderate chilling temperatures [15, 16]. Another example is paneer [17, 18], a dairy product of India, which was developed as Combi-SSP, and will be discussed later. In both cases the thesis work of young scientists was breaking the ground.

1.2.3 Intermediate moisture foods (IMF)

Intermediate moisture foods (IMF) are in the a_w-range of 0.90–0.60, and are often stabilized by additional hurdles, such as heating, preservatives, pH, and E_h [19]. These foods are easy to prepare and storable without refrigeration, i.e. they are cost and energy efficient. Traditional IMF based on meat, fish, fruits and vegetables are common and much liked in different parts of the world, because they are tasty, nutritious, and in general safe. However, the novel IMF have not achieved the expected breakthrough in human nutrition. Some reasons for this disappointing performance are the poor palatability of novel IMF due to the high concentration of humectants, and the need to introduce often high amounts of antimicrobial additives ('chemical overloading of foods') which may cause health concerns and pose legal problems.

In previous communications [2–4] the properties, processing, stability, and safety of some traditional IMF meat products, i.e. Chinese sausage, Chinese dried meats, South African biltong, and Turkish pastirma, have been described in detail and guidelines for their manufacturing were given [4]. These published facts will not be repeated here. The stability and safety of such traditional meats, which are all in the intermediate moisture range, is based on empiric application of hurdle technology [20].

Other examples of empiric applied hurdle technology revealed an extensive study done in 10 Latin America countries (Argentina, Brazil, Costa Rica, Cuba, Chile, Mexico, Nicaragua, Puerto Rico, Uruguay, and Venezuela) on traditional foods of the region, which are storable without refrigeration. In the course of this study (CYTED-D AHI), which was sponsored by Spain, about 260 food items, representing fruits and vegetables, and foods derived from fish, milk, meat, cereals, as well as miscellaneous products were identified, which are stable and safe without refrigeration [21–23]. The majority of these food items were IMF, however some vegetable, meat, fish, cheese, and bakery products had a higher a_w, sometimes as high as 0.97–0.98. It turned out that the stability of these high moisture foods was based on empirically applied combined methods ('CM technology'), i.e. hurdle technology. The understanding of this fact has opened a new avenue of food preservation in Latin America, which is now applied especially to tropical and subtropical fruits in small and bulk containers, as will be discussed later (see section 1.5.1). In the opinion of Latin American scientists [22, 23] the technological achievements of Ibero-America in this area deserve a closer look, in particular by developing countries where refrigeration is scarce. The Latin American approach, that is first to obtain an inventory on foods available in the region which have no need for refrigeration, secondly to reveal the underlying principles behind their stability and finally to improve the stability and quality of such foods by application of hurdle technology, is now pursued in India [18] and China [24, 25]. Since IMF are often not satisfactory from the sensory point and contain high levels of additives, the application of CM technology (hurdle technology) to stabilize high moisture foods, which also need no refrigeration, seems to have great potential [23].

1.3 Behaviour of microorganisms during food preservation

Food preservation implies putting microorganisms in a hostile environment, in order to inhibit their growth or shorten their survival or cause their death. The feasible responses of microorganisms to such a hostile environment determines whether they may grow or die. Related to these responses more basic research and its application are needed, because this might lead to new dimensions of food preservation. At present there are more questions than answers available, as the following discussion will indicate.

1.3.1 Homeostasis of microorganisms

An important phenomenon, which deserves attention in food preservation is the interference by the food with the homeostasis of microorganisms [26]. Homeostasis is the tendency to uniformity or stability in the normal status (internal environment) of the organisms. For instance, the maintenance of a defined pH within narrow limits is a prerequisite and feature of living organisms [27], and this applies to higher organisms as well as to microorganisms. Much is already known about the homeostasis in higher organisms at the molecular, subcellular, cellular and systematic levels in the field of molecular biology, biochemistry, physiology and medicine [27]. This knowledge should now be transferred to microorganisms important for the preservation of foods. If the homeostasis of microorganisms, i.e. their internal equilibrium, is disturbed by preservative factors (hurdles) in foods, they will not multiply, i.e. remain in the lag-phase or even die, before their homeostasis is re-established ('repaired'). Thus, food preservation is achieved by disturbing the homeostasis of the microorganisms in a food temporarily or permanently.

Since food stability is often achieved by water activity reduction, the osmoregulation of microorganisms has been studied already extensively with respect to food preservation. Considerable knowledge is available about the osmoregulatory mechanisms of microorganisms [28–32], and the role of potassium as well as the accumulation of amino acids or polyols (as compatible solutes), in order to reverse plasmolysis and re-establish a metabolically suitable water content inside the cells living in an environment with low a_w. Gould and co-workers pointed out that cell osmoregulation mechanisms involve a considerable energy cost for the synthesis or accumulation of 'compatible solutes'. Thus, any restriction of energy supply will tend to be especially synergistic with lowered a_w when used as method of food preservation [30]. The principles of other known 'repair mechanisms' of the microbial cells under stress should be applied to food preservation too, i.e. could be related to hurdle technology.

1.3.2 Multi-target preservation of foods

For foods preserved by hurdle technology, it has been suspected for some time that different hurdles in a food could not just have an additive effect on stability, but may act synergistically [1]. Example 8 in Figure 1.1 illustrates this. A synergistic effect could become true if the hurdles in a food hit, at the same time, different targets (e.g. cell membrane, DNA, enzyme systems, pH, a_w, E_h) within the microbial cell, and thus disturb the homeostasis of the microorganisms present in several respects. Therefore, employing different hurdles in the preservation of a particular food should have advantages, because microbial stability could be achieved with an intelligent combination of gentle hurdles.

In practical terms, this could mean that it is more effective to use different preservatives in small amounts in a food than only one preservative in larger amounts, because different preservatives might hit different targets within the bacterial cell, and thus act synergistically [6, 7]. This 'multi-target preservation' of foods could become a promising research area, because if small hurdles with different targets are selected, a minimal but most effective preservation of foods could be accomplished. It is anticipated that the targets in microorganisms of the different preservative factors (or hurdles) for foods will be elucidated, and then the hurdles could be grouped into target-classes. A mild and effective preservation of foods, i.e. a synergistic effect of hurdles, is likely if the preservation measures are based on an intelligent selection and composition of multi-target hurdles taken from different target-classes.

1.3.3 Stress reactions and metabolic exhaustion

Some bacteria become more resistant (e.g. toward heat) under stress (because the synthesis of protective stress proteins is induced by heat, a_w, ethanol, etc.), or they are less heat-resistant under stress (and this induced by low pH or catalase inactivation in heat-injured cells). These responses might influence the preservation of foods, because increased resistance under stress could turn out to be problematic in the application of hurdle technology [33]. However, benefits of the application of hurdle technology come primarily from the inhibition of growth, and are less related to survival and death of microorganisms. Nevertheless, further research in stress proteins, the time factor for their synthesis, and the different mechanisms which could switch them on, seems warranted in relation to the application of hurdle technology for food preservation.

Another phenomenon of certainly practical importance is the 'auto-sterilization' of stable hurdle technology foods during storage. This was first observed by us and initially not believed, many years ago [34], with mildly heated (95°C core temperature) liver sausage adjusted to different water activities by the addition of salt and fat, and the product was inoculated with *Clostridium sporogenes* PA 3679 and stored at 37°C. Clostridial spores which survived the heat treatment vanished in the product during storage, if the products were stable. Later this behaviour of clostridia as well as *Bacillus* spores was regularly observed during storage of SSP meat products (see section 1.2.2), especially with F-SSP. The most likely explanation is that bacterial spores which survive the heat treatment are able to germinate in these foods under conditions that are less favourable than those under which vegetative cells of bacilli and clostridia are able to multiply [5]. Therefore, during storage of these products some viable spores germinate, but the germinated spores or vegetative cells deriving from these spores die. Thus the spore counts in stable SSP actually decrease during storage at ambient temperatures. Also during studies in our laboratory, with Chinese dried meat products, we observed the same behaviour [35]. If these meats were recontaminated after processing with staphylococci, salmonellae or yeasts, the counts of these microorganisms on stable products decreased fast during unrefrigerated storage, especially on meats with a water activity close to the threshold for microbial growth. Again the same phenomenon was observed by Latin American researchers [36, 49] in their studies with high moisture fruit products (HMFP), because the counts of a variety of bacteria, yeasts, and moulds which survived the mild heat treatment, decrease quite fast in the products during unrefrigerated storage, because the hurdles applied (see section 1.5.1) did not allow growth. A general explanation for this behaviour might be that vegetative microorganisms which cannot grow will die, and they die more quickly if the stability is close to the threshold for growth, storage temperature is elevated, antimicrobial substances are present, and the organisms are heat-injured. Apparently, the microorganisms in stable hurdles technology foods strain every possible repair mechanism to overcome the hostile environment, by doing this they completely use up their energy and die, if they become metabolically exhausted. Thus, due to autosterilization, the hurdle technology foods, which are microbiologically stable, become more safe during storage, especially at ambient temperatures. So, for example, salmonellae which survived the ripening process in fermented sausages, will vanish more quickly if the products are stored at ambient temperatures, and they will survive longer in products stored under refrigeration.

1.4 Total quality of foods

Stanley [37] indicated that the hurdle technology approach seems to be applicable to a wider concept of food preservation than just microbial stability, but that, in order for it to work, a precise knowledge of the effectiveness of each hurdle for a given commodity is required. Furthermore, he suggested distinguishing between positive and negative hurdles for the quality of foods [37]. Certainly, hurdle technology is not only applicable to safety, but also to quality aspects of foods [7]. However, McKenna [38] emphasized that while hurdle technology is appropriate for securing the microbial stability and safety of foods, the total quality of foods is a much broader field and encompasses a wide range of physical, biological and chemical attributes. The concept of combined processes should work towards the total quality of foods rather than the narrow but important aspects of microbial stability and safety. But at present the tools for applying hurdle technology to total food quality are still not adequate, and this is equally true for predicting food quality by modelling. As an initial step towards improving total food quality, the HACCP (Hazard Analysis Critical Control Points) procedure should be converted from safety targets to other quality targets. The application of combined processes philosophy to physical product quality should be a future development. Researchers should appreciate the wider power of the hurdle technology concept and food industry should use the available tools of combined processes for as many quality enhancements as is possible [38]. The principles of HACCP have been already well established, however, in the original sense HACCP had to be used exclusively to avoid biological, chemical and physical hazards [39], whereas the application of the HACCP concept for quality aspects of foods seems sensible it is still disputed.

1.4.1 Optimal range of hurdles

Some hurdles (e.g. Maillard reaction products) influence the safety as well as the quality of foods, because they have antimicrobial properties and at the same time improve the flavour of the product; and this also applies to nitrite used in the curing of meat. The possible hurdles in foods might influence the stability and safety, as well as the sensory, nutritive, technological, and economic properties of a product, and the hurdles present might be negative or positive for securing the desired total quality of a food. Moreover, the same hurdle could have a positive or a negative effect on foods, depending on its intensity. For instance, chilling to an unsuitable low temperature will be detrimental to fruit quality ('chilling injury'), whereas moderate chilling is beneficial. Another example is the pH of fermented sausages, which should be low enough to inhibit pathogenic bacteria, but not so low as to impair taste. In order to secure the total quality of a food, the safety and quality hurdles should be kept in the optimal range [6]. If the intensity of a particular hurdle in a food is too small it should be strengthened; on the other hand, if it is detrimental to the food quality it should be lowered. By this adjustment, the hurdles in foods should be kept in the optimal range, considering safety as well as quality [7].

1.4.2 Potential safety and quality hurdles

The most important hurdles commonly used in food preservation, either applied as 'process' or 'additive' hurdles, are high temperature (F value), low temperature (t value), water activity (a_w), acidity (pH), redox potential (E_h), competitive microorganisms (e.g. lactic acid bacteria), and preservatives (e.g. nitrite, sorbate, sulfite). However, in addition, more than 50 hurdles of potential use for foods of animal or plant origin, which improve the stability and/or the quality of these products, have hitherto been identified (Table 1.1). In the Final Report of a FLAIR Concerted Action No. 7, Subgroup B [40], these hurdles have been described by Bøgh-Sørensen [41], and other members of the group have briefly reviewed in this Report some of the emerging hurdles (ultrahigh pressure, mano-thermo-sonication, photodynamic inactivation of microorganisms, modified atmosphere packaging for non-respiring products as well as respiring produce, edible coatings, ethanol vapour, Maillard reaction products, and bacteriocins). Although about 50 hurdles have been already identified, the list of possible hurdles for the preservation of foods is by no means closed. But not all of these hurdles will be commonly applied, and certainly not all of them for the same food product.

1.4.3 User guide to food design

Hurdle technology as a concept has proved useful in the optimization of traditional foods as well as in the development of novel products. However, it should be combined, if possible, with the HACCP concept and predictive microbiology. These three concepts are related. However, hurdle technology is primarily used for food design, the HACCP concept for process control, and predictive microbiology for process refinement. By considering these different approaches, an overall strategy for securing stable, safe and high quality foods should now be accomplished. This strategy could be applied in an effective food design, for which a user guide was tentatively suggested [42].

The Federal Centre for Meat Research, Kulmbach, demonstrated the efficiency of hurdle technology for food preservation in a study (supported by the Medical Corps of the German Army) on meat products with fresh-product characteristics which nevertheless are storable without refrigeration [14]. Eight categories of meat products were selected and optimized. Since these meats should be suitable for army provisions, even if produced by large or small enterprises, the manufacturing processes must be standardized and reproducible. Therefore, for the first time, a

Table 1.1 Incomplete list of potential hurdles for foods of animal or plant origin, which improve the stability and/or quality of these products [6, 41]

Temperature (low or high)
pH (low or high)
a_w (low or high)
E_h (low or high)
Modified atmosphere (carbon dioxide, oxygen, nitrogen, etc.)
Packaging (vacuum packaging, active packaging, aseptic packaging, new edible coatings, etc.)
Pressure (high or low)
Radiation (microwaves, UV, ionizing, irradiation, etc.)
Other physical processes (ohmic heating, high electric field pulses, radiofrequency energy, oscillating magnetic fields, photodynamic inactivation, ultrasonication, etc.)
Microstructure (emulsions, solid-state-fermentation, etc.)
Competitive flora (lactic acid bacteria, etc.)
Preservatives (organic acids, lactate, acetate, sorbate, ascorbate, isoascorbates, glucono-delta-lactone, polyphosphates, propylene glycol, diphenyl, chitosan, free fatty acids, phenols, monolaurin, chelators, Maillard reaction products, ethanol, spices, nitrite, nitrate, sulfite, smoke, ozone, hypochlorite, pimaricin, lysozyme, lactoperoxidase, nisin, other bacteriocins, etc.)

linkage between hurdle technology and the HACCP concept was introduced [14, 43]. In the manufacturing plants processing the recommended meats, no microbiological tests have to be carried out; however, other process parameters have to be strictly controlled, and these are: time, temperature, pH, and a_w. These measurements should be done on-line, or at least close to the line. A new instrument became available [44, 45], which allows a_w determinations of meat products within 10–20 minutes. The mentioned army study could be used as a model for other instances, where hurdle technology and HACCP should be linked [13].

The army project also raised the question of how food design should be done in general, by applying hurdle technology combined with HACCP, and possibly predictive microbiology too. Predictive microbiology [46–48] is a promising concept which allows computer-based and quantitative predictions of microbial growth, survival and death in foods. However, the predictive models constructed so far handle only up to four different factors (hurdles) simultaneously. Factors considered to date are temperature, pH, a_w (due to salt or humectants), preservatives (e.g. nitrite, lactic acid) and CO_2. As outlined in section 1.4.2, there are numerous other relevant hurdles to be considered, which are important for the stability, safety, and quality of particular foods. It is unlikely that all, or even a majority of these hurdles, could be covered by predictive modelling. Thus predictive microbiology cannot be a quantitative approach to the totality of hurdle technology. However, it does allow quite reliable predictions of the fate of microorganisms in food systems, while considering few but the most important factors (hurdles). Because several hurdles are not taken into account, the predicted results are fortunately often on the safe side, i.e. the limits indicated for growth of pathogens in foods by the models available are in general more prudent ('fail-safe') than the limits in real foods [6, 7]. But predictive microbiology will be an important tool in future food design, because it can narrow down considerably the range over which challenge tests with relevant microorganisms need to be performed. Although it will never render challenge testing obsolete, it may greatly reduce both time and costs spent in product development. Thus predictive microbiology should be an integral part of advanced food design [7].

For the design of foods 10 steps have been suggested [13, 42] which proved appropriate when solving real product development tasks in the food industry. These steps are listed in Table 1.2, but still should be considered as tentative, until further practical experiences with the application of this user guide have accumulated in the food industry.

In food design different disciplines, including technologists and microbiologists, must work together. The technologist should determine which processes or additives are proper for the enhancement of hurdles in a food, by taking the legal, technological, sensory, and nutritive limitations into

Table 1.2 Steps for food design using an integrated concept, comprising hurdle technology as well as predictive microbiology and HACCP [42]

1. First, for the modified or novel food product the desired sensory properties and the desired shelf-life must be defined.
2. Secondly, a tenative technology for the production of the food should be suggested.
3. The food is now manufactured according to this technology, and the resulting product is analysed for pH, a_w, preservatives or other inhibitory factors, and the temperatures for heating (if intended) and storage as well as the expected shelf-life are defined.
4. For preliminary stability testing of the suggested food product, predictive microbiology could be employed.
5. The product is now challenged with food-poisoning and spoilage microorganisms, using somewhat higher inocula and storage temperatures than 'normal'.
6. If necessary, the hurdles in the product are modified, taking the homeostasis of the microorganisms and the sensory quality of the food (i.e. 'total quality') into consideration.
7. The modified product is again challenged with relevant microorganisms, and if necessary the hurdles are modified once more. Predictive microbiology for assessing the safety of the food might be helpful at this stage too.
8. Now that the established hurdles of the modified or novel food are exactly defined, including tolerances, the methods for monitoring the process are defined (preferably physical methods should be used).
9. Thereafter, the designed food should be produced under industrial conditions, because the possibilities for a scale-up of the proposed process must be validated.
10. Finally, for the industrial process the critical control points (CCPs) and their monitoring has to be established, and thus the manufacturing process should be controlled by HACCP.

account. The microbiologist should determine which types and intensity of hurdles in a particular food are needed for the desired safety and stability of the product. Because the engineering, economic, and marketing aspects have to be taken into consideration too, food design is indeed a multidisciplinary endeavour.

1.5 Application of hurdle technology in less developed countries

Foods based on hurdle technology are prevalent in industrialized as well as in developing countries. In the past and often still today hurdle technology was applied empirically without knowing the governing principles. But now with a better understanding of these principles and improved monitoring devices the intentional use increases. Recent examples of an intelligent application of hurdle technology in developing countries are the preservation of various fruits in Latin America, a dairy product in India, and meat products in China. Therefore, these developments will be briefly discussed here.

For developing countries, foods storable without refrigeration are of special interest, because refrigeration (energy) is costly and not continuously available. Furthermore, in less developed countries food preservation procedures should be inexpensive and simple, but reliable. Over centuries a treasure of knowledge of food preservation methods, which fulfil these requirements, has been accumulated in different parts of the world. Most of the foods which remain stable, safe and tasty during prolonged storage without refrigeration, even under the difficult climatic conditions prevalent in many developing countries, are intermediate moisture foods (see section 1.2.3), with an a_w-range from 0.90 to 0.60. However, such foods are often not sufficiently palatable (too sweet or salty, and/or too tough). Therefore, research is now directed toward high moisture foods (with a water activity as high as 0.95–0.98), which are stabilized by additional hurdles and thus can be stored without refrigeration too. In order to illustrate which improvements in traditional foods have already been achieved in different developing countries by using hurdle technology for high moisture foods, the following examples will be presented.

1.5.1 *Fruits of Latin America*

Recently, processes have been developed for the preservation of high moisture fruit products (HMFP: a_w higher than 0.92) in seven Latin American countries (Argentina, Costa Rica, Cuba, Mexico, Nicaragua, Puerto Rico, and Venezuela), and have been applied to peach halves, pineapple slices, mango slices and purée, papaya slices, chicozapote slices, purée of banana, plum, passion fruit, and tamarind, as well as whole figs, strawberries, and pomalaca [49, 50]. The new technologies were based on the combination of a mild heat treatment (blanching for 1–2 min with saturated steam), slight reduction in water activity (to 0.98–0.93, by the addition of glucose or sucrose), lowering of pH (to 4.1–3.0, by the addition of citric or phosphoric acids) and the addition of antimicrobials (potassium sorbate or sodium benzoate, and sodium sulfite or sodium bisulfite) to the syrup of the products. Thus, 'combined methods technology' (hurdle technology) was applied in these novel processes [36, 49, 50–52]. Hurdle combinations often used for HMFP have been: $a_w = 0.97$, pH = 3.5, and the addition of potassium sorbate = 1000 ppm and sodium bisulfite = 150 ppm for the syrup; during storage of HMFP the sorbate and in particular the sulfite levels decreased, whereas the a_w fell (the a_w-hurdle increased) due to the hydrolysis of sucrose [50, 51].

These minimal processes proved energy efficient, simple to carry out (little capital investment), and were satisfactory to preserve fruits *in situ*. The resulting fresh-like products were still scored highly by 30–50 member consumer panels after 3 months of storage at 35°C for taste, flavour, colour, and especially for texture, which is often problematic for canned fruits. Thus, according to Latin American researchers [49–51], the combined methods applied allow storage of fruits, without losses between seasonal harvest peaks, for direct domestic consumption and for further processing to confectionery, bakery and dairy products, or for preserves, jams, and jellies. Fruit pieces can also be utilized as ingredients in salads, barbecues, pizzas and fruit drink formulations. Moreover, these novel HMFP will open new possibilities for export markets. In general, they provide a better utilization of Latin American indigenous tropical and subtropical fruits, many of them having exotic and quite distinctive flavours, textures and appearances.

The high moisture fruit products stabilized by hurdle technology proved shelf-stable during 3–8 months storage at 25–35°C. Due to the blanching process the initial microbial counts were substantially reduced, and during the storage of the stabilized HMFP the number of surviving bacteria, yeasts and moulds further decreased, often to below the detection limits [36, 49–52]. Banana purée challenged with yeasts, moulds, clostridia and bacilli, known to spoil fruits, and stored without refrigeration for 120 days remained stable if proper hurdles were applied (mild heat treatment, adjustment of $a_w = 0.97$ and pH = 3.4, addition of 100 ppm ascorbic acid and 400 ppm sodium bisulfite). The inoculated microorganisms declined and often vanished below the detection limit. These favourable microbiological results obtained with HMFP are probably due to 'metabolic exhaustion' of the microorganisms present in the stabilized products (see section 1.3.3).

Alzamora and co-workers expressed the opinion that HMFP-technologies will attract much attention in many developing countries, because they are easy to implement and will improve considerably the quality of stored fruits [49]. They even believe that the usefulness of combined methods (hurdle technology) for HMFP may give rise to an 'explosion' of research on minimally processed fruits, and the application of this novel process by the food industry [50]. However, the preservation of HMFP must certainly be based on guidelines for good manufacturing practice (GMP) or preferably on hazard analysis critical control points (HACCP), in order to be successful under artisan or industrial conditions (see section 1.4.3). For instance, the reuse of syrup may become a risk in relation to a build-up of the spoilage flora (e.g. *Zygosaccharomyces bailii*, which could be sorbate-resistant), and therefore the reuse of syrup in HMFP processes should only be recommended after pasteurization.

1.5.2 Dairy product of India

Paneer is a traditional, cottage-cheese type product in fried cubes, with tomato sauce, onions, and spices, which is frequently consumed and much liked in India, because of its nutritive value and characteristic taste. However, paneer generally spoils at room temperature (which in India can often be 35°C) within 2 days, and this is an immense drawback for its industrial production.

Sterilized paneer in cans has severe sensory limitations with regard to flavour, texture and colour. Therefore, together with a visiting scientist from India, Dr K. Jayaraj Rao, in the German Federal Centre for Meat Research, Kulmbach, a mildly heated paneer in cans, with the desired flavour (like prepared fresh), colour (little browning) and texture (not too hard) was developed. This product was stabilized by hurdle technology, and thus is stable and safe for several weeks without refrigeration. The following combinations of hurdles proved effective with this product: $a_w = 0.97$, heating to F value of 0.8, pH = 5.0 or alternatively $a_w = 0.96$, F = 0.4, pH = 5.0 [17].

After his return to India, K.J. Rao continued his work with application of hurdle technology to fried paneer in cubes made from buffalo milk. The product with gravy was packed either in tins or flexible retort pouches, and a set of hurdles, i.e. F = 0.8, $a_w = 0.95$, pH = 5.0, and 0.1% potassium sorbate, was chosen, which had maximum lethal and inhibitory effect on microorganisms and minimum effects on textural and chemical characteristics [18]. The water activity of paneer and gravy was lowered by using humectants, such as *dahi*, skim milk powder, salt and glycerol. The pH was adjusted by changing the *dahi* : skim milk powder ratio. The resulting product had a keeping quality of 1 month at 30°C or over 3 months at 15°C. The product was compared with fresh samples from restaurants and was found to be equally acceptable. In the opinion of

Rao [18], this method of preservation has a large scope for alterations in product formulations, depending on regional taste preferences, without affecting the keeping quality of the product.

Via paneer the hurdle technology was introduced into food science of India, and its application to a variety of indigenous foods is anticipated.

1.5.3 Meat products of China

About 15% of the meat supply of the Peoples Republic of China, which in total uses more meat than any other country in the world, are processed into meat products, and dried meats are very popular. There are three technologies in use for Chinese dried meat production [4], but one variety (rou gan) constitutes more than 95% of the dried meats of China [24]. The technology used for the production of rou gan has not changed for hundreds of years, but improvements are possible and desirable.

The Chinese consumer now prefers products with a softer texture, lighter colour and less sweet taste, i.e. lower sugar addition. Shafu is an improved rou gan which fulfils these expectations, and therefore has been well accepted in the Chinese market. Whereas traditional rou gan has a water activity below 0.70, the novel product shafu, with much superior sensory properties, has an a_w of about 0.79, and nevertheless is storable without refrigeration [24]. Compared with rou gan the moisture content of shafu is higher, and salt and sugar contents are lower. Whereas nitrate is added to rou gan, shafu is produced with nitrite curing salt, and the finished product is vacuum packaged. Both products have low residual levels of nitrite and nitrate, contain few microorganisms including in general no pathogenic or toxigenic bacteria, and may be stored for several months without refrigeration.

The superior sensory properties of shafu, compared with the traditional rou gan, are due to an intelligent application of hurdle technology, and it may be expected that similar modification in the future will be used for other Chinese meats [24].

Traditional meat products of China are quite simple to produce, have a characteristic taste, and they are storable for an extended time without refrigeration. An inventory of the processing and the properties of traditional Chinese meats has been established [25], which covers Chinese bacon (la rou), pressed duck (ban ya), silk rabbit (chan si tu), cured chicken (yuan bao ji), dried meat (rou gan), sweet dried meat (rou pu), pork floss (rou song), Chinese sausage (la chang), and Chinese raw ham (yü nam ho tui and chin hua ho tui). As a second step, the principles of the stability and quality of these meats were described, which are all produced by empirically applied hurdle technology. Now, with a deliberate use of hurdle technology the sensory properties of these meats might be optimized, without sacrificing their microbial stability and safety. For other developing countries, where the storage of meat products without refrigeration is desirable too, the traditional and improved Chinese technologies should be of interest, as well as for industrialized countries as a source of innovation.

1.6 Future potential

The stability and quality of most foods is based on empirical and more recently on knowingly employed hurdle technology. To follow this successful route, further basic as well as applied research is needed.

The basic research in this context should centre around the interrelation of homeostasis and hurdle technology. Results could lead to a better understanding of stress reactions and the metabolic exhaustion of microorganisms, and finally to a materialization of the multi-target preservation of foods, as a new concept.

More on the applied side is the exploration and further applications of hurdle technology in order to improve quality and stability, i.e. total quality of foods. The validation of the suggested user guide to food design, which should include predictive microbiology as well as HACCP, seems timely. Furthermore, the coupling of hurdle technology with modern preservation methods for foods, such as irradiation, ultrahigh pressure or ultrasonic radiation, should be strengthened, because it might well turn out that it is essential for these processes too. Related to more traditional preservation methods, such as fermented foods, shelf-stable products (SSP), and intermediate moisture foods (IMF), many more applications of hurdle technology probably are in store than have been explored until now.

Hurdle technology has advantages for chilled foods too, e.g. as a back-up measure ('invisible technology') in the case of temperature abuse in retail display or in the home of the consumer. However, the most impressive strides in the application of hurdle technology have been made with foods which remain stable, safe and tasty even if stored without refrigeration. A case in point is the admirable achievements of Latin America in respect of the stabilization and quality improvement of a variety of minimally processed high moisture fruits, which are storable without refrigeration. Similar ideas have caught on in China as well as in India, and might lead to impressive results there too. It may be expected that hurdle technology foods with a relatively high water activity will partially replace intermediate moisture foods, because lower amounts of humectants and less drying are required, and this would be desirable from the sensory and nutritive point of view. However, deliberate hurdle technology should be applied without sacrificing the microbial stability and safety of foods, especially those which are stored without refrigeration. Therefore, if the hurdle technology foods are to become more sophisticated, they will require a thorough understanding of the principles involved as well as more back-up by good manufacturing practice (GMP) and possibly by the HACCP concept.

References

1. Leistner, L. (1978) In *Food Quality and Nutrition*, ed Downey, W.K. Applied Science Publishers, London, p. 553.
2. Leistner, L., Rödel, W. and Krispien, K. (1981) In *Water Activity: Influences on Food Quality*, eds Rockland, L.B. and Stewart, G.F. Academic Press, New York, p. 855.
3. Leistner, L. (1985) In *Properties of Water in Foods in Relation to Quality and Stability*, eds Simatos, D. and Multon, J.L. Martinus Nijhoff Publishers, Dordrecht, p. 309.
4. Leistner, L. (1987) In *Water Activity: Theory and Applications to Food*, eds Rockland, L.B. and Beuchat, L.R. Marcel Dekker, New York, p. 295.
5. Leistner, L. (1992) *Food Res. Internat.*, **25**, 151.
6. Leistner, L. (1994) *J. Food Engineer.*, **22**, 4211.
7. Leistner, L. (1994) In *Food Preservation by Combined Processes*, eds Leistner, L. and Gorris, L.G.M. Final Report of FLAIR Concerted Action No. 7, Sub-group B, EUR 15776 EN p.1.
8. Leistner, L. (1992) In *New Technologies for Meat and Meat Products*, eds Smulders, F.J.M., Toldrá, F., Flores, J. and Prieto, M. (1992) Audet Tijdschriften p.1.
9. Leistner, L. (1986) *Fleischwirtschaft*, **66**, 496.
10. Katsaras, K. and Leistner, L. (1991) *Biofouling*, **5**, 115.
11. Robins, M., Brocklehurst, T. and Wilson, P. (1994) *Food Technology International Europe* (in press).
12. Hechelmann, H. and Leistner, L. (1984) *Mitteilungsblatt Bundesanst. Fleischforsch. Kulmbach*, **84**, 5894.
13. Leistner, L. and Hechelmann, H. (1994) In *Proc. Food Pres. 2000 Conf., US Army Research*, Natick (in press).
14. Hechelmann, H., Kasprowiak, R., Reil, S., Bergmann, A. and Leistner, L. (1991) *Stabile Fleischerzeugnisse mit Frischprodukt-Charakter für die Truppe*, Bundesminster der Verteidigung, FBWM 91-11, Bonn.
15. Giavedoni, P. (1994) *Azioni combinante nella stabilizzazione degli alimenti*, PhD Thesis, Udine, Italy.
16. Giavedoni, P., Rödel, W. and Dresel, J. (1994) *Fleischwirtschaft*, **74**, 639.
17. Rao, K.J., Dresel, J. and Leistner, L. (1992) *Mitteilungsblatt Bundesanst. Fleischforsch. Kulmbach*, **31**, 293.
18. Rao, K.J. (1993) *Application of Hurdle Technology in the Development of Long Life Paneer-Based Convenience Food.* PhD Thesis, Karnal, India.
19. Leistner, L. and Rödel, W. (1976) In *Intermediate Moisture Foods*, eds Davies, R., Birch, G.G. and Parker, K.J. Applied Science Publishers, London, p. 120.
20. Leistner, L. (1990) In *Proc. 26th Int. Congr. Meat Sci. Technol.*, Vol. III, Havana, p. 842.
21. Aguilera Radic, J.M., Chirife, J., Tapia de Daza, M.S., Welti Chanes, J. and Parada Arias, E. (1990) *Inventario de alimentos de humedad intermedia tradicionales de Ibero-america*, Instituto Politecnico Nacional, Mexico.
22. Aguilera, J.M., Chirife, J., Parada Arias, E. and Barbosa Cánovas, G.V. (1993) In *Food Dehydration*, eds Barbosa Cánovas, G.V. and Okos, M.R. American Institute of Chemical Engineers, New York, p. 72.
23. Welti, J., Tapia de Daza, M.S., Aguilera, J.M., Chrife, J., Parada, E., López Mals, A., Lopez, L.C. and Corte, P. (1994) *Revista Espanola de Ciencia y Tecnología de Alimentos*, **34**, 53.
24. Wang, W. and Leistner, L. (1993) *Fleischwirtschaft*, **73**, 854.
25. Wang, W. and Leistner, L. (1994) *Fleischwirtschaft*, **74** (in press).
26. Gould, G.W. (1988) In *Homeostatic Mechanisms in Micro-organisms*, eds Whittenbury, R., Gould, G.W., Banks, J.G. and Board, R.G. Bath University Press, Bath, p. 220.

27. Häussinger, D. (ed.) (1988) *pH Homeostasis – Mechanisms and Control*, Academic Press, London.
28. Christian, J.H.B. (1955) *Aust. J. Biol. Sci.*, **8**, 490.
29. Brown, A.D. (1978) *Adv. Microb. Physiol.*, **17**, 181.
30. Gould, G.W., Brown, M.H. and Fletcher, B.C. (1983) In *Food Microbiology: Advances and Prospects*, eds Roberts T.A. and Skinner, F.A. Academic Press, London, p. 67.
31. Gould, G.W. (1985) In *Properties of Water in Foods in Relation to Quality and Stability*, eds Simatos D. and Multon, J.L. Martinus Nijhoff Publishers, Dordrecht, p. 229.
32. Troller, J.A. (1987) In *Water Activity: Theory and Applications to Food*, eds Rockland L.B. and Beuchat, L.R. Marcel Dekker, New York, p. 101.
33. Xavier, I.J., Ugborogho, T.O. and Ingham, S.C. (1994) In *Proc. Food Pres. 2000 Confer., US Army Research*, Natick (in press).
34. Leistner, L. and Karan-Djurdjic, S. (1970) *Fleischwirtschaft*, **50**, 1547.
35. Shin, H.K. (1984) *Energiesparende Konservierungsmethoden für Fleischerzeugnisse, abgeleitet von traditionellen Intermediate Moisture Foods*. PhD Thesis, Stuttgart-Hohenheim, Germany.
36. Sajur, S. (1985) *Preconservación de duraznos por métodos combinados*. MS Thesis, Mar del Plata, Argentina.
37. Stanley, D.W. (1991) *Crit. Rev. Fd. Sci, Nut.*, **30**, 487.
38. McKenna, B.M. (1994) In *Food Preservation by Combined Processes*, eds Leistner, L. and Gorris, L.G.M. Final Report of FLAIR Concerted Action No. 7, Subgroup B, EUR 15776 EN, p.99.
39. Pierson, M.D. and Corlett, D. jr. (1992) *HACCP: Principles and Applications*, Van Nostrand Reinhold, a Division of Wadsworth, USA.
40. Leistner, L. and Gorris, L.G.M. eds (1994) *Food Preservation by Combined Processes*, Final Report of FLAIR Concerted Action No. 7, Subgroup B, EUR 15776 EN.
41. Bøgh-Sørensen, L. (1994) In *Food Preservation by Combined Processes*, eds Leistner, L. and Gorris, L.G.M. Final Report of FLAIR Concerted Action No. 7, Subgroup B, EUR 15776 EN, p. 7.
42. Leistner, L. (1994) In *Food Preservation by Combined Processes*, eds Leistner, L. and Gorris, L.G.M. Final Report of FLAIR Concerted Action No. 7, Subgroup B, EUR 15776 EN, p. 25.
43. Leistner, L. (1993) In *Proc. 45th Ann. Reciprocal Meat Conf.*, p. 1.
44. Rödel, W., Scheuer, R. and Wagner, H. (1989) *Fleischwirtschaft*, **69**, 1396.
45. Rödel, W., Scheuer, R. and Wagner, H. (1990) *Fleischwirtschaft*, **70**, 905.
46. Gould, G.W. (1989) *Food Science and Technology Today*, **3**, 89.
47. McClure, P.J., Baranyi, J., Boogard, E., Kelly, T.M. and Roberts, T.A. (1993) *Int. J. Fd. Microbiol.*, **19**, 161.
48. McMeekin, T.A., Olley, J.N., Ross, T. and Ratkowsky, D.A. (1993) *Predictive Microbiology: Theory and Application*, Research Studies Press, Taunton.
49. Alzamora, S.M., Tapia, M.S., Argaíz, A. and Welti, J. (1993) *Food Res. Internat.*, **26**, 125.
50. Alzamora, S.M., Cerrutti, P., Guerrero, S. and López Malo, A. (1994) In *Proc. ISOPOW – Practicum II*, Puebla (in press).
51. López Malo, A., Palou, E., Welti, J., Corte, P. and Argaiz, A. (1994) *Food Res. Internat.* (in press).
52. Guerrero, S., Alzamora, S.M. and Gerscheson, L.N. (1994) *J. Food Protec.* (in press).

2 Bacteriocins: natural antimicrobials from microorganisms

C. HILL

2.1 Introduction

Microorganisms can grow in and spoil valuable foodstuffs or prove a safety hazard to consumers. In most foods many microorganisms compete for the highly nutritious food components. However, no single food can act as a perfect medium for all microorganisms and growth will be limited to those which can overcome the specific conditions prevailing in the particular food. Those which succeed in exploiting a particular niche can alter the physical and chemical parameters in the food in a positive or negative fashion. Historically, chemical preservatives have been used as additional barriers to limit the number of microorganisms capable of growing within foods, but consumer attitudes have led researchers to examine the possibility of using 'natural' inhibitors from plant and microbial sources.

Bacteria produce many inhibitory compounds which can inhibit the growth of potential spoilage or pathogenic microorganisms. These include fermentation end products such as organic acids, hydrogen peroxide, and diacetyl, in addition to bacteriocins and other antagonistic compounds such as reuterin (Daeschel, 1989). This chapter will focus on one of the most intriguing of these inhibitory compounds; the class of proteins termed bacteriocins. Bacteria possess specific genes to produce, modify and export these proteins, and also encode immunity proteins to protect the production organism from attack. Yet the actual role of bacteriocins in nature is unclear. They may provide a competitive advantage over non-producing neighbours, but this remains to be convincingly demonstrated in the natural world. In any event, bacteriocin production can be exploited by food processors to provide an additional barrier to undesirable bacterial growth in foods. While much fundamental work has focused on the colicins and microcins (produced by *Escherichia coli* and Enterobacteriaceae, respectively), the food industry is more immediately concerned with those bacteriocins produced by food grade organisms such as members of the lactic acid bacteria (LAB). The potential for these compounds lies in the addition of the production organism, or a bacteriocin preparation, to the at-risk food. This chapter explores the production and applications, both real and potential, of bacteriocins in foods.

2.1.1 *Historical*

The term 'bacteriocins' was initially suggested by Jacob *et al.* (1953) to describe the inhibitory activity displayed by certain bacteria to closely related species. Many bacteriocins or bacteriocin-like compounds have been characterized in the last few decades. Considerable variation in size, stability, genetic location, post-translational modifications, and modes of action have been encountered, and defy encapsulation into a single definition. However, bacteriocins can be most simply described as bacterial proteins which exhibit an antimicrobial effect against (usually) related species. It is well established that many of the LAB produce these inhibitory compounds. Most of these bacteriocins only inhibit closely related species, limiting their application in precluding the growth of pathogenic organisms, though not in the prevention of spoilage due to other members of the LAB. However, exceptions can be found, and inhibition of *Listeria monocytogenes* and other food-borne pathogens has been reported. A number of schemes for differentiating between different bacteriocin types have been proposed, but it is generally agreed that at least three definable classes exist, which are listed in Table 1. Class I encompasses the small, post-translationally modified, broad host range lantibiotics of which nisin is the best known example. Class II includes the small heat stable unmodified peptides, while class III contains larger, heat labile molecules such as helveticin J. A further grouping of modified bacteriocins has been proposed by some workers, but these can be dealt with using the previous three-tiered system.

2.2 Bacteriocin structure and function

Mechanistic studies have focused on just a few of the bacteriocins described in Table 2.1. Most attention has been devoted to nisin and the lactococcins, and valuable data have resulted which have provided a detailed picture of the respective mechanisms of action.

2.2.1 *Lantibiotics*

Lantibiotics are also produced by genera other than the LAB, and molecules such as subtilin, epidermin and gallidermin have been characterized to the sequence and structure level. Of the three main LAB lantibiotics which have been characterized to date, nisin is the most well known, and will be dealt with in most detail (De Vuyst and Vandamme, 1994).

Nisin is encoded by the 174-bp *nis* A gene (also termed *spa*N). The pre-nisin molecule, immediately after translation, is a 57 amino acid peptide, with no unusual amino acids or secondary structure. Pro-nisin is

Table 2.1 Bacteriocins characterized from lactic acid bacteria (Adapted from de Vuyst and Vandamme, 1993)

Bacteriocin	Producer	Inhibitory spectrum*	Size (aa)	Location
Class I: Lantibiotics				
Nisin (A and Z)	*Lactococcus lactis*	Broad	34	Transposon
Lacticin 481	*Lactococcus lactis*	Broad	27	Plasmid
Lactocin S	*Lactobacillus sake*	Broad	37	Plasmid
Carnocin U149	*Carnobacterium piscicola*	Broad	35–37	Unknown
Class II: Non-lantibiotic, small heat stable				
Diplococcin	*Lactococcus lactis*	Narrow	51	Plasmid
Lactococcin A	*Lactococcus lactis*	Narrow	54	Plasmid
Lactococcin B	*Lactococcus lactis*	Narrow	47	Plasmid
Lactococcin M	*Lactococcus lactis*	Narrow	48	Plasmid
Lactacin F	*Lactobacillua johsonii*	Narrow	57 & 48	Plasmid
Sakacin A	*Lactobacillus sake*	Broad	41	Plasmid
Curvacin A	*Lactobacillus curvatus*	Broad	38–41	Unknown
Sakacin P	*Lactobacillus sake*	Broad	41	Unknown
Carnobacteriocin A & B	*Carnobacterium piscicola*	Broad	53 & 48	Plasmid
Pediocin AcH	*Pediococcus acidilactici*	Broad	44	Plasmid
Pediocin PA-1	*Pediococcus acidilactici*	Broad	44	Plasmid
Leucocin A-UAL-187	*Leuconostoc gelidum*	Broad	37	Plasmid
Enterocin 1146	*Enterococcus faecium*	Broad	–	Chromosome
Class III: Large, heat-labile				
Helveticin J	*Lactobacillus helveticus*	Narrow	333	Chromosome

*Narrow spectrum indicates those bacteriocins which only affect the producer genus.

formed after cleavage of the initial 23 amino acid leader peptide. The mature, active molecule is formed after a number of dehydration steps (converting serine and threonine to dehydroalanine and dehydrobutyrine, respectively). Sulphide ring formation between these modified residues and cysteine residues results in the formation of one lanthionine and four β-methyllanthionines as shown in Figure 2.1. The molecule is cationic and hydrophobic in character. Active, mature nisin is small (molecular mass of 3353), though it normally exists as stable dimers.

Nisin has a bactericidal effect on sensitive cells, effecting a rapid death within a minute of addition. The target of nisin and related molecules has been shown to be the energy-transducing cytoplasmic membrane of sensitive cells. Adsorption does not appear to involve specific protein receptors (as is the case for colicins), but involves direct interaction between the nisin molecule and the membrane, perhaps facilitated by negatively charged cell-wall components. The action of nisin proceeds through insertion and pore formation, leading to a rapid and aspecific efflux of low molecular weight compounds and the depolarization of the membrane. Membrane insertion relies upon a charged membrane, and does not occur in non-energized liposomes. The current dogma assumes that the membrane-associated nisin molecules aggregate to form amphiphilic pores, through which the low molecular weight compounds can exit. Evidence has also been presented which suggests that nisin may interact directly with components of the electron transport chain, inhibiting oxygen uptake. The affected cell is unable to derive energy with which to synthesize macromolecules such as proteins or nucleic acids, and is rapidly killed. It has been observed that Gram-negative cells, normally insensitive to the

PRE-NISIN

MetSerThrLysAspPheAsnLeuAspLeuValSerValSerLysLysAspSerGlyAlaSerProArg
-23 -20 -10 -1

↑ **IleThrSerIleSerLeuCysThrProGlyCysLysThrGlyAlaLeuMetGlyCysAsnMetLysThr**
1 10 20 23

AlaThrCysHisCysSerIleHisValSerLys
24 30 34

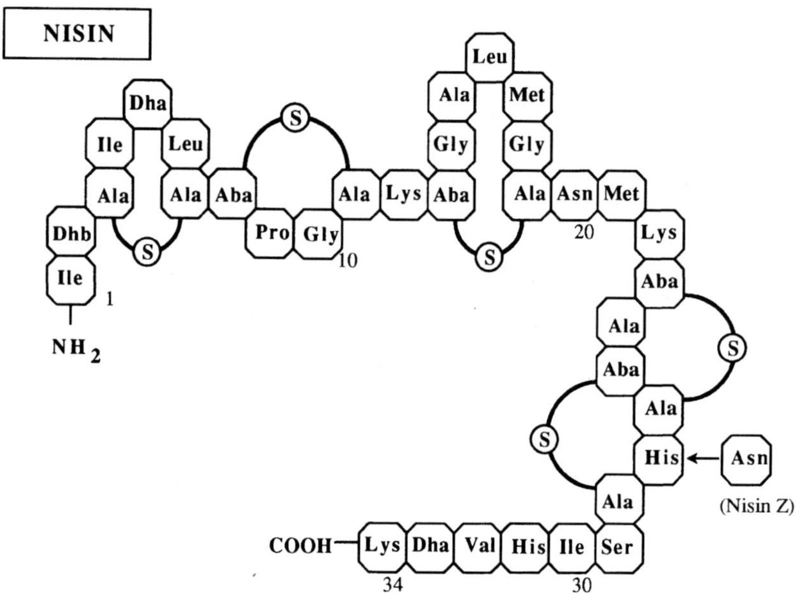

Figure 2.1 Structures of pre-nisin and nisin. Pre-nisin shows the amino acid sequence immediately post-translation. The cleavage site is indicated by an arrow. The residues which will be modified are underlined. Nisin shows the pentacyclic structure of the mature inhibitor. The molecule shown is nisin A, but the substitution of Asn for His_{27} in nisin Z is indicated. Dha is dehydroalanine, Dhb is dehydrobutyrine, Ala-S-Ala is lanthionine and Aba-S-Ala is β-methyllanthionine.

action of nisin, can be sensitized by the addition of chelating agents which disrupt the integrity of the outer membrane and allow the bacteriocin access to the cytoplasmic membrane.

Protein engineering of nisin has been used to define the roles of specific residues in the functional properties of the molecule (Rauch *et al.*, 1994). A natural variant of nisin, nisin Z, has been shown to have improved diffusion properties in agar, but displays similar biological activity. Nisin Z differs from nisin A at a single residue, with a histidine replacing an aspartic acid at position 27 (Figure 2.1). Man-made alterations in the nisin A sequence have also increased our understanding of the roles of individual amino acids in the biosynthesis of active nisin. For the most part, these changes are effected by altering the coding sequence of the *nis*A gene by site-directed mutagenesis, and introducing the altered gene into a nisin-producing strain for expression and modification. For example, one such experiment involved exchanging the serine at position 5 for threonine as a result of a single change in the coding sequence at the appropriate codon. Serine is normally dehydrated to dehydroalanine during post-translational modification of nisin. The resultant engineered derivative now contains a dehydrobutyrine in place of the original dehydroalanine and displays a reduced inhibitory effect against indicator organisms. Introducing an alanine at position 33, in place of dehydroalanine, also results in a diminution of activity. These, and similar studies, hold great promise for the construction of 'custom-made' inhibitory proteins with optimal inhibitory spectra and stability profiles.

Lactocin S is a lantibiotic produced by *Lactobacillus sake* (Nes *et al.*, 1994). The gene encoding lactocin S has been cloned and sequenced and indicates, together with amino acid analysis and mass spectroscopy results, that the mature molecule is a 37 aa peptide with two lanthionine groups and at least four dehydrated residues. Lacticin 481, also produced by *Lactococcus lactis*, has been characterized to the sequence level and shown to belong to the lantibiotic class of bacteriocins (Piard, 1994). Lacticin 481 is unrelated to nisin at the level of the primary sequence, but the mature 27 amino acid peptide also contains characteristic lanthionine residues. Both lactocin S and lacticin 481 are broad host range inhibitors.

2.2.2 Small heat-stable bacteriocins

None of the bacteriocins in this class display any post-translational modification beyond the cleavage of an 18–21 amino acid leader region from the pro-bacteriocin molecule. All are small, between 36 and 57 amino acids after loss of the leader peptide, and to date, no site of action other than the cytoplasmic membrane has been identified.

Lactococcin A is perhaps the most well studied of all the class II bacteriocins, and has been the subject of analysis in a number of different laboratories, who have identified this bacteriocin in at least four different strain backgrounds (van Belkum, 1994). The initial report indicated that the location of the bacteriocin gene was on the conjugative plasmid p9B4, and this plasmid was subsequently shown to encode two other lactococcins, B and M. In one other strain these three bacteriocins have also been associated with a single plasmid, though the plasmid in this instance is non-conjugative and somewhat larger than p9B4. Mature lactococcin A is a 54 amino acid hydrophobic peptide which is bactericidal for other lactococcal strains. The mode of action has been studied in some detail and it has been concluded that, like nisin, lactococcin A acts at the cytoplasmic membrane of sensitive cells, leading to the disruption of the membrane potential and the leakage of small molecular weight compounds. The molecule most probably interacts with a specific membrane receptor (liposomes are insensitive) and aggregates to form pores. Unlike nisin, lactococcin A pore formation is independent of the voltage across the target membrane.

Mature lactococcin B is a 47 amino acid peptide with little homology to the lactococcin A molecule. However, the leader peptides share significant homology (90%) and it is entirely possible that the maturation of both lactococcin A and B is performed by the same processing enzyme. Lactococcin B has a similar mechanism of action to lactococcin A, acting at the level of voltage-independent pore formation in the cytoplasmic membrane. The mechanism of action of lactoccin M remains to be determined.

Another group of small heat-stable inhibitors has been referred to as the pediocin-like bacteriocins. This class of bacteriocins share a number of properties with the identical prototypes pediocin PA-1 and pediocin AcH. One of the features of this group is the high degree of activity against the foodborne pathogen *Listeria monocytogenes*. The pediocin-like group includes leucocin A-UAL, mesenterocin Y105, sakacin P, sakacin A and curvacin A. All are linear peptides (36–44 amino acids in length), and all share a consensus region at the amino terminus of the mature bacteriocin (NH_2–Xaa–Xaa–Tyr–Gly–Asn–Gly–Val–Xaa–Cys– . . .). It has been shown that four cysteine residues in pediocin PA-1 form two disulphide bonds, giving the molecule a degree of secondary structure. When these bonds are disrupted by the use of dithiothreitol (DTT), all activity is lost. Only two of the four cysteine residues of pediocin PA-1 are conserved in the other pediocin-like bacteriocins leucocin A-UAL, mesenterocin Y105, sakacin P and sakacin A, and at least in the case of sakacins A and P, these are not believed to be involved in bridge formation since DTT has no effect on their activity. Another bacteriocin which may fall into the pediocin-like grouping is enterocin 1146, which also appears from genetic evidence to possess the consensus region, and is also highly active against *Listeria*. Many of the pediocin-like bacteriocins show cross-immunity, suggesting similarities at the level of both action and immunity. In a recent study it has been confirmed that Pediocin PA-1 also forms pores in the cytoplasmic membrane of sensitive strains (Chikindas *et al.*, 1993). The data suggest that the activity of pediocin is also voltage independent, and relies upon a specific, though unidentified, membrane receptor.

Lactacin F is a small heat stable bacteriocin produced by *Lactobacillus johnsonii* VPI11088 (formerly *Lb. acidophilus* 11088). This bacteriocin is active against a number of *Lactobacillus* and *Enterococcus* spp., and has been purified and sequenced at the amino acid and nucleotide levels (Klaenhammer *et al.*, 1994). Lactacin F activity was long thought to result from the action of a single peptide of 57 amino acids in length. It has recently been shown that, while this molecule of itself is sufficient to kill *Lb. helveticus* NCK388, the full inhibitory range is only apparent when a second gene encoding another small peptide molecule is present in the production host. When a genetic construct producing only Laf A is placed next to a construct producing the second peptide (Laf X), inhibition of *Lb. delbreuckii* and *E. faecalis* is only seen in the area between the two producers. This strongly indicates that activity against these species is dependent on the presence of both Laf A and X. This type of bacteriocin can be placed in a sub-group, Class IIb, of the small heat stable bacteriocins. The active bacteriocin complex has been shown to induce membrane permeability in sensitive cells.

Lactococcin G is another bacteriocin in which activity relies upon the action of two peptides (Nissen-Meyer *et al.*, 1992). Both peptides, termed α and β, belong to class II bacteriocins, with an estimated length of 39 and 35 amino acids, respectively. Both are highly charged with features consistent with the other pore-forming inhibitors, and are both required for activity. Lactococcin M may also be a class IIb bacteriocin, on the basis of the genetic organization of the lactococcin M operon which suggests that two peptides are required for activity.

2.2.3 Large heat-labile bacteriocins

Only one of the so-called large heat-labile bacteriocins has been characterized in any great detail, helveticin J (Fremaux and Klaenhammer, 1994). This bacteriocin, produced by *Lb. helveticus* 481, is a 37 kDa protein with narrow host range. Most of the information relevant to this molecule has been inferred from the nucleotide sequence of the chromosomally located *hly*J gene, and thus little is known about possible post-translational modifications. The mechanism of action is potentially different from the class I and II bacteriocins, but this remains to be elucidated. The class III group almost certainly includes a number of additional bacteriocins such as caseicin 80, lacticin A and B, and acidophulin A, but these inhibitors await further characterization.

2.3 Genetics of bacteriocins from LAB

This topic has been recently covered in detail in reviews by Klaenhammer (1993) and Klaenhammer *et al.* (1992), and so only a brief overview will be provided here. Genes encoding bacteriocins of classes I, II and III have been cloned and sequenced, and the genetic location, and spread of similar bacteriocins throughout different genera has been noted. Many bacteriocins have been cloned through a process of reverse genetics, in which a portion of the active bacteriocin is sequenced and an oligonucleotide synthesized on the basis of this information. These oligonucleotides can be employed as genetic probes to 'fish' the original coding sequence from a genebank of DNA from the production organism. Perhaps the most elegant example of this approach was used to clone the nisin genes. The deduced structure of mature nisin was used to predict the probable sequence of pre-nisin, and thus to construct a degenerate oligonucleotide probe. In some instances, particularly in those in which the bacteriocin gene(s) is plasmid-located, direct cloning into a suitable expression host has been preferred.

2.3.1 Genetic organization of bacteriocin operons

The organization of five bacteriocin operons is presented in Figure 2.2. The bacteriocins chosen are representative of the main classes and include, where available, the processing genes in addition to the production and immunity determinants. Potential promoters and terminators have also been indicated.

The entire 10-kb nisin coding region includes seven genes, the first of which is *nis*A, the structural gene encoding pre-nisin. This is followed by *nis*B, which is thought to be involved in the modification of pre-nisin, possibly in the dehydration steps. *nis*B is homologous to genes found in the epidermin and subtilin operons. *nis*T encodes a 69 kDa protein thought to play a role in the translocation of nisin across the membrane of the producer cell. The fourth gene, *nis*C, is again most probably involved in post-translational modification of pre-nisin, and has analogues in the epidermin and subtilin operons. A role in immunity has been proposed for the lipoprotein encoded by *nis*I, whereas *nis*P removes the leader sequence to produce mature nisin. The final gene, *nis*R, most probably directs the synthesis of a regulatory protein, and expression of this gene has been shown to be essential for precursor nisin synthesis.

The lactococcin A and B genes are located in close proximity on the plasmid p9B4. Production (*lcn*) and immunity (*lci*) genes are tightly linked in both cases, and the two systems are separated by less than 300-bp. Both systems are under the control of independent promoters. Two additional genes are located immediately upstream of the *lcn*A gene. These genes,

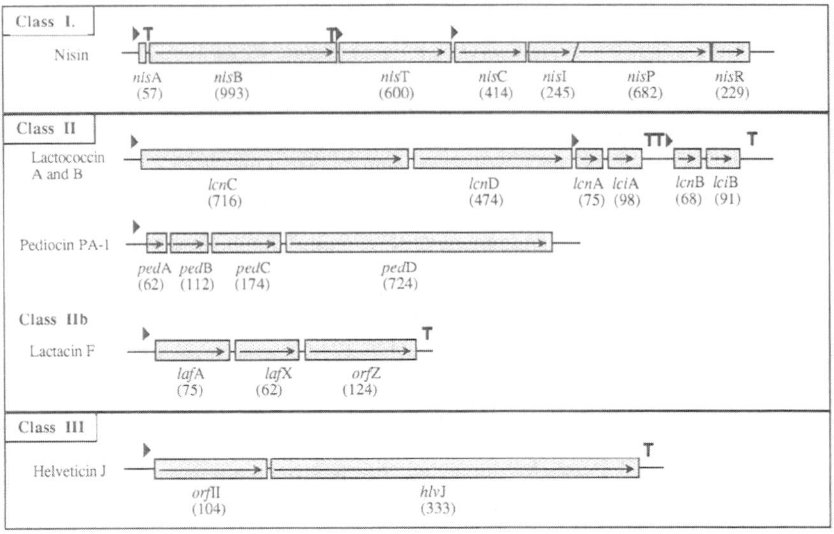

Figure 2.2 Genetic organization of genes involved in bacteriocin biosynthesis and immunity. Representatives of the different classes of bacteriocins are shown. Open reading frames are indicated as shaded boxes, with the gene name underneath. The direction of transcription is shown by the arrow. The size of the gene product in amino acids is given in parentheses. Putative promoters (arrowheads) and terminators (T) are also shown. The different operons are not drawn to a single scale.

*lcn*C and *lcn*D, are essential for the production and secretion of lactococcin A (and B). The *lcn*C gene shows homology to a family of ATP-binding transport proteins, suggesting a role in bacteriocin export, while *lcn*D also shows homology to inner membrane transport proteins.

The pediocin gene cluster encompasses four genes, *ped*A, B, C and D. Pre-pediocin is encoded by the *ped*A gene, whereas *ped*B is most probably involved in pediocin immunity. *ped*D encodes a membrane protein which may perform a similar role to the product of *lcn*C in bacteriocin export. The role of the *ped*C gene product remains to be determined although homology to the *spa*D gene involved in subtilin biosynthesis has been reported.

Lactacin F is one of the class IIb bacteriocins, which require two peptides for the total activity. The *laf*A and *laf*X gene products are required for activity against *E. faecalis* and *Lb. delbreuckii*, whereas only the *laf*A product is required for activity against *Lb. helveticus*. The role of the downstream *orf*Z has not been elucidated, but it does not appear to be essential for bacteriocin expression or immunity.

Finally, the class III helveticin J gene has also been identified and cloned. The *hly*J gene encodes the structural bacteriocin molecule, and this determinant is preceded by *orf*II, a 315 bp coding region which directs the synthesis of a 104 amino acid protein. The role of this protein is as yet undefined, but it may be involved in immunity and/or secretion of helveticin J.

2.3.2 Genetic location of bacteriocin genes

The original definition of bacteriocins required a plasmid location for the genetic determinants. However, this aspect of the original definition has not been strictly adhered to for the bacteriocins of LAB, and some of the bacteriocins analysed to date have chromosomal locations. The location of the bacteriocins described in this chapter are indicated in Table 2.1. The genetic location of the nisin biosynthesis genes was initially the subject of some controversy, but it has been firmly established that the genes are located on a chromosomally located 70-kb conjugative transposon, variously termed Tn*5276* or Tn*5301*, which also encodes sucrose utilization (Dodd and Gasson, 1994). The fact that nisin production, processing and immunity genes are located on a conjugative element has been exploited to move nisin production and immunity phenotypes to other lactococcal strains, and thereby generate nisin producing strains suitable for dairy fermentations (Broadbent and Kondo, 1991).

It can be seen that most, if not all, of the class II bacteriocins are plasmid-borne, though this remains to be verified in some instances. An exception to this rule is enterocin 1146, which is a class II bacteriocin and is chromosomally located. A chromosomal location has also been reported for a number of pediocins, though not for PA-1 or AcH listed in Table 2.1. In the case of lactococcins A, B and M, all three bacteriocin operons are located on a single conjugative plasmid, p9B4. The same three operons have also been identified by use of the polymerase chain reaction on a larger, non-conjugative plasmid in another lactococcal strain DPC938 (Morgan and Hill, unpublished results), and all three are also present in strain LMG2130, but it is not clear why this association of the independent operons should be conserved among different strains. The class III bacteriocin, helveticin J, is chromosomally located.

2.4 Application of bacteriocins in food systems

While bacteriocin characterization, mode of action, and genetic analysis has proceeded at an extraordinary pace, it must be conceded that most applications remain in the developmental rather than commercial stage. However, the field is beginning to yield very encouraging results, and we can confidently expect that bacteriocins will confirm at least some of their much-touted potential in the near future. This section will focus on some of the practical studies and commercial applications of bacteriocins of the LAB. As ever, nisin is in the vanguard of this exciting research, but the pediocins have also been receiving increased attention in recent years.

Nisin is the only one of the LAB bacteriocins that is being commercially produced and used (De Vuyst and Vandamme, 1994; Delves Broughton, 1990). It is produced by fermentation of milk and nisin producing lactococci, and is sold as a commercial preparation under the tradename of Nisaplin® by Aplin and Barret Ltd (UK). Nisin has been approved by the World Health Organization for use in foods since 1968, but it is only since 1988 that it has gained approval from the US Food and Drug Administration. It has been estimated that nisin is currently approved in over 40 countries worldwide and has found a use in many different food systems. This topic has been covered in some detail in a recent review (De Vuyst and Vandamme, 1994). A number of examples, by no means exhaustive, are given in this section.

2.4.1 Dairy industry

In cheese fermentations it is difficult to ensure even distribution of an inhibitor in the final cheese, unless it is added while the milk is still liquid. One of the obvious problems in using a broad host range bacteriocin such as nisin at this stage is that the starter bacteria are as likely to be inhibited as the potential spoilage or pathogenic organism. This can be circumvented by utilizing nisin resistant cultures if nisin is to be added as an exogenous preservative, or by using nisin producers, with innate immunity, as starters for the fermentation. Notwithstanding this problem, the beneficial effects of including nisin in cheese fermentations have been noted by a number of workers. One of the most striking effects is the prevention of 'late gas blowing' due to outgrowth of clostridial spores. This has been reported as a particular problem in a number of hard and semi-hard Dutch and Swiss cheeses, mainly due to the outgrowth of spores of *Clostridium butyricum* and *Clostridium tyrobutyricum*. The addition of exogenous nisin and the use of nisin producing cultures have both proven effective in eliminating this gas defect, but the method has not as yet gained widespread acceptance, probably due to the effect of nisin on the natural flora of the cheese.

An alternative approach which could be used to control specific pathogens or spoilage organisms in dairy foods is to utilize bacteriocins with a highly specific activity range. For example, enterocin 1146, like many of the pediocin-like bacteriocins, is extremely active against *Listeria monocytogenes* at levels which have no effect on lactococcal starters. In our laboratory we have utilized this specificity and the heat stability of enterocin to good effect. Milk is first fermented with *Enterococcus faecium* DPC1146, and subsequently pasteurized. The heat stable bacteriocin is produced in milk and is unaffected by the heat treatment, which effectively renders the enterocin-seeded milk sterile. This milk may be mixed with fresh milk and used for cheesemaking. The lactococcal starters grow and produce acid normally in this mix, but *L. monocytogenes* introduced at the same time is rapidly killed. This inhibitory effect is not seen when milk is fermented with a variant of DPC1146 which no longer produces bacteriocin.

In dairy foods which do not require starter bacteria, broad host range bacteriocins are to be preferred, and so inhibitors can be added without regard for the effect on beneficial flora. Nisin has been included in processed cheese and cheese spreads for many years to prevent the outgrowth of clostridial spores. This is a successful commercial operation, and nisin has been used to prevent the outgrowth of clostridial spores, including those of *Cl. botulinum*, for over 40 years. The nisin is added exogenously (as Nisaplin®), and allows manufacturers to make a higher moisture product without risk of spoilage or health concerns for the consumer.

It is also possible to imagine applications for bacteriocins in the dairy industry which do not rely upon the inhibition of spoilage of pathogenic organisms, but rather depend upon an effect on the starter strains to improve the final product quality. In our laboratory we have noted that *Lactococcus lactis* DPC938 (which contains a plasmid encoding lactococcins A, B and M) has a bacteriolytic effect on other lactococcal strains, including those used in cheese fermentations. The degree of lysis can be measured as an increase in the level of the intracellular enzyme lactate dehydrogenase in the milk after the addition of a bacteriocin solution (Figure 2.3). Since lysis of the starter culture after acid production is one

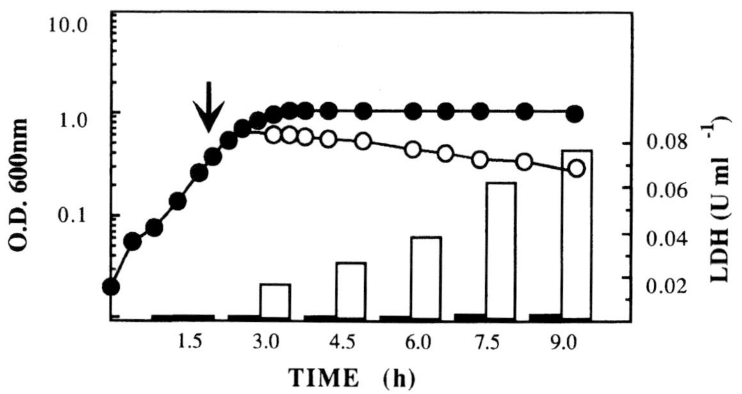

Figure 2.3 Bacteriolytic effect of DPC938 bacteriocins on a sensitive lactococcal strain. The arrow indicates the time of addition of cell-free supernatant from DPC938 (open circles) or DPC3286 (closed circles) to an actively growing cheese starter strain. The levels of LDH released are shown as a histogram.

of the controlling factors in the rate of cheese ripening, we have conducted a series of experiments in which we follow the fate of cheese starters during co-growth with either DPC938 or a bacteriocin negative variant, DPC3286. If the correct ratios of producer and 'sacrificial' strain are used, acid production is not affected, but the subsequent lysis of the starter and release of endogenous enzymes is dramatically improved. Pilot trials are currently in progress to assess the prospects for commercial production of fast-ripening cheeses using this paired system.

2.4.2 Canning industry

Nisin has been used to good effect in both low acid ($pH > 4.6$) and high acid ($pH < 4.5$) canned foods to prevent the outgrowth of clostridial spores. Nisin is used in a number of low acid canned products, including carrot puree, mushrooms, peas and beans, soups, pasteurized hams, potatoes, sausages, fish, rice spaghetti and macaroni (De Vuyst and Vandamme, 1994). In some instances the addition of nisin can allow a more gentle heat process, and thus improve the appearance and taste of the final product.

2.4.3 Meat industry

Meat has a number of intrinsic properties which make the use of bacteriocins more problematic than liquid foods. In particular, it is difficult to achieve an even distribution, there may be solubility problems, bacteriocins can partition to the fat phase, or simply bind to meat proteins. With the exception of nisin in canned hams mentioned above, bacteriocins have not found many applications in fresh or cooked meats. However, one area in which some progress has been reported is in fermented meats and frankfurters. For example, it has been reported that a combination of nisin and low levels of nitrite inhibited the outgrowth of *Clostridium perfringens* spores in frankfurters, whereas nitrite alone, at the levels normally used, was ineffective.

There have been a number of reports concerning the ability of pediocins or pediocin producing cultures to control the growth and survival of *L. monocytogenes* in fermented meats such as summer sausage and in frankfurters. One example will serve to illustrate the potential in this area. Luchansky *et al.* (1992) used a pediocin producer and an isogenic non-producer to inoculate sausage batter which was subsequently spiked with a 4-strain cocktail of *L. monocytogenes*. The numbers of listeriae remained high (10^5/gram) in the sausage containing the non-producer but fell to less than 10^2/gram when the producer strain was used. Significantly, the pediocin produced survived the cooking process at the end of the fermentation and was detectable throughout 60 days storage at 4°C.

Sakacin A, one of the pediocin-like bacteriocins, has also been used to control the growth of *L. monocytogenes* in pasteurized ground beef (Schillinger, 1994). The presence of a sakacin A producer prevented growth of a high inoculum of *L. monocytogenes* at 8°C, whereas in the absence of the sakacin A producer levels increased from 10^5 to 10^9/gram within 10 days. Similar results, including a killing effect against the deliberately added cocktail of listeriae, were reported for sakacin A in comminuted pork.

2.4.4 *Wine and beer*

The malolactate fermentation is an important feature of some wines. This fermentation must be controlled to achieve a high quality product. In many red wines the conversion of dicarboxylic malate to monocarboxylic lactate is a desirable consequence of the growth of lactic acid bacteria such as *Leuconostoc oenos* or *Pediococcus damnosus*. This conversion can lead to a rise of between 0.1 and 0.3 of a pH unit, and also has an effect on the sensory characteristics of the wine. Where the malolactate fermentation is desired the wine can be inoculated with selected strains of *Leuc. oenos*. However, in certain white wines this same fermentation has a negative effect on the wine quality and is undesirable. Nisin can be used to promote the quality of both wine types (Daeschel *et al.*, 1991). For example, nisin and a nisin resistant *Leuc. oenos* can be added to the wine after the yeast fermentation. The nisin will prevent the growth of all LAB other than the deliberately added nisin resistant starter and a highly reliable malolactate fermentation will ensue. Alternatively, nisin alone can be added to prevent all LAB growth, and thus ensure the absence of the malolactate fermentation in the developing wine. An additional benefit to both instances is that fermentation of the wine by wild LAB is prevented, since this unwanted fermentation can be a cause of defects such as ropiness and excess acetate.

The growth of lactobacilli and pediococci in beer is a major cause of spoilage, leading to a beer with too much developed acid and with off-flavours due to the production of diacetyl. It has been demonstrated that nisin can be used to control this problem without having an adverse effect on the beer quality.

2.4.5 *Sauerkraut*

A potential application for nisin producing strains may be in the production of sauerkraut. Two nisin producing lactococcal strains have been isolated from a commercial sauerkraut fermentation (Harris *et al.*, 1992a). The nisin producing strains were used in model sauerkraut fermentations, paired with a naturally nisin resistant *Leuconostoc mesenteroides* (Harris *et al.*, 1992b). The sauerkraut fermentation is sequential and relies upon the initial growth of the heterofermentative *L. mesenteroides* followed by the succession of lactobacilli and further acid production. The nisin producer was successful in delaying the onset of growth of *Lactobacillus* spp., without affecting the growth of *L. mesenteroides*, and this is a potential application which would appear to be worthy of continued research efforts.

2.5 Future prospects for bacteriocins

Many of the problems facing bacteriocin applications and further research efforts concern the perception that bacteriocins have not lived up to their early promise. This may be due to the overenthusiastic promotion of bacteriocin potential by those involved in seeking research funds, or to unrealistic expectations by sponsors of a 'magic bullet' approach to controlling all spoilage and pathogenic problems associated with foods. It is timely, therefore, to reassess the real potential of bacteriocins in the light of the results to date, and to candidly admit to the problems as well as the possibilities for these inhibitors.

The future for bacteriocins certainly does not lie in discovering, or in engineering, the perfect bacteriocin for all applications. Just as is the case for antibiotics in medical therapy, it will be a case of specific bacteriocins for specific tasks, and therefore it is of immense value to continue to identify novel bacteriocins with unique stabilities, target ranges, and modes of action. It is inevitable that some of the bacteriocins which will be found, or are already under characterization, will not find an industrial application. However, there is also academic merit to continuing basic research on these extremely interesting proteins, for which we have not yet identified a role in natural systems.

What then is the future for those bacteriocins with a suitable target range, whether natural or engineered? Such bacteriocins do have a role to play in the control of undesirable bacterial growth, and perhaps also in manipulating the properties of fermented products (e.g. cheese and sauerkraut). However, it is likely that the types of food and the application will be determined by practicalities such as the economics of bacteriocin production on an industrial scale. For example, it is not difficult to imagine an application for LAB bacteriocins in a fermented food. The bacteriocin could be introduced to the product at little or no cost through the use of a bacteriocin producing starter bacterium. The even distribution of the inhibitor throughout the product would be ensured, and the resultant product is likely to be below pH 5.5, a requirement for full bacteriocin activity in many instances. It is more difficult to imagine instances in which the protective effect of the bacteriocin will be of such value that the addition of purified inhibitor will be cost effective, aside from the difficulties in ensuring homogeneous distribution throughout the food. None the less, such applications have been identified for nisin, and so are also possible for other purified bacteriocins. Perhaps a more likely means of addition of bacteriocin will mirror the current use of Microgard®. Microgard® is produced by Wesman Foods, Inc., Oregon, by the fermentation of skim milk with *Propionibacterium shermanii*. The resultant product is added to 30% of the cottage cheese produced in the USA as an inhibitor against psychrotrophic spoilage bacteria (Al-Zoreky *et al.*, 1991). The inhibitory activity almost certainly depends primarily on the presence of propionic acid, but there has also been a role proposed for a bacteriocin-like protein produced during the fermentation (Daeschel, 1989). This use of milk fermented by a bacteriocin producer as an ingredient in milk-based foods may be a useful approach for introducing bacteriocins into foods at little cost.

Additional difficulties will be faced by those wishing to utilize bacteriocins in complex food systems, especially in particulate materials such as meats and vegetables. These products can limit the usefulness of proteinaceous inhibitors because of the propensity of bacteriocins to partition to the fat phase, or to bind to food components. In addition, almost all the LAB bacteriocins are most active at low pH (< pH 5.5), and this may also limit their inhibitory effect in many foods with pH values close to neutral.

Resistance to bacteriocins is another area which has not received sufficient attention. All who work with bacteriocins will have witnessed the appearance of bacteriocin resistant strains at low frequency. This should not be a major problem in the food industry. Firstly, it is not suggested that bacteriocins should be used as the primary barrier to control large numbers of undesirable microorganisms, but rather to eliminate those few which may survive other processing steps. Secondly, given the low frequencies at which resistant mutants appear, they are unlikely to pose a major problem in properly processed foods. In any event, it is more desirable to have the bacteriocin as an additional barrier rather than have no additional barrier at all. This argument becomes less convincing when it is suggested that bacteriocins may replace current safeguards, rather than augment them. It may be that the future for bacteriocins lies simply in culture improvement and in the provision of additional safety barriers, and not as primary inhibitors. For example, a cheese manufacturer may choose a bacteriocin-producing strain over a non-producer because of the no-cost effect on ripening (see section 2.4.1) or an additional barrier against clostridia or listeriae, but this should not be seen as replacing developed acidity or added salt as the primary safety factors in the cheese.

The genetic analysis of bacteriocin operons and regulation must continue, since the ability to overproduce the inhibitor will certainly impact upon the cost effectiveness. Similarly, recombinant strains which produce a range of bacteriocins, or produce bacteriocins in response to specific signals, may be more commercially viable than those produced under the control of the native promoters. Lastly, protein engineering, such as that already initiated with nisin, promises much in the way of additional inhibitors with altered target ranges, stability, and applications. It is to be hoped that bacteriocin research continues to attract the attention of scientists of high calibre, and also of their far-seeing colleagues in industry who will be ultimately responsible for the role of these fascinating inhibitors in modern food-processing industries.

References

Al-Zoreky, N., Ayres, J.W. and Sandine, W. (1991) Antimicrobial activity of Microgard™ against food spoilage and pathogenic microorganisms. *Journal of Dairy Science*, **74**, 758–63.

Broadbent, J.R. and Kondo, J.K. (1991) Genetic construction of nisin-producing *Lactococcus lactis* subsp. *cremoris* and analysis of a rapid method for conjugation. *Applied and Environmental Microbiology*, **57**, 517–24.

Chikindas, M.I., Garcia-Garcera, M.J., Driessen, A.J.M., Ledeboer, A.M., Nissen-Meyer, J., Nes, I.F., Abee, T., Konings, W.N. and Venema, G. (1993) Pediocin PA-1, a bacteriocin from *Pediococcus acidilactici* PAC1.0, forms hydrophilic pores in the cytoplasmic membrane of target cells. *Applied and Environmental Microbiology*, **59**, 3577–84.

Daeschel, M.A. (1989) Antimicrobial substances from lactic acid bacteria for use as food preservatives. *Food Technology*, **43**(1), 164–66.

Daeschel, M.A., Jung, D.-S. and Watson, B.T. (1991) Controlling wine malolactic fermentation with nisin and nisin-resistant strains of *Leuconostoc oenos*. *Applied and Environmental Microbiology*, **57**, 601–3.

Delves-Broughton, J. (1990) Nisin and its uses as a food preservative. *Food Technology*, **44**, 100–17.

De Vuyst, L. and Vandamme, E.J. (1994) Nisin, a lantibiotic produced by *Lactococcus lactis* subsp. *lactis*: properties, biosynthesis, fermentation and applications. In *Bacteriocins of Lactic Acid Bacteria* (eds L. De Vuyst and E.J. Vandamme), Blackie Academic and Professional, Glasgow, pp. 151–221.

Dodd, H.M. and Gasson, M.J. (1994) Bacteriocins of lactic acid bacteria. In *Genetics and Biotechnology of Lactic Acid Bacteria* (eds M.J. Gasson and W.M. de Vos), Blackie Academic and Professional, Glasgow, pp. 211–51.

Fremaux, C. and Klaenhammer, T.R. (1994) Helveticin J, a large heat-labile bacteriocin from *Lactobacillus helveticus*. In *Bacteriocins of Lactic Acid Bacteria* (eds L. De Vuyst and E.J. Vandamme), Blackie Academic and Professional, Glasgow, pp. 397–418.

Harris, L.J., Fleming, H.P. and Klaenhammer, T.R. (1992a) Characterization of two nisin-producing *Lactococcus lactis* strains isolated from a commercial sauerkraut fermentation. *Applied and Environmental Microbiology*, **58**, 1484–89.

Harris, L.J., Fleming, H.P. and Klaenhammer, T.R. (1992b) Novel paired starter culture system for sauerkraut, consisting of a nisin-resistant *Leuconostoc mesenteroides* strain and a nisin-producing *Lactococcus lactis* strain. *Applied and Environmental Microbiology*, **58**, 1477–83.

Jacob, F., Lwoff, A., Siminovitch, A. and Wollman, E. (1953) Definition de quelques termes relatifs a la lysogenie. *Annales des Institute Pasteur Paris*, **84**, 222–4.

Klaenhammer, T.R. (1993) Genetics of bacteriocins produced by lactic acid bacteria. *FEMS Microbiological Reviews*, **12**, 39–86.

Klaenhammer, T.R., Ahn, C., Fremaux, C. and Milton, K. (1992) Molecular properties of *Lactobacillus* bacteriocins. In *Bacteriocins, Microcins and Lantibiotics* (eds R. James, C. Ladzunski and F. Pattus), NATA ASI Series H, **65**, Springer-Verlag, Berlin, pp. 37–58.

Klaenhammer, T.R., Ahn, C. and Muriana, P.M. (1944) Lactacin F, a small hydrophobic heat-stable bacteriocin from *Lactobacillius johnsonii*. In *Bacteriocins of Lactic Acid Bacteria* (eds L. De Vuyst and E.J. Vandamme), Blackie Academic and Professional, Glasgow, pp. 377–96.

Luchansky, J.B., Glass, K.A., Harsono, K.K., *et al.* (1992) Genomic analysis of *Pediococcus* starter cultures used to control *Literia monocytogenes* in turkey summer sausage. *Applied and Environmental Microbiology*, **58**, 3053–59.

Nes, I.F., Mørtvedt, C.I., Nissen-Meyer J. and Skaugen, M. (1994) Lactocin S, a lanthionine-containing bacteriocin isolated from *Lactobacillus sake* L45. In *Bacteriocins of Lactic Acid Bacteria* (eds L. De Vuyst and E.J. Vandamme), Blackie Academic and Professional, Glasgow, pp. 435–49.

Nissen-Meyer, J., Holo, H., Havarstein, S., Sletten, K. and Nes, I.F. (1992) A novel lactococcal bacteriocin whose activity depends on the complementary action of two peptides. *Journal of Bacteriology*, **174**, 5686–92.

Piard, J.-C. (1994) Lacticin 481, a lantibiotic produced by *Lactococcus lactis* subsp. *lactis* CNRZ 481. In *Bacteriocins of Lactic Acid Bacteria* (eds L. De Vuyst and E.J. Vandamme), Blackie Academic and Professional, Glasgow, pp. 251–71.

Rauch, P.J.G., Kuipers, O.P., Siezen, R.J. and de Vos, W.M. (1994) Genetics and protein engineering of nisin. In *Bacteriocins of Lactic Acid Bacteria* (eds L. De Vuyst and E.J. Vandamme), Blackie Academic and Professional, Glasgow, pp. 223–49.

Schillinger, U. (1994) Sakacin A produced by *Lactobacillus sake* Lb 706. In *Bacteriocins of Lactic Acid Bacteria* (eds L. De Vuyst and E.J. Vandamme), Blackie Academic and Professional, Glasgow, pp. 419–34.

van Belkum, M.J. (1994) Lactococcins, bacteriocins of *Lactococcus lactis*. In *Bacteriocins of Lactic Acid Bacteria* (eds L. De Vuyst and E.J. Vandamme), Blackie Academic and Professional, Glasgow, pp. 301–18.

3 Natural antimicrobials from animals

R.G. BOARD

3.1 Introduction

Natural is a word uppermost in the thoughts of forward-looking food microbiologists. They consider the immune systems of animals and ponder on the possibilities of identifying and exploiting antimicrobial agents of such systems in the day-to-day preservation of food. Over the past 20 years or so molecular biologists have added enormously to our knowledge of the antimicrobial systems of mammals, insects and amphibians. Their efforts have led to a large literature of primary and review articles. It would be impossible to review fully this literature in the present chapter. Instead a commentary is offered on the nature of various antimicrobial systems. It draws attention to antimicrobial agents that have attributes which ought to commend them to food microbiologists who seek improvements in one or other of the following:

- The compilation of a cocktail of antimicrobial agents, including one or more from natural sources, in order to extend the shelf-life and assure the safety of perishable foods.
- An amendment of a process, such as thermal processing or high-pressure treatment, so that the keeping qualities of foods are not compromised whilst organoleptic properties suffer little damage.

During the past century or so a large arsenal of antimicrobial agents have been identified during studies of the immunity of mammals, fish and invertebrates (Table 3.1). In what may be considered to be the most evolutionary advanced (the induced) systems, the efficacy of antibodies, phagosomes and the complement system depends on prior exposure to alien cells, particularly specialist pathogens, and a cascade of biochemical events triggered by at least two signals. This level of sophistication would appear to offer little, if any, prospect of the immediate use of such systems in food preservation. One key attribute of the above, however – self/non-self recognition – is found in the antimicrobial agents which make up the constitutive defence systems. An agent active against prokaryotic but not mammalian cells that functions without the need for elaborate transport or other back-up systems is of obvious interest to food microbiologists. Several such agents are listed in Table 3.1. One example is the much studied lysozymes, muramidases that hydrolyse a key chemical bond

Table 3.1 Antimicrobial systems in animals

System based on	Examples	Occurs in: Mammals						
		Blood	Phagosomes	Mouth	Milk	Avian eggs	Fish	Invertebates
Self/non-self recognition	Antibodies	+	(+)	+	+	+*	+	–
	Complement	+	(+)	+	+	–	+	–
	Induced factors other than antibodies	+	+	–	–	–	–	+
Nutritional immunity	Metal sequestrants	+	+	+	+	+	+	×
Inhibition of physiology or perturbation of organization of the alien cell	Lysozyme	+	+	+	+	+	+	+
	Peroxidases	–	+	+	+	–	×	–
	Phenoloxidases	–	–	–	–	–	×	+
Biological structure	Skin, membranes shell, integument or rind	+	–	+	–	+	+	+
References		1	2,3	4	5,6	7	8	9,10,11,12

+, Major importance; –, no importance; ×, no available data; (+), present but of secondary importance.
*Present but of minor importance.

Table 3.2 Occurrence of selected antibiotic peptides in animals*

General Class	References
Defensins	
Polymorphonucleated neutrophiles of man, rabbit, rat, guinea pig	13
Small bowel of mice and pigs	14, 15
Airway epithelial cells – bovine	16
Insects	9, 10
Cecropins	
Insects	9, 11
Small bowel – bovines	17, 18, 19
Seminal fluid – *Drosophila melanogaster*	20
Attacins	
Insects	9
Magainins	
Skin e.g. *Xenopus laevis*	21, 22
Others	
Seminal plasmin – bovine	23
Antimicrobial domains in human neutrophil Cathepsin G	24
Bactericidal domain of lactoferrin	25, 26

*For an up to date list of antibiotics of insects, see [9]. For a perspective on the role of such peptides in immunity, see [27, 28].

(section 3.6) in the peptidoglycan – the unique exoskeleton of the eubacterial cell. Others are the peptides (Table 3.2) that impair the physiological functions of the cell envelope of eubacteria and fungi as well as eukaryotic cells in certain instances. Antimicrobial peptides are common throughout the animal kingdom (Tables 3.1 and 3.2). Obviously they were selected very early in biological evolution and have been conserved because they confer a selective advantage to a wide range of animals. Thus, such peptides – or more probably man-synthesized analogues – ought to be of immediate interest to food microbiologists.

Metal sequestrants are components of the constitutive defence system. In the opinion of some, these cause nutritional immunity – alien cells are deprived of essential elements. Table 3.1 shows that sequestrants are of common occurrence in the defence systems of a wide range of animals. Such ubiquity ought to commend sequestrants to the attention of food microbiologists particularly as some, lactoferrin for example, have both a bacteriostatic and a bactericidal action (section 3.4).

Perusal of Table 3.1 leads to a generalization – the antimicrobial defence systems of animals are based on a *cocktail* of agents. It is most unlikely, therefore, that any one component of such systems would provide food microbiologists with a 'golden bullet' – a sovereign remedy for the control of microbial spoilage of a particular type or class of food. This ought to cause little surprise. Animals have evolved systems that confer protection against a phylogenetically and hence physiologically diverse range of bacteria and microfungi. In the day-to-day survival of an animal, the majority of micro-organisms – the general contaminants in the environment or on or in the animal body – will not be specialist pathogens. Such a diversity is a feature also of the general contaminants acquired by most foods in their passage from field to fork.

It was noted above that at least two signals are needed to activate phagocytes. Such sophistication ought not preclude consideration of events down-stream of activation. The phagosome formed following phagocytosis is of particular interest. This membrane-bound vesicle within a specialist white blood cell can be considered to be both a killing ground and a charnel house. Some of the agents that kill and some that digest dead or moribund bacteria are listed in Table 3.3.

Table 3.3 A generalized overview* of the antimicrobial events and agents associated with the killing and decomposition of eubacteria in the phagosome of professional phagocytes (PM)†

Event	Agent(s)	Outcome
1	Chemical attractants, antibodies, complement	Contact of alien cell with surface of PM PM changes from quiescent to active state Alien cell ungulfed in phagosome within PM
2	Activation of plasma membrane-bound NAD (P) H – oxidase	Production of reactive oxygen intermediates – O_2^- H_2O_2 OH^- O_2'
3	L-Arginine oxidation	Production of NO^- NO_2^- NO_3^-
4	Alkalisation followed by acidification of phagosome	Optimal pI for various lytic enzymes and other agents released from lysosomes
	Phagosome – lysosome fusion with release of ‡:	
	Myeloperoxidase	May catalyse microbiolocidal reactions using H_2O_2 and halides as substrates
	Lactoferrin	Bacteria denied Fe (III)
	Lysozyme	Hydrolysis of bacterial cell walls
	Cathepsin G	Proteolysis and possibly direct killing of bacteria
	Defensins	Death of susceptible bacteria
	Bactericidal-permeability increasing protein	Increased permeability to hydrophobic antimicrobial agents

*Based on reviews [2, 3, 29].
†Professional phagocytes comprise [3]: polymorphonuclear granulocytes (cells of limited life span, high secretory potential and aggressive nature) and mononuclear phagocytes (major effectors of defence with a life-span of months, extremely heterogenous with respect to antibacterial activity and defence capacity.
‡This list is illustrative not exhaustive.

3.2 The phagosome

Several events are associated with the change from the quiescent to the active state in professional phagocytes (Table 3.3). Closer study reveals that two active states contribute to a host's defence. In one, the oxygen burst associated with the plasma membrane-bound NAD(P)H-oxidase occurs in phagocytes in oxygenated tissues. With the anoxic conditions in highly infected or damaged tissue, antimicrobial activity in phagosomes follows the fusion of the latter with lysosomes: membrane-bound depots of antimicrobial agents (Table 3.3).

Some of the antimicrobial products of the oxygen burst – O_2^-, H_2O_2, OH^-, O_2^- – are already used by food microbiologists to decontaminate wrapping materials used in aseptic packaging; or for that matter, indirectly, to decontaminate food by irradiation. The general cytotoxicity of reactive oxygen intermediates is well known. It is noteworthy that evolution has led to the 'confinement' of the phagosomes to professional phagocytes, the cytoplasm of which contains the neutralizing agents, superoxide dismutase, catalase and glutathione peroxidase [29]. Encapsulation of the reactive nitrogen intermediates (Table 3.3) in phagosomes is presumably the outcome also of selection that seeks to protect an animal from excessive exposure to the carcinogenic *N*-nitrosamines arising from reactions involving NO. At one time nitrate was used more widely in the curing of meats, for example. Its use has now been limited because of fears about *N*-nitrosamine formation. Even now, however, NO_3^- is allowed (reluctantly) to be used to control 'late blowing' in Gouda cheese by *Clostridium tyrobutyricum* [6]. It is evident from the above discussion that food microbiologists have already exploited in particular situations the antimicrobial agents produced during the oxygen burst in professional phagocytes. As yet, however, little attention has been given to the agents that cause the death and/or destruction of bacteria following the fourth event in the activation of a professional phagocyte (Table 3.3). Although phagocytosis has been an area of intensive study over many years [see reviews 2, 3, 30] and the literature can be used to fashion a seductive scenario of the events culminating in the destruction of bacteria, it must be recognized that the variety of agents and the diversity of the reactions they bring about will probably defy a comprehensive interpretation of the exact contribution of every antimicrobial agent to host defence. This point was stressed in a previous review [29]. It need not, however, bar investigations of particular agents listed under event 4 in Table 3.3 and their potential utility in food preservation. A conceptual somersault is all that is required for such investigations. It can be argued that the majority, if not all, of the antimicrobial agents present in the lysosomes of professional phagocytes have their origins in the constitutive antimicrobial defence systems in 'primitive' animals and that selection led to their presence in professional phagocytes. Consequently, counterparts of the antimicrobial agents in the phagosome can be sought in constitutive defence systems that are amenable to critical investigation. For example, the myeloperoxidase system of saliva or the analogous lactoperoxidase system in mammalian milk (section 3.4.1) and lysozyme in avian eggs (section 3.6). The observation that proteases release a bactericidal peptide [24] from a protease, cathepsin G, of the phagosome directs attention to an analogous situation, the presence of a bactericidal domain in the lactoferrin molecule (see page 49). Of the other factors listed under event 4 in Table 3.3, defensins are of particular interest because of their high concentrations in professional phagocytes (5–7% of total cellular proteins and 30–50% of the protein of azurophil granules in human neutrophils [31]) and their ubiquity in the animal kingdom (Table 3.2). In view of this ubiquity it is remarkable that the copious literature generated by food microbiologists who study bacteriocins ([32] see Chapter 2) or yeast killer toxins [33] rarely if ever allude to antimicrobial peptides formed by eukaryotic cells.

3.3 Antibiotic peptides

This term is of common use in the literature on immunity in animals. Through emphasis on mode of action, cytolytic pore-forming proteins and peptides may be a preferable term because it would encompass studies of peptides from animals, bacteria and fungi [34].

3.3.1 Biological role

Recently Boman [35] discussed possible reasons for the ubiquity of antibiotic peptides in the animal kingdom. He surmised that they evolved some 600 million years ago in response to a common need of all organisms – protection from the microflora of the environment. In other words these early members of antimicrobial defence systems were associated with the day to day survival of an animal rather than with its defence against sporadic encounters with specialist pathogens. To be effective in the former role, antibiotic peptides would need to have a broad spectrum of activity against prokaryotic rather than eukaryotic cells (Table 3.4) and to be produced with low energy requirements in non-specialist cells not endowed with 'memory' in the immunological sense but responding rapidly to a simple signal (degradation products of peptidoglycans or free lipopolysaccharides).

'A common need' for an animal to be protected from the indigenous flora of its environment was exemplified in the studies by Zasloff [21]. Over many years he and his colleagues had cut through the skin and muscles of the abdomen in order to operate on the ovaries of the frog, *Xenopus laevis*.

Table 3.4 Some chemical attributes of selected antibiotic peptides of animal origin*

Peptide	Attribute	Reference
Defensins Mammalian	Non-glycosolidated; 29–35 amino acid residues including six cysteines; arginine rich; cationic. Structurally rigid B sheets stabilised by disulfide bonds. No alpha helical domains	13, 36
Cecropins Insects	35–39 amino acid residues; *N*-terminal parts strongly basic and *C*-terminal parts are neutral and contain long hydrophobic stretches. Nearly two perfect amphipathic alpha helices joined by a hinge region containing Gly-Pro.	37
Magainins Frogs	23 amino acid residues; lysine-rich; lacks tyrosine; no cysteine; basic; amphipathic alpha helices	38

*An overview of the structures of the following insect peptides is given in reference [10]: attacin, sarcotoxin II, apidaecin, abaecin, and diptericin.

They noted that infection of wounds was extremely rare despite the use of non-aseptic surgical methods and the return of an animal to contaminated water following an operation. Their observations led to the identification of a new class of antibiotic peptides, the magainins (see Table 3.2), or the 'sterilizing' agents in frogs' skin [22]. Recent studies have demonstrated also that antibiotic peptides may well play a role in the antimicrobial defence of particular organs in an animal's body. Thus, it has been surmised that andropin, a bactericidal peptide which shows some similarity to cecropins, may protect the seminal fluid and male reproductive organs of *Drosophila melanogaster* against microbial contamination [20]. A similar role has been ascribed to seminal plasmin, a 47-residue peptide in the seminal plasma of cattle [23]. A tracheal antibiotic peptide, which is active against Gram-positive and Gram-negative bacteria as well as *Candida albicans*, occurs in the airways of cattle [16]. The relatively germ-free status of the small intestine of mammals has been attributed, at least in part, to defensins (termed cryptdins) in the cells in the small bowel of mice [14]. A similar claim has been made for pigs [19]. The most abundant cryptdin in the mouse gut is active against a mutant but not a wild-type (virulent) strain of *Salmonella typhimurium*.

3.3.2 Chemical attributes and spectrum of action

There is a range in the number of amino acid residues and the configuration of the molecules of the antibiotic peptides of animals (Table 3.4). This Table also shows that particular amino acids are notable for their presence or absence. Thus, in mammalian defensins, six cysteines confer a characteristic structure to the molecule. Magainins lack both cysteine and tyrosine but are rich in lysine. It needs to be stressed that there are subtle differences in the amino acid composition of particular antibiotic peptides both from those in a species as well as between families [37]. In spite of such differences in composition and molecular architecture, antibiotic peptides of animals have a predilection for the cell envelope of bacteria (Table 3.5) – the outer membrane of Gram-negative bacteria (attacin P5, sarcotoxin 11, diptericin, coleoptericin, all attacin-like peptides of insects [10]) or the inner (cytoplasmic) membrane of Gram-positive ones (cecropins and defensins of insects [10]). Boman [35] noted that the cecropins and probably the defensins and magainins act stoichiometrically rather than catalytically. The general cytotoxicity of the antibiotic peptides must not be overlooked. Boman [35] surmised that the storage of defensins in the granules of phagocytes is a means whereby mammals maintain depots of potent antimicrobial agents without the danger of exposing their own tissues to the dangers of cytotoxins. Bevins and Zasloff [22] noted synergism when antibiotic peptides from frogs were mixed, the antibiotic

Table 3.5 Spectrum, site and mode of action of antibiotic peptides of animal origin*

Peptide	*Site of action* Activity spectrum	Mode of action (references)
Defensins Mammalian	*Cell envelope* Gram-positive bacteria and fungi (Gram-negative bacteria)†	Electrostatic adsorption to site near cell membrane; penetration of membrane and disruption of its integrity by pore formation [13, 40]
Insect	*Cell membrane* Gram-positive *but* not Gram-negative bacteria	? [10]
Attacins	Outer membrane of cell envelope (facilitates action of lysozyme and cecropins) *Escherichia coli* and some other Gram-negative bacteria	Blockage of synthesis of proteins in the outer membrane, cell division inhibited and long chains of cells formed [10, 37]
Cecropins	*Cell membranes* Gram-positive and Gram-negative bacteria	Voltage-dependent ion-channels of variable size [10]
Magainins‡	*Cell membranes* Gram-positive and Gram-negative bacteria, fungi and protozoa	Depolarise cell membrane [21, 22, 39, 40]

*Some of the animal peptides have pharmacological properties also [22].
†Gram-negative bacteria less sensitive than Gram-positive ones.
‡Combinations of magainins and other antibiotic peptides of frog increases antimicrobial activity but cytoxic activity induced also [22]. Numerous synthetic analogues have antibiotic action which is enhanced by increasing the alpha helix potential [41, 42].

potency being increased 20–50-fold when two peptides occurred at a 1 : 1 molar stoichiometry.

When considering the mode of action of antibiotic peptides from many sources, Ojcius and Young [34] emphasized their pore-forming propensities. They stressed that a large number of pore-forming toxins are known to produce channels in the cytoplasmic membrane *via* a 'barrel-stave' mechanism. There are three steps in pore formation:

- water-soluble peptides bind to the membrane;
- the peptides penetrate the membrane; and
- the individual peptides, providing they span the membrane completely, become arranged like the staves of a barrel such that a pore is formed.

The pore allows the passive flux of ions and small molecules across the membrane such that a cell dies due to colloid osmotic lysis. To support their hypothesis of pore-formation by antibiotic peptides, the authors cited the following examples: *Staphylococcus* S-toxin, a peptide of 26 amino acid residues which forms a relatively stable alpha-helix; magainins, peptides of 23 amino acid residues which form amphiphilic alpha helices; cecropins, again peptides (37 amino acid residues) which form an amphiphilic alpha helical configuration. In a recent study of lactococcin B, a thiol-activated bacteriocin from *Lactococcus lactis*, it was demonstrated [43] that low concentrations of this small (5–10 kDa) cationic peptide caused the efflux of ions and amino acids taken up by systems driven by proton motive force. Large concentrations caused the efflux of glutamate whose uptake is ATP driven. This evidence can be taken to mean that the size of a pore formed in a cytoplasmic membrane and hence the events leading to cell death are determined by the concentration of an antibiotic peptide, large amounts being required for large pores and the loss of large molecules. In another study of a lactococcal bacteriocin it was demonstrated that the formation of pores in the cytoplasmic membrane of sensitive cells required two complementary peptides [44]. The above discussion shows that there is abundant information on the occurrence and mode of action of antibiotic peptides. Indeed the data-base is such that those who work on the possible utility of bacteriocins ought to extend their interests to peptides other than those of microbial origin and to examine the efficacy of mixtures of such peptides.

Combinations exhibited enhanced cytotoxicity as evidenced by lysis of various eukaroytic cells [22]. The authors suggested that in the latter instance, mixtures of antibiotic peptides may play an important role in amphibia, the antiseptic properties of combinations of peptides being linked with the destruction of eukaryotic cells during wound repair.

3.4 Protein amendment and production of antibiotic peptides

Recently it has been shown that two proteins, cathepsin G and lactoferrin, contribute in two ways to the antimicrobial defence of mammals. Cathepsin G, a chymotrypsin-like protease, occurs in phagosomes (see Table 3.3). One of its functions is presumably the digestion of dead or moribund cells. Another role was suggested by Odeberg and Olsson [45]. They demonstrated antibiotic activity *in vitro* of cathepsin G derived from the granules of phagocytes. This attribute was not lost when the proteolytic activity of cathepsin G was inhibited with the serine protease inhibitor, diisopropylfluorophosphate [24]. When the arginine-rich cathepsin G was treated with the arginine-specific cysteine protease, clostripain, two low molecular weight fragments lacking protease, but not antibacterial activity, were obtained as shown with cultures of *Staphylococcus aureus* and *Neisseria gonorrhoeae*. Synthetic peptides with sequences corresponding to those of the two fragments in the digest of cathepsin G were shown to have a broad-spectrum antimicrobial activity *in vitro* [24]. Thus the evidence suggests that this protease of the phagosome has two antimicrobial domains in its molecule either or both of which may contribute to a host's defence.

Lactoferrin, an iron-chelating glycoprotein, is another component of the phagosome (see Table 3.3). Conventional wisdom would have us believe that it acts as a bacteriostatic agent through depriving bacteria of an essential element (section 3.1). Presumably it could fulfil this role without the need to make intimate contact with a bacterial cell. In practice such contact occurs [46] and early studies suggested that, in the case of Gram-negative bacteria, structural damage occurred, namely the release of lipopolysaccharides [47, 48]. Other studies [e.g. 25] have demonstrated that the death of bacteria exposed to lactoferrin cannot be accounted for in terms of iron-deprivation alone. This topic has been reviewed in detail [35, 51] and recently the bactericidal domain of lactoferrin identified [52]. Enzymic digests of bovine lactoferrin prepared with porcine pepsin, cod pepsin or an acid proteinase derived from *Penicillium duponti* were shown to have bactericidal activity greater than that of the native protein [50]. A single antibacterial peptide from near to the *N*-terminus of lactoferrin – i.e. away from the iron-binding sites – was obtained by hydrolysis with gastric pepsin. It consisted of a loop of 18 amino acid residues formed by a disulfide bond between cysteine residues 20–37 of human (lactoferricin H) and 19 and 36 of lactoferrin from bovine milk (lactoferricin B). When assayed against *Escherichia coli* 0111 (changes in optical density of a broth culture), purified lactoferricins H and B were respectively 2- and 12-fold more active than native lactoferrin. Lactoferricin B also inhibited the growth of *Klebsiella pneumoniae*, *Pseudomonas aeruginosa*, *Listeria monocytogenes* and *Staphylococcus aureus*. Synthetic analogues of both lactoferricins had an antibacterial potency comparable with those of the peptides obtained by proteolysis of lactoferrin. Lactoferricin B causes the release of lipopolysaccharides from the cell envelopes of Gram-negative bacteria [51], the death of a range of bacteria [52] and *Candida albicans* [53] as well as the inhibition of glucose uptake by Dermatophytes, microfungi and yeasts [26].

3.4.1 Iron

Iron is the second most abundant metal and the fourth most plentiful element in the Earth's crust. Its solution chemistry is of particular importance. Under normal physiological conditions of neutral pH and oxygen availability, free Fe(II) is readily oxidized to Fe(III). The latter is hydrolysed and polymerized to insoluble ferric hydroxides and oxyhydroxides:

$$Fe^{3+} + 3OH^- \rightarrow \text{ferric hydroxide(s)}\ K_{sp} \approx 10^{49}\ M^3$$

Iron present in the high molecular weight colloidal aggregates resulting from such hydrolysis and polymerization is unavailable for physiological purposes. Another attribute of iron needs to be noted. If Fe(II) oxidation occurs under certain conditions within a cell, then hydroxyl radicals may well be formed by the Fenton reaction. The cytotoxicity of such radicals was discussed above in section 3.2.

In view of the solution chemistry of iron, it is hardly surprising that organic evolution has selected systems whereby this element is solubilized from colloidal aggregates, transported in the bloodstream and, if need be, stored in a manner that impedes production of hydroxyl radicals. A class of iron binding proteins, the transferrins, have evolved to fulfil the first two functions and ferritin the third [54]. Following the demonstration that transferrin *in vivo* binds two iron atoms and two synergistic anions (HCO_3^- or possibly CO_3^{2-}) with the release of six protons per molecule [55], this class of sequestrants has been the subject of an enormous research effort. The available information has been reviewed by Harris and Aisen [49] and Aisen [56]. Transferrins are widely distributed in the animal kingdom (see Table 3.1). The greatest research effort has been given to those of the serum, neutrophiles, exocrine secretions and milk of mammals and the blood and egg whites of birds [49]. All the transferrins from these sources are glycoproteins (molecular mass *c*.8 kDa), they consist of about 700 amino acid residues which form two symmetrical homologous parts, each of which carries an iron-binding site [54]. Subtle differences in composition have led to the identification of three members of the transferrin group from the sources noted above, namely serotransferrin in mammalian blood, lactoferrin in the neutrophiles, exosecretions and milk of mammals and ovotransferrin in egg white [54]. The amino acid sequences of all three have been determined [49, 56]. The sequence homology between serotransferrin and human lactoferrins or hen ovotransferrin is 59% and 51% respectively and between human lactoferrin and ovotransferrin, 49%. There is a consensus of opinion that transferrins contribute to the antimicrobial defence systems of animals through chelation of iron. What is still in doubt, however, is the actual extent of this contribution in particular situations. Some investigators [57] have concluded that serotransferrin causes bacteriostasis of microorganisms that invade the bloodstream, such stasis being released when the chelating potential of this ligand is quenched with Fe(III) [58].

Mammalian milk contains both transferrin and lactoferrin, the amounts and relative proportions of which vary considerably in the milk of different species [59]. As yet, however, there does not appear to be an unequivocal view on the role of the transferrins in milk [60]. This ought not to come as a great surprise because the possible contribution of these ligands to the defence of milk against spoilage has to be set against the action of other antimicrobial agents, complement-mediated bacteriocidal effect of antibodies, lysozyme and the lactoperoxidase system [60, 61].

Shakespeare did not appear to have entertained equivocal views about the antiseptic properties of egg white – "I'll fetch some flax and whites of eggs to apply to his bleeding face" (*King Lear*, Act III, Scene 1). Over the past 75 years or so, many of the proteins of egg white have been shown to have biological properties that could conceivably contribute to the antimicrobial defence of eggs [7]. The contribution of ovotransferrin is paramount. This particular protein was first isolated from egg white and named conalbumin by Osborne and Campbell [62]. Its chelating action and role as a bacteriostatic agent was demonstrated by Alderton *et al.* [63] and Schade and Caroline [64]. Since that time a great number of studies have confirmed the role of ovotransferrin in the defence of eggs against infection with both saprophytic and pathogenic bacteria [7].

3.4.2 Avidin

This protein constitutes about 0.05% of the total proteins in the albumen of hens' eggs [7]. It can be recovered in relatively high yields (75%) from albumen by a cation exchange method [65]. Avidin is a basic tetrameric glycoprotein, 128×4 amino acid residues; M_r 62 400) [66]. A related compound, streptavidin, having a more acid isoelectric point than avidin is produced by *Streptomyces* spp. [67]. Avidin combines with the vitamin biotin (four molecules of the latter to one of the former). The combination is remarkable for its extraordinary affinity, $K_a = 10^{15}\,M^{-1}$ a value that corresponds to a free energy of association of about 21 kcal/mol [68]. The crystal structure of both holo-avidin [69] and apo-avidin [70] is known. The apo-avidin has a marked resistance to denaturation by heat or breakdown by proteolytic enzymes [71]. The original assays for avidin concentrations in materials were based on methods using biotin-requiring strains of yeasts

[72] or *Lactobacillus* spp. [73]. Today spectrophotometric methods are used [74]. The bioassay methods for determining avidin concentrations 'spiked' the literature on egg microbiology with an obvious contention; namely that this protein contributes to the antimicrobial defence of eggs. Proof in support of such claims is still awaited [7]. Indeed the possible biological role of avidin in egg white has attracted little if any attention.

3.5 The lactoperoxidase system (LPS)

Peroxidases bring about the following reaction:

$$AH_2 + H_2O_2 \rightarrow A + 2H_2O$$

where AH_2 and A are the reduced and oxidized forms of suitable electron donors. A wide variety of organic and inorganic compounds can serve as electron donors. Animal peroxidases utilize the pseudohalide (thiocyanate, SCN^-) and halides as donors. The following peroxidases occur in mammals: thyroid peroxidase, eosinophil peroxidase, myeloperoxidase (see Table 3.1) and lactoperoxidase [75]. The last mentioned, which has been considered in several recent reviews [61, 76, 77], will be discussed in this section.

Lactoperoxidase, a glycoprotein (MW 78 000) which contains one haem group (protohaem 9) and about 10% carbohydrate, occurs in a soluble form in the milk, saliva and lachrymal secretions of mammals [75]. The amounts of the enzyme in the milk of different species varies to a marked extent, e.g. 1.3–19.4 units per ml of cows' milk, 0.06–0.97 in human milk and 22 in that of guinea pigs [61].

The thiocyanate anion is the principal electron donor in cows' milk, its actual concentration (0.1–1.5 mg/100 ml) being determined by the breed of cow and type of feed [61]. Thiocyanate is derived from two classes of compounds in the feed, the glucosinolates (commonest in the genus *Brassica* – viz. kale, cabbage, turnips) and the cyanogenic glucosides (found in cassava, potatoes, peas, beans, etc.). The hydrolysis of glucosides releases cyanide which is detoxified via a rhodonase-mediated reaction with thiosulphate derived from the metabolism of sulphur-containing amino acids [75]. There are several sources of H_2O_2 for the reaction noted above. It can be formed by catalase-negative lactic acid bacteria, lactobacilli, lactococci and streptococci, which contaminate milk containing dissolved oxygen. Alternatively H_2O_2 may be generated in milk by the action of ascorbic acid oxidase, xanthine oxidase (substrate, hypoxanthine) or glucose oxidase [61]. Of course H_2O_2 may be added directly (British patent 1468 403). The following mechanism of action has been proposed for lactoperoxidase-catalysed oxidation of thiocyanate [see reviews 61, 75]:

$$2SCN^- + H_2O_2 \rightarrow (SCN)_2 + 2H_2O$$
$$(SCN)_2 + H_2O \rightarrow HOSCN + SCN^- + H^+$$
$$HOSCN \rightarrow OSCN^- + H^+$$

Thiocyanate may be oxidized directly, viz.:

$$SCN^- + H_2O_2 \rightarrow OSCN^- + H_2O$$

There is general agreement that $OSCN^-$ is the major intermediate oxidation product of SCN^- and the principal antimicrobial agent generated by lactoperoxidase in milk [77]. Hypothiocyanite ($OSCN^-$) exists in an acid–base equilibrium with hypothiocyanous acid (HOSCN), K_a 5.3 [78]. In terms of antimicrobial activity, there is a general consensus that the lactoperoxidase system causes oxidation of thiol groups on enzymes and affects the cytoplasmic membrane of sensitive organisms such that ions, amino acids and even polypeptides are lost from the cell contents [76]. The reaction or otherwise of specific organisms under specific conditions to the lactoperoxidase system was reviewed by Wolfson and Sumner [61]. The lactoperoxidase system is obviously of interest to dairy microbiologists who seek to extend the storage life of milk particularly in countries lacking refrigeration on the farm. Judging from a recent review [61], this still remains a theoretical possibility only.

3.6 Lysozymes

Lysozymes (EC 3.2.1.17) are defined as 1,4-β-*N*-acetylmuramidases. They cleave the glycosidic bond between C-1 of *N*-acetylmuramic acid and the C-4 of *N*-acetyl glucosamine. These two compounds form the backbone of the polymer in the peptidoglycan in the cell walls of eubacteria [7]. The peptidoglycan is an exoskeleton. It determines the morphology of a bacterial cell and, through curtailing expansion in the underlying cytoplasmic membrane, prevents unfettered water uptake by the cell contents. A simple interpretation of the action of lysozyme can be offered. The enzyme makes contact with the peptidoglycan in the cell wall, the polymer is cleaved, osmotic uptake of water leads to the expansion and eventual rupture of the cytoplasmic membrane. The last event causes cell death [7]. This course of events occurs most often with Gram-positive bacteria. Such organisms have relatively simple cell walls containing up to 90% peptidoglycan. Consequently there is little impediment to the diffusion of lysozyme to its substrate. In some Gram-positive bacteria, such as the staphylococci, teichoic acids and other cell wall accessory materials bind lysozyme and prevent its diffusion. In other Gram-positive organisms such as *Bacillus cereus*, the absence of *N*-acetyl groups on the glucosamine residues confers resistance to lysozyme [79].

In general Gram-negative bacteria are resistant to lysozyme. Their cell walls contain relatively small amounts of peptidoglycan (5–10%) which occurs beneath the outer membrane of the cell envelope. Such an arrangement prevents the diffusion of lysozyme to its substrate. Lysozyme sensitivity can be induced by mechanisms that disrupt the integrity of the cell envelopes of Gram-negative bacteria. Removal of Ca^{2+} and Mg^{2+} by EDTA, rapid shift in pH or temperature are among the many methods that have been devised to induce lysozyme-sensitivity in Gram-negative bacteria [79]. Lysozymes from certain sources e.g. papaya exhibit a pronounced chitinase activity [80].

Lysozymes have a ubiquitous distribution in nature being found in mammals, insects, phage, plants, etc. [80]. Though of ubiquitous occurrence, there is very little evidence of their primary role in antimicrobial defence systems [81]. Indeed it would appear that they work in tandem with other antimicrobial agents to bring about cell lysis. The egg white of the domestic hen contains relatively large amounts of lysozyme, upwards of 3–4% (wet weight) of the albumen proteins [7]. Even in this system, there is no convincing evidence that lysozyme plays a primary role in the egg's defences. Indeed many studies have indicated that it plays a very minor role in a system in which ovotransferrin is of major importance in defending the yolk against microbial attack [7].

The above observations have not prevented intensive studies of lysozymes. It was the first protein to be sequenced containing all 20 amino acids and for which a detailed mechanism of action was defined [80, 81]. Moreover the possible utility of lysozyme in food preservation has attracted many investigators [81]. In Europe, lysozyme obtained from the egg white of domestic hens is widely used to control late blowing in Gouda cheese due to the growth of *Clostridium tyrobutyricum* [6]. In terms of successful patent applications the Japanese have achieved considerable success in adopting lysozyme as a food preservative in a very wide range of applications (see review by Cunningham *et al.* [82]).

3.7 Prospects

This commentary on the natural antimicrobial systems of animals has highlighted three important issues:

(1) The importance of a cocktail of antimicrobial agents in the defence of animals against attack by the commonly occurring organisms in their environment.
(2) The ubiquity and antiquity of particular antimicrobials such as the antibiotic peptides.
(3) The enormous contribution of the molecular biologists to our under-

standing of the molecular basis of the mode of action of many of the antimicrobial agents in nature.

Indeed it would seem that at long last food microbiologists have an adequate data and a battery of apposite techniques to explore systematically the possible use of natural antimicrobial agents in food preservation. This chapter has demonstrated clearly that no one antimicrobial agent has become the paramount member of defence systems that have evolved over millions of years. This situation must be heeded by food microbiologists. The available evidence indicates that at best a natural antimicrobial agent will have to be included in a cocktail in order to contribute to the preservation and safety of a food.

References

1. Taylor, P.W. (1983) *Microbiol Rev.*, **47**, 46.
2. Kaufmann, S.H.E. and Flesch, I.E.A. (1992) In *Molecular Biology of Bacterial Infection: Current Status and Future Perspectives* (eds Cdr. C.E. Hormaeche, C.W. Penn and C.J. Smyth), Cambridge University Press, Cambridge, p. 97.
3. Kaufmann, S.H.E. (1993) *Ann. Rev. Immunol.*, **11**, 129.
4. Tenovuo, J. and Pruit, K.M. (1984) *J. Oral Path.*, 537.
5. Anon. (1986) *Antimicrobial Systems in Milk*, International Dairy Federations, Brussels.
6. Fox, P.F. (1993) *J. Food Biochem.*, **17**, 173.
7. Tranter, H.S. and Board, R.G. (1982) *J. Appl. Biochem.*, **4**, 295.
8. Fletcher, T.C. (1982) *Dev. Comp. Immunol. (Suppl.)*, **2**, 123.
9. Kimbrell, D.A. (1991) *Bioessays*, **13**, 657.
10. Hutmark, D. (1993) *Tig*, **9**, 178.
11. Hoffman, J.A. and Hetru, C. (1992) *Immunol. Today*, **13**, 411.
12. Boman, H.G. (1986) In *Natural Antimicrobial Systems* (eds G.W. Gould, M.E. Rhodes-Roberts, A.K. Charnley, R.M. Cooper and R.G. Board), Bath University Press, Bath, p. 116.
13. Lehrer, R.I., Lichenstein, A.K. and Ganz, T. (1993) *Ann. Rev. Immunol.*, **11**, 105.
14. Selsted, M.E., Miller, S.I., Henschen, A.H. and Oullette, A.J. (1992) *J. Cell Biol.*, **118**, 929.
15. Jones, D.E. and Bevins, C.L. (1993) *FEBS Lett.*, **315**, 187.
16. Diamond, G., Jones, D.E. and Bevins, C.L. (1993) *Proc. Natl. Acad. Sci. USA*, **90**, 4596.
17. Lee, J-Y., Boman, A., Sun, C., Anderson, M., Jörnvall, H., Mutt, V. and Boman, H.G. (1989) *Proc. Natl. Acad. Sci. USA*, **86**, 9159.
18. Agerberth, B., Lee, J-Y., Bergman, T., Calquist, M., Boman, H.G., Mutt, V. and Jörnvall, H. (1991) *Eur. J. Biochem.*, **202**, 849.
19. Agerberth, B., Boman, A., Anderson, M., Jörnvall, H., Mutt, V. and Boman, H.G. (1993) *Eur. J. Biochem.*, **216**, 623.
20. Samakoulis, C., Kylsten, P., Kimbrell, D.A., Engström, A. and Hutmark, D. (1991) *EMBO J.*, **10**, 167.
21. Zasloff, M. (1987) *Proc. Natl. Acad. Sci. USA*, **84**, 5449.
22. Bevins, C.L. and Zasloff, M. (1990) *Ann. Rev. Biochem.*, **59**, 395.
23. Sitaram, S., Subbalakshmi and Nagaraj, R. (1993) *FEBS Lett.*, **328**, 239.
24. Bangapore, N., Travis, J., Onunka, V.C., Pohe, J. and Shafer, W.M. (1990) *J. Biol. Chem.*, **265**, 13584.
25. Bellamy, W., Takase, M., Yamanchi, K., Wakabayashi, H., Kawase, K. and Tomita, M. (1992) *Biochem. Biophys. Acta*, **1121**, 130.
26. Bellamy, W., Yamauchi, K., Wakabayashi, H., Takase, M., Takamura, N., Shimamura, S. and Tomita, M. (1994) *Letts Appl. Microbiol.*, **18**, 230.

27. Boman, H.G. (1991) *Cell*, **65**, 205.
28. Lehrer, R.I., Ganz, T. and Selsted, M.E. (1991) *Cell*, **64**, 229.
29. Banks, J.G., Board, R.G. and Sparks, N.H.C. (1986) *Biotech. Appl. Biochem.*, **8**, 103.
30. Root, R.K. and Cohen, M.S. (1981) *Rev. Infect. Dis.*, **3**, 565.
31. Lehrer, R.I., Ganz, T. and Selsted, M.E. (1991) *Cell*, **64**, 229.
32. Nettles, C.G. and Barefoot, S.F. (1993) *J. Food Prot.*, **56**, 338.
33. Hildago, P. and Flores, M. (1994) *Food Microbiol.*, **11**, 161.
34. Ojcius, D.M. and Young, J.D-E. (1991) *Trends Biochem. Sci.*, **16**, 225.
35. Boman, H.G. (1991) *Cell*, **65**, 205.
36. Hill, C.R., Lee, J. and Eisenberg, M.E. (1991) *Science*, **251**, 1481.
37. Boman, H.G. and Hultmark, D. (1987) *Ann. Rev. Microbiol.*, **41**, 103.
38. Poulter, L., Terry, A.S., Williams, D.H., Giovannini, M.G., Moore, C.H. and Gibson, B.W. (1988) *J. Biol. Chem.*, **263**, 3279.
39. Zasloff, M., Martin, B. and Chen, H-C. (1988) *Proc. Natl. Acad. Sci. USA*, **85**, 910.
40. Soravia, E., Martini, G. and Zasloff, M. (1988) *FEBS Letts*, **228**, 337.
41. Chen, H-C., Brown, J.H., Morell, J.L. and Huang, C.M. (1988) *FEBS Letts*, **236**, 462.
42. Cuervo, J.H., Rodriquez, B. and Houghten, R.A. (1988) *Peptide Res.*, **1**, 81.
43. Venema, K., Abee, T., Haandrikman, A.J., Leenhouts, K.J., Kok, J., Konings, W.N. and Venema, G. (1993) *Appl. Environ. Microbiol.*, **59**, 1041.
44. Nissen-Meyer, J., Holo, H., Håvarstein,, L.S., Sletten, K. and Nes. I.F. (1992) *J. Bact.*, **174**, 5686.
45. Odeberg, H. and Olsson, I. (1975) *J. Clin. Invest.*, **56**, 1118.
46. Arnold, R.R., Cole, M.F. and McGhee, J.R. (1977) *Science*, **197**, 263.
47. Ellison, R.T., Giehe, T.J. and La Force, F.M. (1988) *Infect. Immun.*, **56**, 2774.
48. Ellison, R.T., La Force, F.M., Giehe, T.J., Boose, D.S. and Dunn, B.E. (1990) *J. Gen. Microbiol.*, **136**, 1437.
49. Harris, D.C. and Aisen, P. (1989) In *Iron Carriers and Iron Proteins* (ed. T.M. Loehr), VCH Publishers Inc., New York, p. 241.
50. Tomita, M., Bellamy, W.R., Takase, M., Yamauchi, K., Wakabayashi, H. and Kawase, K. (1991) *J. Dairy Sci.*, **74**, 4137.
51. Bellamy, W.R., Wakabayashi, H., Takase, M., Kawase, K., Shimamura, S. and Tomita, M. (1993) *J. Appl. Bact.*, **75**, 478.
52. Bellamy, W., Takase, M., Wakabayashi, H., Kawase, K. and Tomita, M. (1992) *J. Appl. Bact.*, **73**, 472.
53. Bellamy, W., Wakabayashi, H., Takase, M., Kawase, K., Shimamura, S. and Tomita, M. (1993) *Med. Microbiol. Immunol.*, **182**, 97.
54. Crichton, R.R. and Charloteaux-Wauters, M. (1987) *Eur. J. Biochem.*, **164**, 485.
55. Schade, A.L., Reinhart, R.W. and Levy, H. (1949) *Arch. Biochem. Biophys.*, **20**, 70.
56. Aisen, P. (1989) In *Iron Carriers and Iron Proteins* (ed. T.M. Loehr), VCH Publishers Inc., New York, p. 355.
57. Kochan, I. (1977) In *Advances in Chemistry Series*, No. 162 (ed. K.N. Raymond), American Chemical Society, Washington, p. 55.
58. Weinberg, E.D. (1984) *Physiol. Rev.*, **64**, 65.
59. Ribadeau-Dumas, B. (1986) In *Antimicrobial Systems in Milk* (ed. Anon.), International Dairy Federation, Brussels, p. 31.
60. Reiter, B. (1986) In *Antimicrobial Systems in Milk* (ed. Anon.), International Dairy Federation, Brussels, p. 59.
61. Wolfson, L.M. and Sumner, S.A. (1993) *J. Food Prot.*, **56**, 887.
62. Osborne, I.B. and Campbell, G.F. (1900) *J. Am. Chem. Society*, **22**, 422.
63. Alderton, G., Ward, W.H. and Fevold, H.L. (1946) *Arch. Biochem. Biophys.*, **11**, 9.
64. Schade, A.L. and Caroline, L. (1944) *Science*, **100**, 14.
65. Durance, T.D. and Nakai, S. (1988) *J. Food Sci.*, **53**, 1096.
66. Wilchek, M. and Bayer, E.A. (eds) (1990) *Methods in Enzymology*, Vol. 184, Avidin-Biotin Technology, Academic Press Inc., London, p. 5.
67. Chaiet, L. and Wolf, F.J. (1964) *Arch. Biochem. Biophys.*, **106**, 1.
68. Green, N.M. (1975) *Adv. Protein Chem.*, **29**, 84.
69. Pugliere, L., Coda, A., Malcovati, B. and Bolognesi, M. (1993) *J. Mol. Biol.*, **231**, 698.
70. Pugliese, L., Malcovati, M., Codq, A. and Bolognesi, M. (1994) *J. Mol. Biol.*, **235**, 42.

71. Györgi, P. and Rose, C.S. (1943) *Proc. Soc. Exp. Biol. Med.*, **53**, 55.
72. Hertz, R. (1943) *Proc. Soc. Esp. Biol. Med.*, **52**, 15.
73. Landy, M., Dicken, D.M., Bicking, M.M. and Mitchell, W.R. (1942) *J. Bact.*, **43**, 5.
74. Green, N.M. (1965) *Biochem. J.*, **94**, 230.
75. Pruitt, K.M. and Reiter, B. (1989) In *The Lactoperoxidase System: Chemistry and Biological Significance* (eds K.M. Pruitt and J. Tenovuo), Marcel Dekker, New York, p. 143.
76. Björck, L. (1986) In *Antimicrobial Systems in Milk* (ed. Anon.), International Dairy Federation, Brussels, p. 18.
77. Reiter, B. and Harnülv, B.G. (1984) *J. Food Prot.*, **47**, 724.
78. Thomas, E.L. (1989) In *The Lactoperoxidase System: Chemistry and Biological Significance* (eds K.M. Pruitt and J. Tenovuo), Marcel Dekker, New York, p. 31.
79. Banks, J.G. and Tranter, H.S. (1986) In *Antimicrobial Systems in Milk* (ed. Anon.), International Dairy Federation, Brussels, p. 39.
80. Jollès, P. and Jollès, J. (1984) *Mol. Cell. Biochem.*, **63**, 165.
81. Proctor, V.A. and Cunningham, F.E. (1984) *CRC Crit. Rev. Fed. Sci. Nutr.*, **26**, 359.
82. Cunningham, F.E., Proctor, V.A. and Oetsch, S.J. (1991) *World's Poult. Sci.*, **47**, 141.

4 Natural antimicrobials from plants

G.J.E. NYCHAS

4.1 Introduction

Food preservation is becoming more complex. New food products are being introduced onto the market. Generally these require longer shelf-lives and greater assurance of freedom from foodborne pathogenic organisms. The search for new substances to be used in food preservation is hampered by regulatory restrictions. Consequently a great deal of time and money may be required to develop a new chemical preservative and to get it approved especially in view of the public pressure against chemical additives in general. Such obstacles provide new opportunities for those seeking alternative routes in the search for new food preservatives. The excessive use of chemical preservatives, some of which are suspect because of their supposed or potential toxicity, has resulted in increasing pressure on food manufacturers to either completely remove chemical preservatives from their food products or to adopt more 'natural' alternatives for the maintenance or extension of a product's shelf life. There is considerable interest in the possible use of such natural alternatives as food additives either to prevent the growth of foodborne pathogens or to delay the onset of food spoilage. Many naturally occurring compounds, such as phenols (phenolic acid, polyphenols, tannins), and organic acids (acetic, lactic, citric) have been considered in this context. Many spices and herbs and extracts possess antimicrobial activity, almost invariably due to the essential oil fraction (Deans and Ritchie, 1987). Thus the essential oils of citrus fruits exhibit antibacterial activity to foodborne bacteria (Dabbah *et al.*, 1970) and moulds (Akgul and Kivanc, 1989) so too have the essential oils of many other plants such as oregano, thyme (Salmeron *et al.*, 1990; Paster *et al.*, 1990), sage, rosemary, clove, coriander etc. (Farag *et al.*, 1989; Aureli *et al.*, 1992; Stecchini *et al.*, 1993). The antibacterial and antimycotic effects of garlic and onion have been well documented also (Mantis *et al.*, 1978; Sharma *et al.*, 1979; Saleem and Al-Delaimy, 1982; Conner and Beuchat, 1984a,b).

To date phenolics have been used mainly in foods for purposes (e.g. antioxidants) and other than as antimicrobial agents. For this reason they are worthy of study either alone or in combination with processing methods or with chelators in order to establish if they could extend the shelf-life of foods. It needs to be stressed that, although chelators are not considered as preservatives *per se*, they can potentiate other antimicrobial agents (Kabara, 1991). For example such compounds can perturb the structure of the cell membrane of Gram-negative bacteria such that resistance to antimicrobial agents is overcome.

According to Mitscher (1975) plants are poorly exploited sources of alternative antimicrobial agents whose structures and modes of action may well differ from those derived from the more well known sources. In their recent review, Wilkins and Board (1989), reported that over 1389 plants are potential sources of antimicrobial agents. Food microbiologists have investigated the antimicrobial properties of many herbs, spices and food plants, but in many cases they have ignored a number of potentially useful plants listed by Wilkins and Board (1989). For this reason they should keep abreast of current advances in pharmacognosy and related sciences where an increasing interest is being taken in the potential use of plant constituents as drugs or antimicrobials in general.

4.2 Phytoalexins

According to Dixon *et al.* (1983) and Mansfield (1986) phytoalexins may be defined as "... host synthesized, low molecular weight broad spectrum antimicrobial compounds whose synthesis from distant precursors is induced in plants in response to microbial infection or treatment of plant tissues with a range of naturally occurring or synthetic, artificial compounds (biotic or abiotic elicitors)". The molecules that signal plants to begin the process of phytoalexin synthesis are called elicitors. The occurrence of endogenous elicitors, does not exclude the possibility that microbial metabolites or cell wall components may also act directly as elicitors themselves. These compounds are not found in healthy, unaffected tissues. Darvill and Albersheim (1984) presented data demonstrating that the enzymes catalysing the synthesis of phytoalexins are themselves synthesized *de novo* when plant cells are exposed to microbes or other effective stimuli.

The resistance of plants to colonization by bacteria, nematodes and fungi could well be attributed to the accumulation of phytoalexins (Mansfield, 1986). It needs to be stressed that the antimicrobial spectrum of phytoalexins is directed mainly against fungi but activity has also been reported against a variety of bacterial species. Gram-positive bacteria have been found to be more sensitive than Gram-negative ones.

Among plant phenolics, isoflavonoids – with C_6–C_3–C_6 basic skeleton structure (Harbone, 1980) – are the predominant chemical class of phytoalexins. Recently Kubo *et al.* (1985) refer to oleuropein as a probable phytoalexin precursor. In section 4.5 the antimicrobial action of phenolics will be discussed in some detail. When considering the commercial potential of these compounds (phytoalexins), one could recommend the production of such secondary metabolites in cultured plant cells.

4.3 Organic acids

Nowadays organic acids such as acetic, lactic, propionic, benzoic, sorbic and their derivatives are commonly used as antimicrobial agents. Citric, succinic, malic and tartaric are found in a variety of fruits (e.g. citrus, rhubarb, grapes, pineapples) and vegetables (e.g. broccoli, carrots). Through use as acidulants – e.g. malic acid addition to apple juice – some of these contribute to the antimicrobial properties of raw material and products. Lactic and propionic acids are formed during natural processing involving fermentation. The antimicrobial activity of these acids is well documented (Doores, 1983; Eklund, 1989; Beauchat and Golden, 1989) and has been reviewed recently (Eklund, 1989; Kabara and Eklund, 1991). Cell walls, cell membranes, metabolic enzymes, protein synthesis system and genetic material are the main targets of their action against a wide range of microorganisms (Eklund, 1989).

In this chapter the antimicrobial action of naturally occurring organic acids from spices, herbs, fruits and vegetables will be reviewed only. These phenolic compounds are well known for their mainly antioxidant characteristics or their effect on taste and odour.

4.4 Essential oils

True spices are defined as the roots, bark, buds, seeds or fruits of aromatic plants which usually grow in tropical countries (Wilkins and Board, 1989). The antimicrobial spectrum (Table 4.1) of extracts (steam distillation or volatile organic solvents) from several types of plants (Table 4.2) and plant parts have been recognized for many years. The antimicrobial compounds in plant materials are commonly in the essential oil fraction. These compounds are mainly responsible for the characteristic aroma and flavour of the spices. They are recovered from plant materials primarily by steam distillation, although some are expressed cold, by dry or vacuum distillation (Farell, 1985).

According to Hargreaves *et al.* (1975) essential oils are defined as being a group of odorous principles, soluble in alcohol and to a limited extent in water, consisting of a mixture of esters, aldehydes, ketones and terpenes. It needs to be noted that the use of volatile solvents (e.g. acetate, ethanol, ethylene chloride) could provide not only a more complete flavour profile than the essential oil (oleoresins) alone but also a more potential antimicrobial inhibitor (Dziezak, 1989). Among the compounds having wide spectra of antimicrobial effectiveness are thymol from thyme and oregano, cinnamic aldehyde from cinnamon and eugenol from cloves. According to Shelef (1983) phenolic compounds are probably the major antimicrobial components of the essential oils of spices. For this reason, emphasis will be given in this chapter to an assessment of antimicrobial action of essential oils extracted with steam distillation only. Studies referred either to specific compounds, for example phenolics such as eugenol, thymol etc. from essential oil, or volatile solvent was used for the extraction will be reviewed in the section related with phenolics.

The extent of sensitivity of a test organism varies with the strain under study, the environmental (pO_2) conditions imposed (Pasteur *et al.*, 1990; Stecchini *et al.*, 1993), the reaction in the Gram stain (Farag *et al.*, 1989; Tassou and Nychas, unpublished results). It was observed (Paster *et al.*, 1990), for example, that the antimicrobial activity of thymol and oregano oil was enhanced by low pO_2.

In general Gram-positive are more sensitive than Gram-negative bacteria to the antimicrobial compounds in spices (Dabbah *et al.*, 1970; Farag *et al.*, 1989; Shelef, 1983; Tassou and Nychas, unpublished). Variation in the rate or extent of inhibition was also evident among the Gram-negative bacteria. For example *Escherichia coli* was less resistant than *Pseudomonas fluorescens* or *Serratia marcescens* when tested with essential oils from sage, rosemary, cumin, caraway, clove and thyme oils (Farag *et al.*, 1989). Inhibition of growth ranged from 88% with *Aerobacter aerogenes* to 100% with *Alcaligenes faecalis* as test organisms (Dabbah *et al.*, 1970). *Salmonella enteritidis* and *S. typhymurium* were less sensitive than *Pseudomonas fragi* to sage and mastic gum oils (Tassou and Nychas, unpublished). *Salmonella typhimurium* was found to be more sensitive than *Pseudomonas aeruginosa* to essential oil from oregano and thyme (Paster *et al.*, 1990). Deans and Ritchie (1987), who studied the effect of 50 plant essential oils against 25 genera of bacteria, concluded that Gram-positive and Gram-negative organisms were both susceptible to the essential oils and there was no evidence that the degree of sensitivity to the oils was reflected in the Gram reaction of the organism.

Most of the foodborne bacterial pathogens (Table 4.1) examined are sensitive to extracts either from spices or several types of plants and plants part. *Staphylococcus aureus* is probably the most commonly used bacterium in studies of antimicrobial activity of essential oils (Table 4.2). Almost all the essential oils from spices and herbs tested against this organism were found to have an inhibitory activity on its growth as well as on enterotoxin production. Indeed oils from linden flower, orange, lemon, grapefruit, mandarine, sage, rosemary, oregano, thyme, cumin, caraway clove, thyme, allspice, mastic gum (*Pistachia lentiscus* var. *chia*) and onion were found to retard the growth of this organism (Elinma *et al.*, 1983; Shelef *et al.*, 1980; Briozzo *et al.*, 1989; Farag *et al.*, 1989; Aktug and

Table 4.1 Antimicrobial spectrum of essential oils from herbs, spices and various plants

Microorganisms	Reference	Microorganisms	Reference
Acetobacter spp.	1	*Micrococcus luteus*	14
Acinetobacter sp.	14	*Moraxella* sp.	14
calcoacetica		*Mycobacterium* sp.	1
Aeromonas hydrophila	3, 14	*Mycobacterium phlei*	18
Alcaligenes sp.	1	*Mucor* sp.	2
Alcaligenes faecalis	14	*Neisseria* sp.	1
Arthobacter sp.	1	*sicca*	12
Aspergillus niger	2, 20	*Pediococcus* sp.	1
parasiticus	4, 5, 6, 23, 32	*Penicillium* sp.	5,
flavus	4, 5, 6, 8, 20, 27	*chrysogenum*	2, 10
ochraceus	5, 20	*patulum*	5
Bacillus sp.	1	*roquefortii*	5
subtilis	10, 12, 14, 18, 22, 25, 27	*citrinum*	5
cereus	16	*Pityrosporum ovale*	10
Beneckea natriegens	14	*Propionibacterium acnes*	10
Brevibacterium ammoniagenes	10	*Proteus* spp.	1
linens	14	*vulgaris*	14
Brochothrix thermosphacta	14	*Pseudomonas* spp.	1, 21
Campylobacter jejuni	20	*aeruginosa*	10, 11, 14, 20
Candida albicans	10, 11	*fluorescens*	18
Citrobacter sp.	1	*fragi*	13
freundii	14	*clavigerum*	23

Clostridium botulinum	9, 24, 26	*Rhizopus* sp.	2
perfrigenes	11, 15	*Saccharomyces cerevisiae*	10, 18, 27
sporogenes	14, 20	*Salmonella* spp.	1
Corynebacterium sp.	1	*enteriditis*	13
Edwardsiella sp.	1	*senftenberg*	21
Enterobacter sp.	1	*typhimurium*	19, 20
aerogenes	10, 14, 31	*pullorum*	14
Erwinia carotovora	14	*Sarcina* spp.	1, 18
Escherichia coli	1, 10, 11, 12, 14, 18, 28	*Serratia* sp.	1
Flavobacterium sp.	1	*marcencens*	14, 18
suaveolens	14	*Staphylococcus* spp.	1
Klebsiella sp.	1	*aureus*	10, 11, 12, 13, 14, 15, 18, 19, 20, 21, 31
pneumoniae	11, 14	*'Streptococcus' feacalis*	14
Lactobacillus sp.	1	*mutans*	10
plantarum	13, 14, 29	*Trichophyton mentagrophytes*	14
minor	13	*Yersinia enterocolitica*	14
Leuconostoc sp.	1	*Vibrio* sp.	1
cremoris	14	*Vibrio parahaemolyticus*	19
Listeria monocytogenes	7, 17, 30		
Micrococcus sp.	1, 18		

1: Shelef *et al.* (1980); 2: Akgul and Kivanc (1989); 3: Stecchini *et al.* (1993); 4: Salmeron *et al.* (1990); 5: Azzouz and Bullerman (1982); 6: Karapinar (1985); 7: Aureli *et al.* (1992); 8: Conner and Beuchat (1984a,b); 9: DeWit *et al.* (1979); 10: Kubo and Himejima (1991); 11: Briozza *et al.* (1989); 12: Nagy and Tengerdy (1967); 13: Tassou and Nychas (unpublished); 14: Deans and Ritchie (1987); 15: Mantis *et al.* (1978; 1979); 16: Saleem and Al-Delaimy (1982); 17: Ting and Deibel (1992); 18: Farag *et al.* (1989); 19: Gonul and Karapinar (1987); 20: Paster *et al.* (1990); 21: Dabbah *et al.* (1970); 22: Al-Khayat and Blank (1985); 23: Kivanc and Akgul (1990); 24: Ismaiel and Pierson (1990); 25: Blank *et al.* (1987); 26: Hall and Maurer (1896); 27: Sankaran (1976); 28: DeMedici *et al.* (1992); 29: Karaionnoglou *et al.* (1977); 30: Ting and Deibel (1992); 31: Kubo *et al.* (1992); 32: Llewellyn *et al.* (1982).

Table 4.2 Plant essential oils tested for antibacterial properties

Plant	References	Plant	References
Almond (bitter)	1	Mace	20
(sweet)	1	Mandarin	1, 6, 10
Allspice	16	Marjoram	1, 10
Anethole	11	Mastica gum	14
Angelica	1	Melissa	1
Anise	1, 5, 16	Mint (apple)	1
Basil	1, 10	Mustard	16
Bay	1, 20	Neroly	10
Bergamot	1	Nutmeg	1, 8, 10, 20
Calmus	1	Onion	10, 16, 18, 22
Camomile – German	10	Orange	1, 5, 6, 10, 21
Roman	10	(bitter)	1, 6
Caraway	1, 3	Oregano	4, 9, 10, 16, 18
Cardamom	1	Paprika	16
Celery	1, 19	Parsley	1, 5, 10
Cinnamon	1, 19, 16, 18	Pepper	1, 8, 10, 16, 18, 20
Citronella	1	Peppermint	1, 10
Clove	1, 3, 8, 10, 12, 15, 16, 18, 19	Pettigrain	10
Coriander	1, 5, 8, 10	Pimento	1, 10, 18
Cornmint	5	Rose	1

Cortuk*	17	Rosemary	1, 3, 7, 10, 16
Cumin	3, 5, 10	Saffron	10
Dill	1, 5	Sage	1, 3, 5, 7, 10, 16
Estragon	10	Sagebrush	13
Fennel	1, 5, 10	Savory	5
Garlic	10, 16, 18, 22	St. John's Wort	1
Geranium	1	Sassafras	1
Ginger	1, 10	Spike	1
Grapefruit	6	Star Anise	1
Laurel	1, 5, 10	Tarragon	4
Lavender	1	Thuja	1
Lemon	1, 5, 6, 10	Thyme	1, 3, 4, 5, 9, 10, 18
Lime	1, 6	Turmeric	16
Linden flower	2	Valerian	1
Lovage	1	Verbena	1
		Vanilla	10

1: Deans and Ritchie (1987); 2: Aktug and Karapinar (1987); 3: Farag *et al*. (1989); 4: Paster *et al*. (1990); 5: Akgul and Kivanc (1989); 6: Dabbah *et al*. (1970); 7: Shelef *et al*. (1980); 8: Stecchini *et al*. (1993); 9: Salmeron *et al*. (1990); 10: Aureli *et al*. (1992); 11: Kubo and Himejima (1991); 12: Briozzo *et al*. (1989); 13: Nagy and Tengerdy (1967); 14: Tassou and Nychas (unpublished data); 15: Al-Khayat and Blank)1985); 16: Azzouz and Bullerman (1982); 17: Kivanc and Akgul (1990); 18: Ismaiel and Pierson (1990); 19: Blank *et al*. (1987); 20: Hall and Maurer (1986); 21: Sankaran (1976); 22: Elnima *et al*. (1983).
*A Turkish (local) spice.

Karapinar, 1987; Dabbah *et al.*, 1970; Paster *et al.*, 1990; Tassou and Nychas, unpublished).

The effects of essential oils against *Listeria monocytogenes*, the psychrotrophic Gram-positive bacterium responsible for listeriosis, were examined recently by Aureli *et al.* (1992) and Ting and Deibel (1992). The latter workers used 13 while the former 32 spices. Cloves and oregano were found to inhibit this organism by both teams. The activity of cinnamon was disputed by these two teams. Essential oil from clove at 0.5 and 1% concentrations was bactericidal to this organism at 4°C and 24°C (Ting and Deibel, 1992). When these two essential oils were added to a meat slurry inoculated with *L. monocytogenes*, it was found that neither spice at a concentration of 1% inhibited the growth of the organism with incubation at 4°C or 24°C (Ting and Deibel, 1992). In the case of the essential oil from thyme, however, it was found that there was a reduction of the number of viable *L. monocytogenes* cells in minced pork meat supplemented with this material (Aureli *et al.*, 1992). It is worthy of note that both groups of workers concluded that the effectiveness of essential oils decreased when the experiments were conducted *in vivo*. This could be due to high protein and fat content of meat which can mask the antimicrobial effect of essential oils (Shelef, 1983). This subject is discussed in detail in Chapter 3.

Essential oils of clove, coriander and nutmeg were effective in inhibiting the growth of *Aeromonas hydrophila* (Stecchini *et al.*, 1993). A marked reduction of this organism also occurred in inoculated samples of cooked, noncured pork treated with clove or coriander and the meat packaged either under vacuum or air and stored at 2°C and 10°C. The lethal effect of these two oils was more pronounced in the vacuum packaged samples (Stecchini *et al.*, 1993).

Inhibiting the growth of spore forming bacteria by spices and their essential oils has been studied recently also. Inhibition of *Clostridium botulinum*, *Cl. sporogenes* and *Cl. perfringens* has been reported by Mantis *et al.* (1979), Hall and Maurer (1986), Paster *et al.* (1990) and Ismaiel and Pierson (1990). Essential oils from garlic, mace, achiote, onion, cinnamon, thyme, origanum, clove, pimenta and black pepper were found to be the most inhibitory against those organisms. It was also observed that the effect of spice oils on spore germination was reversible (Ismaiel and Pierson, 1990). The oils of black pepper and clove had greater inhibitory effects on vegetative cell growth than the other oils examined. None of the oils had a significant effect on outgrowth from spores. There was a decrease in the rate and extent of germination of *Bacillus subtilis* spores in the presence of clove and eugenol (Blank *et al.*, 1987). It is worth mentioning that the composition of the media used in this study affected the activity of essential oils.

The growth of *Salmonella* spp. was inhibited by essential oils of linden flower, thyme, oregano, orange, lemon, grapefruit, mandarine, almond, bay, clove, coriander, cinnamon, pepper and mastic gum (*Pistachia lentiscus* var. *chia*) (Dabbah *et al.*, 1970; Aktug and Karapinar, 1987; Deans and Ritchie, 1987; Paster *et al.*, 1990; Tassou and Nychas, unpublished).

The antimycotic effect of 26 substances from herbs, spices and plant materials, against *Penicillium* spp. and *Aspergillus* spp. was examined by Azzouz and Bullerman (1982). Clove was the strongest antifungal spice, cinnamon was also quite inhibitory while mustard, garlic, allspice and oregano gave smaller degrees of inhibition.

Contradictory results have been obtained with oregano and thyme oil in tests with *Aspergillus niger*, *Asp. ochraceus*, *Asp. flavus*. Paster *et al.* (1990) reported that these organisms were inhibited by both oils. Akgul and Kivanc (1989) reported also that thyme and wild thyme were the most active oils against *Asp. niger*, *Penicillium chrysogenum*, *Rhizopus* sp. and *Mucor* sp. In contrast Salmeron *et al.* (1990) showed that thyme and oregano oils stimulated the growth of *Asp. flavus* and *Asp. parasiticus* which at the same time acting as antiaflatoxigens. *Aspergillus* and *Penicillium* isolates from black table olives were inhibited with methyl eugenol and the essential oil of the spice *Echinophora sibthorpiana* (Kivanc and Akgul, 1990).

There are limited studies concerning the antimicrobial compounds in spices. Allicin, which is present in garlic oil, is responsible for the antimicrobial action against Gram-positive and Gram-negative bacteria as well as pathogenic fungi (Shelef, 1983). Cinnamon contains 0.5–2.0% volatile oil, mainly cinnamaldehyde, eugenol and *o*-methoxy-cinnamaldehyde. The antimicrobial activity of oils could be attributed to phenols. Indeed, in their study Katayama and Nagai (1960) concluded that the active components of oils consist of eugenol, carvacrol, isoborneol, thymol, vanillin and salicylaldehyde and that these compounds are considered to be phenolics. For this reason the mode of action of essential oils will be discussed in the next section. If the exploitation of essential oils is going to be increased, certain parameters and characteristics should be taken into account. For example the extraction methodology, the variation of composition in the same species which may be influenced by geographical location, climatic condition etc. For this reason the active constituent present in inhibitory oils should be chemically separated and identified. Moreover the mode of action of these compounds should be also examined in more detail.

4.5 Phenolics, pigments and related compounds

The contribution of phenolic compounds in defence mechanisms of plant tissues in response to infections or injuries (Friend, 1979; Harborne, 1980) as well as to sensory (taste, odour, appearance) and nutritional quality of fresh or processed plant foods and their characteristics in taste and odour (Torres *et al.*, 1987; Gutierrez *et al.*, 1989; Ho, 1992; Robishaud and Noble, 1990; Shahidi, 1992; Fisher, 1992; Brenes-Balbuena *et al.*, 1992) is well established.

The term 'phenolic' or 'polyphenol' can be defined chemically as a substance which possesses an aromatic ring bearing a minimum of one but more frequently more than two hydroxy substituents (Harborne, 1980; Ho, 1992), including functional derivatives. Phenolic compounds rarely occur in the free state in living plant tissue due to their caustic properties (Harborne, 1980); they are practically always present in conjugated form. In the simplest instance they are bound to sugars as β-D-glucopyranosides but a wide array of other bound forms are known such as methyl ethers, esters (e.g. hydroxycinnamic acids occur more frequently as esters than glycosides).

The phenolic compounds (Table 4.3) which occur commonly in food material may be classified into three groups, namely, simple phenols and phenolic acids (e.g. *p*-cresol, 3-ethylphenol, hydroquinone, protocatechuic, vanillic, gallic, syringic, ellagic), hydroxycinnamic acid derivatives (e.g. *p*-coumaric, caffeic, ferulic, sinapic) and flavonoids. The third group, is the most important single group of phenolics in food. It consists of catechins, proanthocyanins, anthocyanidins and flavons, flavonols and their glycosides (Davidson, 1983; Ho, 1992). Finally 'tannins' are a diverse group of plant phenolics which are unified by a single common property: the ability to precipitate protein from aqueous solution. Tannins are polymeric phenolics. It needs to be stressed that phenolics of low molecular weight, including phenolic acids and simple flavonoids, may bind protein but cannot crosslink the complexes as is required for precipitation (Hagerman, 1992).

In general the occurrence of phenolics in different plants and parts thereof (Harborne, 1980; Terazawa, 1986; Tsukamoto *et al.*, 1985; Beuchat and Golden, 1989; Bailey *et al.*, 1990; Khurana, 1992; Okuda *et al.*, 1992; Omar, 1992; Shahidi, 1992; Tassou, 1993) as well as their potential antimicrobial action in different broth media have been demonstrated by many workers using a variety of solvents (Table 4.4) and test organisms (Table 4.5) (Davidson, 1983; Rodriguez *et al.*, 1988; Tassou, 1993).

Surprisingly, the antimicrobial action of the antioxidant phenolics, such as butylated hydroxytoluene (BHT), butylated hydroxyanisole (BHA), propylgallate (PG) and 2-tert-butylhydroquinone (TBHQ) as well as parabens (alkyl esters of *p*-hydroxybenzoic acid) have been examined to a limited extent compared to the studies with the naturally occurring ones (Branen *et al.*, 1980). The exception are those coming from olives.

Indeed the antimicrobial action of the naturally occurring phenolics from olives, tea and coffee have been studied in more detail than those from other sources (see Table 4.3) (Toda *et al.*, 1989, 1991; Diker *et al.*, 1991;

Table 4.3 Naturally occurring phenols and phenolic compounds from different plants; Oleaceae (1), skin and seeds of wine grapes (2), avocado (3), tea (4), spices (5), Brassica oilseeds (6), artichoke (7), Berberis (8), hops (9), thymol (10), chicory (11), vanilla (12), sesame oil (13), plum (14), apple (15), carrots (16), boldo (17), herbs (ginger, gentian) (18)

Phenolic compound	Plant	Phenolic compound	Plant
Apigenin-7-glucoside	1	4-hydroxycinnamic acid	1, 16
Benzoic acid	6	Isovanillic	3
Berbamine	8	Linalool	17
Berberine	8	Lupulon	9
Caffeine	4	Luteoline-5-glucoside	1
Caffeic acid	1, 3, 7, 15	Ligustroside	1
3-*o*-Caffeylquinic acid	7, 14, 16	Myricetin	4
4-*o*-Caffeylquinic acid	7, 14, 16	3-methoxybenzoic acid	1
5-*o*-Caffeylquinic acid	14	Oleoside	1
Caryophelene	9	Oleuropein	1
Catechin	2, 4	Paradols	5
Cinnamic acid	6	Protocatechuic acid	1, 3
Chlorogenic acid	6, 7, 15	*o*-pyrocatechuic	3
Chelldonic acid	8	Quercetin-3-rutinoside (rutin)	1, 4
Chicorin	11	Quercetin	4
Columbamine	8	α-resocrylic	3
Coumarin	5	β-resocrylic	3
p-coumaric acid	1, 3, 6, 15	Sesamol	13
o-coumaric	3	Shogoals	5
m-coumaric	3	Syringic acid	6
Cynarine	7	Sinapic	3, 6
Dihydrocaffeic acid	1	Tannins	2, 5, 6, 17, 18
Dimethyloleuropein	1	Tannic acid	2
Esculin	11	Thymol	10
Ferulic acid	1, 3, 6	3,4,5-trimethoxybenzoic acid	1
Gallic acid	1, 2, 3, 6	3,4,5-trihydroxyphenylacetic	1
Gingerols	5	Tyrosol	1
Humulon	9	Verbascoside	1
Hydroxytyrosol	1	Vanillin	5, 12
4-hydroxybenzoic acid	1, 3, 6, 12, 16	Vanillic acid	1, 3, 6, 12

Based on Ragazzi and Veronese (1967); Vazquez Roncero *et al.* (1974); Fedeli and Camurati (1974); Fedeli (1977), Fleuriet *et al.* (1984); Amiot *et al.* (1986); Gariboldi *et al.* (1986); Robichaud and Noble (1990); Torres *et al.* (1987); Liang *et al.* (1990); Terazawa (1986); Shahidi (1992); Omar (1992); Fu *et al.* (1992); Spanos *et al.* (1990); Babic *et al.* (1993).

Tranter *et al.*, 1993). This may be due in part to the high value of the products being processed (debittering of olives, acceleration of olive fermentation) and in part to ecological problems associated with such processing. Olive waste waters, for example, have a high BOD5 and a large content of substances that inhibit the activity of microorganisms that could affect biodegradation.

The bactericidal activities of tea and coffee have been examined only recently. *Campylobacter jejuni* and *C. coli* were found to be killed by phenolic extracts of black and green tea (Diker *et al.*, 1991). Sakanaka *et al.* (1989, 1990) reported the antibacterial effect of extracts of Japanese

Table 4.4 Solvents used for the extraction of phenolic compounds from various plants

Solvent	Plant or parts used
Ethyl acetate	Green olives and waste waters from olive oil mills, tea leaves
Methanol	green olives, bark, waste water from olive oil mills, leaves, stems, tea leaves, avocado
Methanol/chloroform	olive oil, olives and leaves, tea leaves
Ethanol	green olives
Acetone	leaves
n-propanol	waste water from olive oil mills
phosphate buffered Saline	tea leaves, coffee beans

Based on Panizzi *et al.* (1960); Ragazzi and Veronese (1967); Vazquez-Roncero *et al.* (1974); Rodriguez *et al.* (1988); Sakanaka *et al.* (1989, 1990), Diker *et al.* (1991), Toda *et al.* (1989, 1991), Torres, *et al.* (1987).

Table 4.5 Microorganisms tested with phenolic antioxidants or phenolic compounds

Microorganism	BHA	BHT	PG	TBHQ	OLP	Other
Aerobacter aerogenes					+	
cloacae					+	
Aeromonas liquefaciens					+	
Alcaligenes feacalis					+	
Aspergillus flavus	+	+			+	+
parasiticus	+	+			+	
Bacillus subtilis					+	
macerans					+	
pumilus					+	
cereus					+	
megaterium					+	
Candida krusei					+	
Cellulonmmonae flavigena					+	
Clostridium botulinum	+	+	+			
Corynebacterium michiganese					+	
Debaryomyces membranefaciens					+	
Enterobacter aerogenes					+	
cloacae NRRL B-414					+	
Erwinia carotovora tracheiphila					+	
Escherichia coli	+				+	+
intermedia					+	
Hansenula subpelliculosa					+	
Kloeckera apiculata					+	
Kutrhia bessonii					+	
Lactobacillus plantarum W50					+	
brevis 50					+	
bulgaricus					+	
casei					+	
delbrueckii					+	
fermenti					+	
lactis					+	
plantarum					+	

Table 4.5 (*continued*)

Microorganism	BHA	BHT	PG	TBHQ	OLP	Other
Leuconostoc mesenteroides 42					+	
citrovorum					+	
dextranicum					+	
Listeria monocytogenes	+	+		+		+
Micrococcus spp.					+	
Paracolobactrum aerogenoides					+	
Pichia membranefaciens					+	
Pediococcus cerevisiae 39					+	
Penicillium spp.					+	+
Pseudomonas fluorescens	+				+	
solanacearum					+	
fragi	+					+
aeruginosa						+
lachrymans					+	
Rhodotorula glutinis					+	
minuta					+	
rubra					+	
Saccharomyces rosei					+	
cerevisiae var.					+	
ellipsoideus					+	
Salmonella typhimurium	+	+			+	+
enteritidis					+	
seftenberg		+				
Staphylococcus aureus	+	+	+	+	+	+
Streptococcus cremoris					+	
diacetylactis					+	
feacalis					+	
lactis					+	
liquefaciens					+	
thermophilus					+	
mutans						+
sobrinus						+
Xanthomonas vesicatoria					+	
stewartii					+	
Vibrio parahaemoliticus	+					

Based on Ayaz *et al.* (1980); Tassou (1993); Garrido-Fernandex and Vaughn (1978); Fleming *et al.* (1973); Thompson (1991); Robach and Pierson (1979); Robach *et al.* (1977); Payne *et al.* (1989); Davidson (1983); Sakanaka *et al.* (1989, 1990); Ruiz-Barba *et al.* (1990, 1991, 1993) Rico-Munoz *et al.* (1987); Paster *et al.* (1988); Davidson and Branen (1980); Person *et al.* (1980); Ruiz-Barba and Diez (1989); Stern *et al.* (1979).

green tea against the cariogenic bacterium, *Streptococcus mutans*, as well as their perturbing effects on the synthesis of the glucans responsible for bacterial cell adhesion to the tooth surface. Recently it was reported (Toda *et al.*, 1991) that black tea exhibited bactericidal activity against *Vibrio cholerae* O1. It was found also that *Escherichia coli* and *Pseudomonas aeruginosa* were the most resistant of those bacteria tested with extracts from coffee and tea. *Staphylococcus aureus*, *St. epiderimidis*, *V. cholerae* non O-1 and *Plesiomonas shigelloides* were inhibited by extracts from both tea and coffee. *Salmonella typhi*, *S. typhimurium*, *S. enteritidis*,

Shigella flexneri and *Sh. dysenteriae* were inhibited only with tea extracts from black, green or Pu-erh tea. *Aeromonas sobria* was inhibited by extracts from black tea and coffee but *A. hydrophila* by those from coffee. Extracts from coffee but not from tea (black, green and Pu-erh) inhibited *Enterobacter cloacae*. These results illustrate the variation in the inhibitory effect of the same extract against organisms belonging to the same family or even the same species.

Ethyl acetate extracts from tea inhibited the growth of *St. aureus* in a medium and in skim milk (model food system). The inhibition in broth samples was greater than that in the model food system. Enterotoxin B production was diminished in both cases also but again production in broth samples was lower than that in skim milk samples (Tassou and Nychas, unpublished observations). Kubo *et al.* (1992) showed that various non-polar polyphenolic flavour components (indole, linalool, neralidol, 1-octanol, cis-jasmone, geraniol, caryophylene etc.) from green tea inhibited a wide range of microorganisms such as *B. subtilis*, *St. aureus*, *Strep. mutans*, *Ps. auruginosa*, *Ent. aerogenes*, *Esch. coli*, *Sacch. cerevisiae*, *Candida utilis* (see Table 4.3).

Phenolic extracts from olives inhibit or delay the rate of growth of a range of bacteria and microfungi (Table 4.6) (Juven and Henis, 1970; Fleming *et al.*, 1973; Mas and Peinado, 1984; Moreno *et al.*, 1987; Rico-Munoz *et al.*, 1987; Rodriguez *et al.*, 1988; *Paster et al.*, 1988; Gourama *et al.*, 1989; Nychas *et al.*, 1990; Tassou *et al.*, 1991; Ruiz-Barba *et al.*, 1993; Tranter *et al.*, 1993). Indeed it was found that *Pseudomonas fragi*, *Lactobacillus plantarum*, *St. aureus*, *St. carnosus*, *Enterococcus faecalis*, and *Bacillus cereus* were inhibited by ethyl acetate extracts from olives (Tassou, 1993). Nychas *et al.* (1990) demonstrated that this extract inhibited protein secretion by *St. aureus*. It should be noted that there have been limited studies of the effects of phenolic extracts on food spoilage bacteria in broths or in model food systems. With food-borne pathogens the phenolics from olives, when added to a model food system (skim milk),

Table 4.6 Inhibition of lactic acid bacteria by hydrolysis products of oleuropein as indicated by paper disc bioassay

Bacteria	Oleuropein	Aglycone	Elenolic acid	Dihydroxy-phenyl-ethanol	Methyl *o*-methyl elenolate
Lactobacillus plantarum	−	+	+	−	−
Pediococcus cerevisiae	−	+	+	−	−
Lactobacillus brevis	−	+	+	−	−
Leuconostoc mesenteroides	−	+	+	−	−

+ : inhibition.
− : no inhibition.
Based on Fleming *et al.* (1973).

influenced enterotoxin B production, with the initial concentration of extract having an important bearing on the extent of inhibition (Tassou and Nychas, 1994).

Phenolic extracts of callus tissues of olive, which contains mainly caffeic acids and to a lesser extent catechin and coumarins, inhibited by 90% aflatoxin production without inhibiting the growth of *Asp. flavus* on media containing 0.5% or 1.0% (w/v) of extract (Paster *et al.*, 1988). Phenolic substances from either olives or in waste waters (alpechin) from olive oil mills exhibited antimicrobial activity on *Bacillus* spp. in soil and inhibited sporulation or spore germination of *B. megaterium* and *B. cereus* T in the laboratory (Rodriguez *et al.*, 1988; Paredes *et al.*, 1986, 1987; Moreno *et al.*, 1987; Gonzalez *et al.*, 1990; Perez *et al.*, 1992; Tassou *et al.*, 1991). These substances, acting as growth inhibitors in the wastes, are known to delay biodegradation.

Among the phenolic compounds (see Table 4.3) found in olives and related plants – e.g. *Olea* spp., *Fraxinus mandshurica* var. *japonica*, *Syringa vulgaris* – (Inouye *et al.*, 1975; Terazawa, 1986; Tassou, 1993; Tsukamoto *et al.*, 1985; Kuwajima *et al.*, 1988) it has been shown by many that oleuropein, a bitter glucoside, has antimicrobial activity (Fleming and Etchells, 1967; Juven *et al.*, 1968a,b, 1972; Ruiz-Barba *et al.*, 1990, 1991, 1993). Others have failed to demonstrate such activity (Garrido-Fernandez and Vaughn, 1978). Indeed Garrido Fernandez and Vaughn (1978) tested the effect of 0.2% and 0.4% (w/v) oleuropein in combination with 1% (w/v) glucose on microorganisms. All of the cultures of bacteria and yeasts grew in the presence of 0.2 and 0.4% (w/v) oleuropein with or without glucose. Even so there was a difference in the time required for the development of turbidity in the presence of oleuropein. The difference was least with *Bacillus* spp. The cultures of *Rhodotorula* spp. did not exhibit any response. None of the species of lactic acid bacteria grew with oleuropein as the sole carbon source. There was a variation, however, in the time required for development of dense turbidity in the media containing 0.2% (w/v) as compared with those containing 0.4% (w/v) oleuropein. There was an inhibitory effect on lactic acid bacteria but not all cultures were affected to the same extent by an increasing concentration of oleuropein. This led them to conclude that the bitter glucoside is utilized by most, if not all, of the microorganisms associated with the fermentation of olives.

Tassou (1993) found that *Bacillus subtilis*, *B. cereus* and *St. aureus* were inhibited by oleuropein whereas the growth of *Ps. fragi* was not influenced. However, when the paper disc bioassay technique was used with *Ps. fragi*, a clear zone of inhibition was evident. Using impedance measurements such inhibition was not evident (Nychas, unpublished observation). The antibacterial effect of oleuropein was not affected by either the size of inoculum of *Ps. fragi* or the composition of the media study. The use of EDTA in combination with oleuropein affected the growth of this organism significantly (Tassou and Nychas, unpublished observations). When oleuropein was tested against *Salmonella enteritidis*, it was found that the initial pH of the medium, the concentration of oleuropein and the size of the initial inoculum affected the rate of growth of this organism (Nychas, unpublished results). In particular oleuropein was not effective against *S. enteritidis* at low concentrations; at high concentrations (>0.2% w/v), detection times were delayed but the final levels of growth (expressed in microsiemens) did not differ from those of the controls. Tranter *et al.* (1993) found that the enterotoxin B production by *St. aureus* was influenced by the addition of oleuropein to broths.

It needs to be stressed that the ethyl acetate extracts from olives used in many studies, contain not only oleuropein (the phenolic glucoside), but also its hydrolysis products, β-3,4-dihydroxyphenylethyl alcohol, elenolic acid and oleuropein aglycon (Fleming *et al.*, 1973; Walter *et al.*, 1973). The last mentioned, is also bitter and it was found to be much more inhibitory than oleuropein to lactic acid bacteria. Fleming *et al.* (1973) concluded that green olives had an enzyme system which caused the hydrolysis of oleuropein to the antibacterial aglycone when olives were brined. On degradation the aglycone yields elenolic acid, a compound which is also antibacterial (Table 4.6). The inhibitory activity of aglycone and elenolic acid fractionated by HPLC on *Lact. plantarum* was shown also by Federici and Bongi (1983). As the aglycone inhibited *S. cerevisiae* and *B. subtilis*, Kubo and Matsumoto (1984a,b) and Kubo *et al.* (1985) concluded that this is the active moiety of oleuropein. Recently Ruiz Barba *et al.* (1993) confirmed that three glucosidic compounds, verbascoside, oleuropein and luteolin-7-glucoside, are present in untreated but not in the NaOH-treated olive brines. Hydroxytyrosol, tyrosol and vanillic acid were found in both brines. All the above phenolic compounds, except hydroxytyrosol, showed no bactericidal effect against *Lact. plantarum* ATCC 8014 when assayed as single fractions in the concentrations found in brines. They exhibited a remarkable additive antimicrobial effect however when tested in two-by-two combination. Hydroxytyrosol was highly bactericidal when tested alone or in combination with oleuropein or verbascoside. Gourama *et al.* (1989) reported that *Penicillium* spp. and *Aspergillus* spp. were inhibited with tyrosol.

Oleuropein, at a level of 0.2% (w/v), inhibited *Geotrichum candidum*, but not *Saccharomyces cerevisiae*, *S. oviformis*, *S. carlsbergensis*, *Candida albicans*, *C. tropicalis*, *C. krusei* and *Pichia membranefaciens* (Juven and Henis, 1970). Among the fungi tested, *Trichoderma lignorum*, *Rhizopus* sp., *Rhizoctonia solani*, *Aspergillus niger*, *Phoma* sp. and *Penicillium cyclopium*, only the growth of a *Rhizopus* sp. and *Rhizoctonia solani* was partially arrested by 0.2% (w/v) oleuropein. Oleuropein stimulated the growth of *Asp. parasiticus* and *Penicillium* spp. but inhibited the production of aflatoxin (Gourama and Bullerman, 1987; Gourama *et al.*, 1989).

There is little information relating to the antimicrobial action of the other naturally occurring phenolics, caffeic, *p*-coumaric, ferulic, gallic, tannic acids and quercetin. These phenolics were found to inhibit *Listeria monocytogenes* (Payne *et al.*, 1989). Caffeic acid was among the phenolics with the lowest activity against this organism as well as against *Penicillium* spp. and *Aspergilus* spp. (Gourama *et al.*, 1989). Phenolics from spices, such as eugenol, gingeron, zingerone and capsaicin, have been found to be sporostatic for *Bacillus subtilis* (Al-Khayat and Blank, 1985).

Among 54 plants tested for their antimicrobial activity, it was found that *Celastrus scandens* (root bark), *Juglans nigra* (fruit husks), *Kalmia latifolia* (leaves), *Pelargonium xhortorum* (leaves) and *Rhus glabra* (root bark) were the most active when *Streptococcus mutans*, *Candida albicans* and *Trichophyton rubrum* were used on the test organisms (Heisey and Gorham, 1992). Although this particular study did not give information about the nature of the inhibitory compounds, they probably contained phenolics, because the extraction solvent was methanol/dichloromethane.

4.5.1 Factors affecting antimicrobial action

Proteins, lipids, salts, pH and temperature are factors which affect the antimicrobial activity of phenolics (Tassou, 1993). For example Rico-Munoz and Davidson (1983) found that the antimicrobial effects of BHA and TBHQ were influenced by the presence of different amounts of casein and corn oil. They reported that an increase in protein content of the media influenced the inhibitory effect of BHA and TBHQ against *St. aureus*, *Pseudomonas fluorescens* and *Saccharomyces cerevisiae*. In the presence of 3% casein, BHA caused less than one log cycle decrease in the growth of *St. aureus* while the same amount of BHA together with higher concentrations of casein (6% and 9%) did not inhibit growth of this organism. Cornell *et al.* (1971) reported that BHA was bound reversibly to casein, probably through hydrophobic interactions, and that its antimicrobial action in nonfat dry milk system was thereby decreased. They found that, at low concentrations, the amount of BHA bound was a linear function of the concentration of free antioxidant.

The loss in antimicrobial activity (expressed as final viable counts) of phenolic extract from olives, against *St. aureus* was attributed to binding of the phenolics by the milk proteins (Tassou and Nychas, 1994) especially in samples with low concentrations of the inhibitor. Although the final extent of growth of *St. aureus* in these samples did not differ significantly, the amount of staphylococcal enterotoxin B produced in the samples differed significantly and was found to be influenced even by the addition of the lowest concentration of extract in the nonfat milk. Tranter *et al.* (1993) reported similar findings for staphylococcal enterotoxin B production in BHI medium. Higher concentrations of extracted compounds in milk not only inhibited significantly the final growth of this organism but also affected drastically the amount of enterotoxin B production. The effect of fats has been examined also. Rico-Munoz and Davidson (1983) observed that the effectiveness of BHA was reduced by small amounts (1.5–3.0%) of corn oil but that larger amounts (4.5%) had a smaller effect on TBHQ.

The synergistic effect of salt in the inhibitory effectiveness of phenolics is disputed. It has been reported that 100 ppm BHA inhibited the growth of *St. aureus* completely in broths containing 3.0% (w/v) salt, whereas only 50 ppm BHA gave the same level of inhibition when the salt concentration was raised to 7.0% (Robach and Stateler, 1980; Kabara, 1991). In the case of phenolics from olives, however, Fleming *et al.* (1973) reported that the combination of oleuropein and 5% salt accentuated the inhibitory effect. Other research workers have not found this synergistic effect with *Salmonella enteritidis* (Tassou, 1993) or *Lactobacillus* spp. (Ruiz-Barba *et al.*, 1993).

The effect of pH on the preservation capacity of phenolics is slightly more complex, as it depends on the microorganism tested. In studies on the effect of conidial outgrowth of *Asp. flavus*, BHA was found to be more active at pH values of 5.5 and 8.5, but growth of *St. aureus* was inhibited most markedly at neutral pH. Similar results have been obtained with oleuropein against *St. aureus* (Tassou and Nychas, 1994). Indeed they found that the commercially 'pure' oleuropein and phenolics extracted from olives inhibited the growth and enterotoxin production by *St. aureus* S-6 in broth as well as in reconstituted milk (model food system). It was found that the inhibition of this organism in N-Z Amine A broth was influenced by the initial inoculum size, the pH of a medium and the concentration of additive. In particular growth and enterotoxin B production by *St. aureus* was inhibited in broth with a high concentration of oleuropein (0.6%). The inhibition was more pronounced in samples with low pH (pH 4–5) and low inoculum size.

Temperature plays an important role in the activity of some phenolics. Indeed in the case of oleuropein it has been reported recently that this compound retains activity even after heating at 121°C/15 min (Ruiz-Barba *et al.*, 1991). The effectiveness of this compound was increased by freezing olives before extraction (Table 4.7). Freezing probably caused a release of the inhibitor due to physical changes in the olive tissue or some chemical alteration. Cruess and Alsberg (1934) contended that the olive contains an emulsin-like enzyme that is liberated by freezing and possibly another enzyme that splits or oxidizes the bitter ester remaining after hydrolysis of the glucoside. Both these two thermal characteristics would suggest its potential use in heated or frozen processed foods.

It needs to be noted that the composition of phenolics is affected by the processing and storage of the plants (Spanos *et al.*, 1990; Spanos and

Table 4.7 Antimicrobial spectrum of extract from olives by paper disc bioassay

Microorganism	Extract containing oleuropein	Acid-hydrolysate of oleuropein	Extract from frozen olives
Lactobacillus			
plantarum W50	−	+	+
brevis 50	−	+	+
Pediococcus cerevisiae 39	−	+	+
Leuconostoc mesenteroides 42	−	+	+
Staphylococcus aureus	+	+	+
Bacillus subtilis	+	+	+
Enterobacter			
aerogenes	−	−	−
cloacae NRRL B-414	−	−	−
Escherichia coli	−	−	+
Salmonella typhimurium	−	+	+
Pseudomonas			
fluorescens	−	−	+
solanacearum	+	+	+
lachrymans	−	+	+
Erwinia			
carotovora	−	+	+
tracheiphila	−	−	+
Xanthomonas vesicatoria	−	+	+
Corynebacterium michiganese	−	−	+
Saccharomyces			
rosei	−	−	−
cerevisiae var. *ellipsoideus*	−	−	−
Hansenula subpelliculosa	−	−	−
Kloeckera apiculata	−	−	−
Debaryomyces membranefaciens	−	−	−
Pichia membranefaciens	−	−	−
Candida krusei	−	−	−

+ : inhibition.
− : no inhibition.
Based on Fleming *et al.* (1973).

Wrolstad 1992; Fu *et al.*, 1992; Babic *et al.*, 1993; Tassou, 1993). Moreover, it has been reported that nonenzymatic autoxidative reaction of phenolic compounds (polyphenolic) could occur under certain conditions of storage (Cilliers and Singleton, 1990). These two characteristics should be taken into account by those who wish to study these compounds in detail.

4.6 Modes of action

Although the antimicrobial activity of phenolic compounds, phytoalexins as well as of the essential oils, is established, the mechanism of the action of these compounds on microorganisms has not been elucidated. Conner and Beuchat (1984a,b) suggested that the antimicrobial action of essential oils may be due to impairment of a variety of enzyme systems including those involved in energy production and structural component synthesis.

Spore-forming bacteria (e.g. *Bacillus*) may be more sensitive than non spore-forming ones (Moreno *et al.*, 1987; Paredes *et al.*, 1986, 1987) to phenolic compounds. It has been found that the addition of phenols to bacteria, whether sporulated or not, perturbs the cytoplasmic membrane of the microorganisms (Fogg and Lodge, 1945; Davidson and Branen, 1981; Blank *et al.*, 1987; Rico-Munoz *et al.*, 1987; Denyer and Hugo, 1991). Indeed Vas (1953) proposed that phenol attacked the cytoplasmic membrane releasing intracellular constituents. Judis (1963) found that up to 50% of intracellular Na glutamate-3,4-^{14}C and 12% of the $NaH_2{}^{32}PO_4$ were lost by *E. coli* in the presence of phenolics. It was concluded that the loss was due to a weakening or destruction of the permeability barrier of the cell membrane. In 1972 Juven *et al.*, studying the mode of action of oleuropein, concluded that it is a surface active agent affecting the cell membranes. It caused haemolysis of human erythrocytes and induced significant leakage of glutamate, potassium, and inorganic phosphate from *Lact. plantarum*. It had no effect on the rate of glycolysis, but caused a decrease in the ATP content of the cells of the latter.

Bernheim (1972) and Davidson and Branen (1980) proposed that phenolics could react with the phospholipid component of the cell membrane of *Pseudomonas aeruginosa* thereby causing an increase in the permeability of the cell membrane or could cause significant changes in the fatty acid composition and phospholipid content of these organisms. Phenol was later shown to cause rapid swelling of *P. aeruginosa* cells (Bernheim, 1974). *Pseudomonas fluorescens* and *P. fragi* were also found to leak intracellular UV-absorbing material and ^{14}C labelled compounds in the presence of butylated hydroxy-anisole, BHA. According to Hugo (1991) leakage is known to be a general phenomenon induced by many antibacterial substances.

Degre and Sylvestre (1983) found that the mode of action of BHA against *Staph. aureus* was related to the leakage of nucleotides. BHA being a highly hydrophobic molecule, the cytoplasmic membrane is likely to be the main site of adsorption since the cell wall of *Staph. aureus* contains as little as 1% or 2% lipid material. The low water solubility of many phenolics (e.g. BHA, oleuropein, etc.) probably precludes any significant diffusion into the cytoplasm. Moreover, low temperature decreases the solubility and hence the concentration of the phenol in the cell membrane lipid (Wanda *et al.*, 1976). While this is one possibility, the effect of temperature on rates of reaction is probably more important. It has been reported that no inactivation of cells occurred at 0°C with 0.02 M BHT, whereas a 1000-fold reduction in bacterial numbers was observed at 25°C

(Wanda *et al.*, 1976). Lysis of bacterial protoplasts by BHA occurred at concentrations lower than those causing leakage of nucleotides from whole cells (Degre and Sylvestre, 1983). This is an indication that the cell wall might play an important role in the relative resistance of whole cells lysis by low concentration of phenolics. Similar conclusions have been drawn by Prindle and Wright (1977) who noted that the mode of action of phenols is concentration dependent. Low concentrations affect the activity of enzymes associated with energy production whilst amounts caused a precipitation of proteins. In contrast Judis (1963) speculated as to whether or not the alleged damage caused to the cell membrane is quantitatively related to the amount of phenol derivative to which the cell is exposed or whether the effect is such that, once small damages are caused, the lesions enlarge and leakage proceeds continuously.

Tassou (1993) also suggested that phenolic compounds from olives are membrane-active agents possibly because they affect permeability. Indeed the different protein patterns of broths inoculated with *St. aureus*, as shown by either HPLC or SDS-page electrophoresis, could be due to either significant changes in the exoprotein excretion of this organism in the presence of the various concentrations of oleuropein or to protein (e.g. cytoplasmic) precipitation by oleuropein. Indeed it has been shown that phenolic antioxidants are able to bind proteins, most probably through hydrophobic interactions (Rico-Munoz *et al.*, 1987; Cornell, 1979; Cornell *et al.*, 1971). However, extracellular material, as demonstrated with HPLC, is only a measure of a generalized loss of membrane function and it is more likely that these compounds interfere with energy metabolism, synthesis of macromolecules or the cell membranes of actively growing cells. With the last mentioned view, the use of scanning electron microscopy in the studies by Ruiz-Barba *et al.* (1990) and Tassou (1993) showed that the whole cells of untreated *Lact. plantarum*, *B. cereus* and *St. aureus* were smooth compared with those treated with phenolics for 24 h, in which case the material surfaces become irregular and rough.

It is well known that BHA, BHT, *p*-coumaric and caffeic acid not only attack the cytoplasmic membrane, thereby destroying its permeability and releasing intracellular constituents, but also cause membrane dysfunction in respect of electron transport, nutrient uptake, nucleic acid synthesis and ATPase activity, etc. (Eklund, 1980; Degre and Sylvestre, 1983; Degre *et al.*, 1983; Rico-Munoz *et al.*, 1987; Ruiz-Barba *et al.*, 1990; Denyer and Hugo, 1991; Hugo, 1991; Kabara and Eklund, 1991; Eklund and Nes, 1991; Martin, 1992). In other words the bactericidal/bacteriostatic effect of phenolic compounds is shown by perturbations at two different levels: cell wall and membrane integrity as discussed above and the physiological status of bacteria (Kabara and Eklund, 1991).

Indeed, as far as it concerns the second level of physiological alteration, it was found (Tranter *et al.*, 1993) that phenolics inhibited *Staphylococcus* enterotoxin B (SEB) and lactate production as well as the rate of glucose assimilation in well-buffered media, despite little or no effect on final cell mass. The decrease in the percentage of glucose utilization as well as the reduction in the formation of L-lactate could be due to the inhibitory effect of oleuropein on substrate uptake, on specific enzyme(s) or on the electron transport chain (Fogg and Lodge, 1945; Degre and Sylvestre, 1983; Degre *et al.*, 1983; Hugo, 1991). Inhibition of growth and enterotoxin A production by *Staph. aureus* strain 100 by BHA was also noted by Ayaz *et al.* (1980).

Rico Munoz *et al.* (1987) found that the phenolic compounds from olives which they tested against *Staph. aureus* LP and A100 did not have any overall effect on the activity of the membrane-bound ATPase. They concluded therefore that phenolic compounds probably do not share a common mechanism of action and that there may not be a single target associated with the inhibition of microorganisms by these compounds. Indeed this could be the case with parabens (esters of *p*-hydroxybenzoic acid). There is clear evidence that parabens inhibit both growth and isolated uptake system for amino acids in parallel (Eklund, 1980). The ability of parabens to act on different targets in bacterial cells, as well as the variation between organisms as to relative sensitivity of these targets (Eklund and Nes, 1991) needs to be noted. As far as the activity of naturally occurring compounds is concerned, Gram positive bacteria were more sensitive, in general, to the above mentioned substances than Gram-negative ones (Tassou, 1993; Tassou and Nychas, unpublished). Even the variation in the rates of inhibition was evident also among the Gram-negative bacteria. A possible explanation for their general reduced activity towards Gram-negative bacteria could be that due to their lipophilic nature – they fail to diffuse across the outer membrane.

The activity of phenolics from olives could be decisive also for spores, since the effectiveness of germination inhibitors may depend upon their ability to permeate the spore coat and block germination-promoting sites (Cook and Pierson, 1983). In general it is well known that the principal changes occurring during endospore outgrowth are the initiation of RNA, protein and membrane syntheses (Vinter, 1970). Moreover, phenolic compounds, such as chlorocresol which have been tested against spore germination, may denature germination enzymes, inhibit the lytic enzyme subtilopeptidase A (Parker and Bradley, 1968), or interfere with the use of L-alanine or other amino acids necessary for the initiation of the germination process (Sierra, 1970).

The ability of phytoalexins and phenolics to affect many cell types could be explained if their mechanisms of action were to cause membrane perturbations thus leading to cell disfunction. This type of mode of action could be beneficial in that it would be difficult for microbes to evolve phytoalexin resistant strains (Weinstein and Albersheim, 1983). For example

Tokutake *et al.* (1992) found that phenolics inhibit the cytochrome bc1 of rat-liver mitochondria. This proposal would also explain the fact that phenolics, essential oils and phytoalexins generally cause static rather than outright toxic effects; cell membranes that leak or function poorly would not necessarily be lethal but would most probably cause a slowing of metabolic process such as cell division (Darvill and Albersheim, 1984). Indeed this is true with oleuropein, the phenolic which has been considered to be a phytoalexin precursor (Kubo *et al.*, 1985).

4.7 Health and legislative aspects

It has been reported that the use of phenolics could affect our nutrition and could be detrimental for our health. However, it was found recently in the area of nutrition and cancer, that many phenols exhibit protective effects. Prochaska and Talalay (1992) cited the work of Wattenberg who is primarily responsible for stimulating scientific interest in these commonly used food additives. He demonstrated that these compounds protected against the toxic and neoplastic effects of a surprisingly wide range of carcinogens in a number of target tissues. Thus polyphenols and their derivatives present in green and black tea have been suggested to have protective effects against cancer and heart disease so that regular intake of vegetables and fruits and the consumption of tea reduce the risk of diverse cancers in humans. It is now recognised that phenols have a protective effect against liver cirrhosis, emphysema and arteriosclerotic heart disease and on cancer (Weisburger, 1992; Newmark, 1992). It is also well known that phenols (e.g. caffeic and ferulic acid) are highly effective against nitrite ion, particularly at acid pH. This results in a major contribution of those plants phenols, commonly found in human foods, to the prevention of the nitrosation of susceptible secondary amines and amides to form potent carcinogenic nitrosamines and nitrosamides *in vitro* in foods and *in vivo*. Sources of nitrites in foods are ubiquitous, particularly in fermented or smoked foods, or are added as preservation aids in cured meat products. In addition, nitrates occurring naturally in foods are readily recycled to the saliva after ingestion and absorption, and then reduced to nitrite by the buccal flora thereby resulting in gastric nitrosation of susceptible amines. The function of dietary plant phenolics in blocking these reactions in foods, in food processing and cooking as well as *in vivo* has probably been underrated as a major cancer prevention process (Newmark, 1992). Pharmacological properties (e.g. as a hypotensive agent) have been attributed to oleuropein (Mazet, 1941; Julius-Bijlsma, 1961). Movsumov *et al.* (1987) suggested that the hypotensive effect of the extracts from olive leaves and fruits is mainly related to oleuropein.

Many other biological properties have been attributed to oleuropein and its derivatives as well as to phenolic substances in general. Molluscicidal activity ($LD_{50} = 250$ ppm) of oleuropein against *Biomphalaria glabratus* was reported by Kubo and Matsumoto (1984b). Degradation products from the hydrolysis of oleuropein inhibit oviposition of *Dacus oleae*, an insect which infests Mediterranean olive trees. This has been attributed to the formation of 3,4-dihydroxyphenylethanol and to other phenolic substances present in the juice trickling from the oviposition wounds of *Dacus oleae*. These substances deter further oviposition on olives already infested (Girolami *et al.*, 1975, 1981). Bongi (1986) noted that oleuropein inhibited flowering in *Kalanchoe blossfeldiana*, stimulated rooting in *Vigna radiata* cuttings, and irreversible stomatal closure in bioassays with *Commelina communis*. It inhibited growth but induced stomatal closure on small olive trees in pots and decreased olive callus growth appreciably.

4.8 Conclusions

As the food industry searches for naturally occurring antimicrobial compounds, further investigations into the effect of phenolic compounds such as oleuropein should be carried out. In particular, the effects of oleuropein and the other compounds *in combination with other antibacterial agents* or hurdles should be examined, in more detail. The existence naturally of phenolics in various food system should also studied. For example the type and the concentration of phenolic compounds found in the vegetable oils differed significantly (Tassou, 1993). These differences presumably reflect the extraction methods as well as the vegetables used for oil extraction (e.g. olive) (Kiritsakis and Makrakis, 1987; Tassou, 1993). These phenolics in olive oil accelerated the rate of death of *Salmonella enteritidis* in fat-rich food systems such as mayonnaise (Radford *et al.*, 1991). This could be due to greatly expanded oil/water interface which exerts a marked influence on the distribution of long-chain free fatty acids and phenolics between the aqueous and oil phases (Cornell, 1979; Collins, 1985). Thermal treatment could play an important role as well. However, it needs to be stressed that these compounds often have a severe effect on the flavour and taste of a food (Galli *et al.*, 1985).

Of major interest is that certain pharmaceutical workers are already considering the possible medical use of phenolics (phytoalexins) due to the increasing problems posed by bacterial resistance to current antibiotics (Haenen, 1985).

References

Akgul, A. and Kivanc, M. (1989) Sensitivity of four foodborne moulds to essential oils from turkish spices, herbs and citrus peel. *J. Sci. Food Agric.*, **47**, 129–132.

Aktug, S.E. and Karapinar, M. (1987) Inhibition of foodborne pathogens by thymol, eugenol, menthol and anethole. *Inter. J. Food Microbiol.*, **4**, 161–166.

Al-Khayat, M.A. and Blank, G. (1985) Phenolic spice components sporostatic to *Bacillus subtilis*. *J. Food Sci.*, **50**, 971–980.

Amiot, M.J., Fleuriet, A. and Macheix, J.J. (1986) Importance and evolution of phenolic compounds in olive during growth and maturation. *J. Agr. Food Chem.*, **34(5)**, 823–826.

Aureli, P., Constantini, A. and Zolea, S. (1992) Antimicrobial activity of some plant essential oils against *Listeria monocytogenes*. *J. Food Protect.*, **55**, 344–348.

Ayaz, M., Luedecke, L.O. and Branen, A.L. (1980) Antimicrobial effect of butylated hydroxyanisole and butylated hydroxytoluene on *Staphylococcus aureus*. *J. Food Protect.*, **43(1)**, 4–6.

Azzouz, M.A. and Bullerman, L.B. (1982) Comparative antimycotic effects of selected herbs, spices, plant components and commercial antifugal agents. *J. Food Protect.*, **45**, 1298–1301.

Babic, I., Amiot, M.J., Nguyen-The, C. and Aubert, S. (1993) Changes in phenolic content in fresh ready-to-use shredded carrots during storage. *J. Food Sci.*, **58**, 351–355.

Bailey, R.G., McDowell, I. and Nursten, H.E. (1990) Use of an HPLC photodiode-Array detector in a study of the nature of a black tea liquor. *J. Sci. Food Agric.*, **552**, 509–525.

Bernheim, F. (1972) The effect of chloroform, phenols, alcohols and cyanogen iodide on the swelling of *Pseudomonas aeruginosa* in various salts. *Microbios.*, **5**, 143.

Bernheim, F. (1974) Effect of aromatic hydrocarbons, alcohols, ketones and aliphatic alcohols on cell swelling and potassium efflux in *Pseudomonas aeruginosa*. *Cytobios.*, **11**, 9.

Beuchat, L.R. and Golden, D.A. (1989) Antimicrobials occurring naturally in foods. *Food Technol.*, **43**, 134–142.

Blank, G., Al-Khayat, M. and Ismond, M.A.H. (1987) Germination and heat resistance of *Bacillus subtilis* spores produced on clove and eugenol based media. *Food Microbiol.*, **4**, 35–42.

Bongi, G. (1986) Oleuropein: an *Olea europaea* secoiridoid biologically active on growth regulation. *Acta Horticult.*, **179**, 245–249.

Branen, A.L., Davidson, P.M. and Katz, B. (1980) Antimicrobial properties of phenolic antioxidants and lipids. *Food Technol.*, **34**, 42–53.

Brenes Balbuena, M., Garcia Garcia, P. and Garrido Fernandez, A. (1992) Phenolic compounds related to the black color formed during the processing of ripe olives. *J. Agric. Food Chem.*, **40**, 1192–1196.

Briozzo, J., Nunez, L., Chirife, J. and Herszage, L. (1989) Antimicrobial activity of clove oil dispersed in a concentrated sugar solution. *J. Appl. Bacteriol.*, **66**, 69–75.

Cilliers, J.J.L. and Singleton, V.L. (1990) Autoxidative phenolic ring opening under alkaline conditions as a model for natural polyphenols in food. *J. Agric. Food Chem.*, **38**, 1797–1798.

Collins, M.A. (1985) Effect of pH and acidulant type on the survival of some food poisoning bacteria in mayonnaise. *Microbiol. Aliments Nutr.*, **3**, 215–221.

Conner, D.E. and Beuchat, L.R. (1984a) Sensitivity of heat-stressed yeasts to essential oils of plants. *Appl. Environ. Microbiol.*, **47**, 229–233.

Conner, D.E. and Beuchat, L.R. (1984b) Effects of essential oils from plants on growth of food spoilage yeasts. *J. Food Sci.*, **49**, 429–434.

Cook, F.K. and Pierson, M.D. (1983) Inhibition of bacterial spores by antimicrobials. *Food Technol.*, **36**, 115–126.

Cornell, D.G. (1979) Distribution of some antioxidants in dairy products. *J. Dairy Sci.*, **62**, 861–868.

Cornell, D.G., Devilbiss, E.D. and Pallansch, M.J. (1971) Binding of antioxidants by milk proteins. *J. Dairy Sci.*, **54**, 634–639.

Cruess, W.V. and Alsberg, C.L. (1934) The bitter glucoside of the olive. *J. Am. Chem. Soc.*, **56**, 2115–2117.

Dabbah, R., Edwards, V.M. and Moats, W.A. (1970) Antimicrobial action of some citrus fruit oils on selected food-borne bacteria. *Appl. Microbiol.*, **19**, 27–31.

Darvill, A.G. and Albersheim, P. (1984) Phytoalexins and their elicitors: a defence against microbial infection in plants. *Ann. Rev. Plant Physiol.*, **35**, 243–275.

Davidson, M.P. (1983) Phenolic compounds. In *Antimicrobial in Foods*. Eds A.L. Branen and P.M. Davidson, pp. 37–74; Marcel Dekker, New York.

Davidson, P.M. and Branen, A.L. (1981) Antimicrobial activity of non-halogenated phenolic compounds. *J. Food Prot.*, **44(8)**, 623–632.

Davidson, P.M. and Branen, A.L. (1980) Antimicrobial mechanisms of BHA against two *Pseudomonas* species. *J. Food Sci.*, **45**, 1607–1613.

Deans, S.G. and Ritchie, G. (1987) Antibacterial properties of plant essential oils. *Int. J. Food Microbiol.*, **5**, 165–180.

Degre, R. and Sylvestre, M. (1983) Effect of butylated hydroxyanisole on the cytoplasmic membrane of *Staphylococcus aureus* Wood 46. *J. Food Prot.*, **46(3)**, 206–209.

Degre, R., Ishaque, M. and Sylvestre, M. (1983) Effect of butylated hydroxyanisole on the electron transport system of *Staphylococcus aureus* Wood 46. *Microbios.*, **37**, 7–13.

De Medici, D., Pieretti, S. and Salvatore, G. (1992) Chemical analysis of essential oils of malagasy medicinal plants by gas chromatography and nmr spectroscopy. *Flavour and Fragrance J.*, **7**, 275–281.

Denyer, S.P. and Hugo, W.B. (1991) Biocide induced damage to the bacterial cytoplasmic membrane. In *Mechanisms of Action of Chemical Biocides; Their Study and Exploitation*. Eds Denyer, S.P. and Hugo, W.B. The Society for Applied Bacteriology, Technical Series No. 27. Oxford, Blackwell Scientific Publications.

DeWit, J.C., Notermans, S., Gorin, N. and Kampelmacher, E.H. (1979) Effect of garlicoil or onion oil on toxin production by *Clostridium botulinum* in meat slurry. *J. Food Protect.*, **42**, 222–224.

Diker, K.S., Akan, M., Hascelik, G. and Yaradkok, M. (1991) The bactericidal activity of tea against *Campylobacter jejuni* and *Campylobacter coli*. *Ltr Appl. Microbiol.*, **12**, 34–35.

Dixon, R.A., Dey, P.M. and Lamb, C.-J. (1983) Phytoalexins: enzymology and molecular biology. *Adv. Enzym.*, **55**, 1–136.

Doores, S. (1983) Organic acids. In *Antimicrobials in Food*, Eds A.L. Branen and P.M. Davidson, Marcel Dekker, New York, pp. 75–108.

Dziezak, J.D. (1989) Spices. *Food Technol.*, **43**, 102–116.

Eklund, T. (1980) Inhibition of growth and uptake processes in bacteria by some chemical food preservatives. *J. Appl. Bacteriol.*, **48**, 423–432.

Eklund, T. (1989) In *Mechanisms of Action of Food Preservation Procedures*, Ed G.W. Gould, Elsevier, London.

Eklund, T. and Nes, I.F. (1991) Effect of biocides on DNA, RNA and Protein synthesis. In *Mechanisms of Action of Chemical Biocides, Their Study and Exploitation*, pp. 225–234. Eds Denyer, S.P. and Hugo, W.B. Technical Series No. 27, Blackwell Scientific Publications, Oxford.

Elnima, N.I., Syed, A.A., Mekkawi, A.G. and Mossa, J.S. (1983) Antimicrobial activity of garlic and onion extracts. *Pharmazie*, **38**, 743–748.

Farag, R.S., Daw, Z.Y., Hewedi, F.M. and El-Baroty, G.S.A. (1989) Antimicrobial activity of some egyptian spice essential oils. *J. Food Protect.*, **52**, 665–667.

Farell, K.T. (1985) *Spices, Condiments and Seasonings*. AVI, Westport, Conn.

Fedeli, E. (1977) Lipids of olives. *Prog. Chem. Fats and other Lipids*, **15**, 57.

Fedeli, E. and Camurati, F. (1974) *12th World Congress, ISF*, Paper 75, Milan, Sept. 2–7.

Federici, F. and Bongi, G. (1983) Improved method for isolation of bacterial inhibitors from oleuropein hydrolysis. *Appl. Environ. Microbiol.*, **46(3)**, 509–510.

Fisher, C. (1992) Phenolic compounds in spices. In *Phenolic Compounds in Food and their Effect on Health I, Analysis, Occurrence, and Chemistry* (eds Chi-Tang Ho, Chang Y. Lee, Mou-Tuan Huang). ACS symposium (No. 506) Series, Washington, DC.

Fleming, H.P. and Etchells, J.L. (1967) Occurrence of an inhibitor of lactic acid bacteria in green olives. *Appl. Microbiol.*, **15**, 1178–1184.

Fleming, H.P., Walter, W.M. and Etchells, J.L. (1973) Antimicrobial properties of oleuropein and products of its hydrolysis from green olives. *Appl. Microbiol.*, **26**, 777–782.

Fleuriet, A., Macheix, J.J., Andary, C. and Villemur, P. (1984) Identification and quantita-

tive determination of verbascoside by HPLC in the fruit of six cultivars of *Olea europaea* L. *C.R. Acad. Sci. Ser. III*, **299**, 253–256.
Fogg, A.H. and Lodge, R.M. (1945) The mode of antibacterial action of phenols in relation to drug-fastness. *Trans. Faraday Soc.*, **41**, 359–365.
Friend, J. (1979) Phenolic substances and plant disease. In *Biochemistry of Plant Phenolics*, pp. 557–588, Eds Swain, T., Harborne, J.B. and Van Sumere, C.F., New York, Plenum Press.
Fu, H-Y., Huang, T-C. and Ho, C-T. (1992) Changes in phenolic compounds during plum processing. In *Phenolic Compounds in Food and their Effect on Health I, Analysis, Occurrence, and Chemistry* (eds Chi-Tang Ho, Chang Y. Lee, Mou-Tuan Huang) pp. 287–295. ACS symposium (No. 506) Series, Washington, DC.
Galli, A., Franzetti, L. and Briguglio, D. (1985) Antimicrobial properties *in vitro* of essential oils and extract of spices used for food. *Industr. Aliment.*, **24**, 463–466.
Gariboldi, P., Jommi, G. and Verotta, L. (1986) Secoiridoids from *Olea europaea*. *Phytochem.*, **25(4)**, 865–869.
Garrido-Fernandez, A. and Vaughn, R.H. (1978) Utilization of oleuropein by microorganisms associated with olive fermentations. *Can. J. Microbiol.*, **24**, 680–684.
Girolami, V., Pellizzari, G., Ragazzi, E. and Veronese, G. (1975) Prospects of increased egg production in the rearing of *Dacus oleae* Gmelin by the use of chemical stimuli. Meeting on Sterility Principle for Insect Control, 1974, Vienna, pp. 209–217.
Girolami, V., Vianello, A., Strapazzon, A., Ragazzi, E. and Veronese, G. (1981) Ovipositional deterrents in *Dacus oleae*. *Ent. Exp. Appl.*, **29**, 177–188.
Gonul, S.E. and Karapinar, M. (1987) Inhibitory effect of linden flower (Tilia flower) on the growth of foodborne pathogens. *Food Microbiol.*, **4**, 97–100.
Gonzalez, M.D., Moreno, E., Quevedo Sarmiento, J. and Ramos Cormezana, A. (1990) Studies on antibacterial activity of waste waters from olive oil mills: Inhibitory activity of phenolic and fatty acids. *Chemosphere*, **20(3/4)**, 423–432.
Gourama, H. and Bullerman, L.B. (1987) Effect of oleuropein on growth and aflatoxin production by *Aspergillus parasiticus*. *Lebensm.-Wiss.u.-Technol.*, **20**, 226–228.
Gourama, H., Letutour, B., Tantaoui-Elaraki, A., Benbya, M. and Bullerman, L.B. (1989) Effects of oleuropein, tyrosol and caffeic acid on the growth of mold isolated from olives. *J. Food Protect.*, **52(4)**, 264–266.
Gutierrez, F., Albi, M.A., Palma, R., Rios, J.J. and Olias, J.M. (1989) Bitter taste of virgin olive oil: Correlation of sensory evaluation and instrumental HPLC analysis. *J. Food Sci.*, **54(1)**, 68–70.
Haenen (1985) Phytoalexins: antibiotic substances from higher plants. *Pharm. Int.*, **Aug.**, 194–196.
Hagerman, A.E. (1992) Tannin–Protein Interactions. In *Phenolic Compounds in Food and their Effect on Health I, Analysis, Occurrence, and Chemistry*, pp. 236–247 (eds Chi-Tang Ho, Chang Y. Lee, Mou-Tuan Huang). ACS symposium Series No. 506, Washington, DC.
Hall, M.A. and Maurer, A.J. (1986) Spice extracts and propylene Glycols as inhibitors of *Clostridium botulinum* in turkey frankfurter slurries. *Poult. Sci.*, **65**, 1167–1171.
Harborne, J.B. (1980) Plant phenolics. In *Encyclopedia of Plant Physiology*. New series V.8 (eds E.A. Bell and B.V. Chartwood), pp. 329–402.
Hargreaves, L.L., Jarvis, B., Rawlinson, A.P. and Wood, J.M. (1975) *The Antimicrobial Effects of Spices, Herbs and Extracts from these and Other Food Plants*. The British Food Manufacturing Industries Research Association Scientific and Technical Surveys No. 88.
Heisy, R.M. and Gorham, B.K. (1992) Antimicrobial effects of plant extracts on *Streptococcus mutans*, *Candida albicans*, *Trichophyton subrum* and other microorganisms. *Ltr Appl. Microbiol.*, **14**, 136–139.
Ho, C-T. (1992) Phenolic compounds in Food. In *Phenolic Compounds in Food and their Effect on Health I, Analysis, Occurrence, and Chemistry* (eds Chi-Tang Ho, Chang Y. Lee, Mou-Tuan Huang). ACS symposium Series, Washington, DC.
Hugo, W.B. (1991) A review: A brief history of heat and chemical preservation and disinfection. *J. Appl. Bacteriol.*, **71**, 9–18.
Inouye, H., Nishioka, T. and Kaniwa, M. (1975) Glucosides of *Fraxinus japonica*. *Phytochemistry*, **4**, 304.
Ismaiel, A.A. and Pierson, M.D. (1990) Inhibition of germination, outgrowth and vegetative growth of *Clostridium botulinum* 67B by spice oils. *J. Food Protect.*, **53**, 755–758.

Judis, J. (1963) Studies on the mechanism of action of phenolic disinfectants II. *J. Pharm. Sci.*, **52(2)**, 126–131.
Julius-Bijlsma, J.A. (1961) A review of the investigations of hypotensive substances from the leaves of the olive tree, *Olea europaea* L. *Pharm. Weekblad*, **96**, 417–434.
Juven, B. and Henis, Y. (1970) Studies on the antimicrobial activity of olive phenolic compounds. *J. App. Bacteriol.*, **33**, 721–732.
Juven, B., Samish, Z. and Henis, Y. (1968a) Identification of oleuropein as a natural inhibitor of lactic fermentation of green olives. *Isr. J. Agric. Res.*, **18**, 137–138.
Juven, B., Samish, Z., Henis, Y. and Jacoby, B. (1968b) Mechanism of enhancement of lactic acid fermentation of green olives by alkali and heat treatments. *J. Appl. Bacteriol.*, **31**, 200–207.
Juven, B., Henis, Y. and Jacoby, B. (1972) Studies on the mechanism of antimicrobial action of oleuropein. *J. Appl. Bacteriol.*, **35**, 559–567.
Kabara, J.J. (1991) Phenols and chelators. In *Food Preservatives*, Eds Russell, N.J. and Gould, G.W., pp. 200–214. Blackie, Glasgow.
Kabara, J.J. and Eklund, T. (1991) Organic acids and esters. In *Food Preservatives*, Eds Russell, N.J. and Gould, G.W., pp. 44–71. Blackie, Glasgow.
Karaioannoglou, P.G., Mantis, A.J. and Panetsos, A.G. (1977) The effect of garlic extract on lactic acid bacteria (*Lactobacillus plantarum*) in culture media. *Lebensm.-Wiss.u.-Technol.*, **10**, 148–150.
Karapinar, M. (1985) The effect of citrus oils and some spices on growth and aflatoxin production by *Aspergillus parasiticus* NRRL 2999. *Inter. J. Food Microbiol.*, **2**, 239–245.
Katayama, T. and Nagai, I. (1960) Chemical significance of the volatile components of spices in the food preservative viewpoint. VI. Structure and antibacterial activity of terpenes. *Bull. Japanese Soc. Scient. Fisher.*, **26**, 29–32.
Khurana, A.L. (1992) High-Performance Liquid chromatographic analysis of phenolic compounds in foods. In *Phenolic Compounds in Food and their Effect on Health I, Analysis, Occurrence, and Chemistry* (eds Chi-Tang Ho, Chang Y. Lee, Mou-Tuan Huang), pp. 77–84, ACS symposium Series, Washington, DC.
Kiritsakis, A. and Markakis, P. (1987) Olive oil: A review. In *Advances in Food Research* Vol. 31, eds Chichester, C.O., Mrak, E.M. and Schweigert, B.S. Academic Press.
Kivanc, M. and Akgul, A. (1990) Mould growth on black table olives and prevention by sorbic acid, methyl-eugenol and spice essential oil. *Die Nahrung*, **34**, 369–373.
Kubo, I. and Matsumoto, A. (1984a) Secreted oleanolic acid on the cuticle of *Olea europaea*; a chemical barrier to fungus attack. *Experientia*, **40**, 937–938.
Kubo, I. and Matsumoto, A. (1984b) Molluscicides from *Olea europaea* and their efficient isolation by countercurrent chromatographies. *J. Agric. Food Chem.*, **32**, 687–688.
Kubo, I., Matsumoto, A. and Takase, I. (1985) A multichemical defense mechanism of bitter olive *Olea europaea* (Oleaceae). Is oleuropein a phytoalexin precursor? *J. Chem. Ecol.*, **11(2)**, 251–263.
Kubo, I., Muroi, H. and Himejima, M. (1992) Antimicrobial activity of green tea flavour components and their combination effects. *J. Agric. Food Chem.*, **40**, 245–248.
Kubo, I. and Himejima, M. (1991) Anethole, a synergist of polygodial against filamentous microorganisms. *J. Agric. Food Chem.*, **39**, 2290–2292.
Kuwajima, H., Uemura, T., Takaishi, K., Inoue, K. and Inouye, H. (1988) A secoiridoid glucoside from *Olea europaea*. *Phytochemistry*, **27(6)**, 1757–1759.
Liang, Y.R., Liu, Z.S., Xu, Y.R. and Hu, Y.L. (1990) A study on chemical composition of two special green teas (*Camellia sinensis*). *J. Sci. Food Agric.*, **53**, 541–548.
Llewellyn, G.C., Burkett, M.L. and Eadie, T. (1981) Potential mold growth, aflatoxin production and antimycotic activity of selected natural spices and herbs. *J. Assoc. Off. Anal. Chem.*, **64**, 955–960.
Mansfield, J.W. (1986) Induced antimicrobial systems in plants. In *Natural Antimicrobial Systems*. Eds G.W. Gould, M.E. Rhodes-Roberts, A.K. Charnley, R.M. Cooper and R.G. Board. Bath University Press, Bath, pp. 191–205.
Mantis, A.J., Karaioannoglou, P.G., Spanos. G.P. and Panetsos, A.G. (1978) The effect of garlic extract on Food Poisoning bacteria in culture media I. *Staphylococcus aureus*. *Lebensm.-Wiss.u.-Technol.*, **11**, 26–29.
Mantis, A.J., Koidis, P.A., Karaioannoglou, P.G. and Panetsos, A.G. (1979) Effect of garlic extract on food poisoning bacteria. *Lebensm.-Wiss.u.-Technol.*, **12**, 330–332.

Martin, S.A. (1992) Effects of extracellular pH and phenolic monomers on glucose uptake by *Fibrobacter succinogenes* S85. *Ltr Appl. Microbiol.*, **15**, 26–28.
Mas, M. and Peinado, J.M. (1984) Phenol effect on the growth of yeasts isolated from 'alpechin'. *Cienc. Biolog.*, **9**, 205–209.
Mazet, G. (1941) Gaz. Med. de France 1 Jan. 1938, *door Rev. Phytotherap.*, **5**, 27.
Mitcher, L.A. (1975) Antimicrobial agents from higher plants. In *Recent Advances in Phytochemistry*, Vol. 9. Ed. V.C. Runeckles. Plenum Press, London, pp. 243–282.
Moreno, E., Perez, J., Ramos Cormezana, A. and Martinez, J. (1987) Antimicrobial effect of waste waters from olive oil extraction plants selecting soil bacteria by incubation with diluted wastes. *Microbios.*, **51**, 169–174.
Movsumov, I.S., Aliev, A.M. and Tagieva, Z.D. (1987) Pharmacochemical studies of the *Olea* in the Azerbaijan S.S.R. *Farmatsiya Mosc.*, **36(2)**, 32–34.
Nagy, J.G. and Tengerdy, R.P. (1967) Antibacterial action of essential oils of Artemisis as an ecological factor. *Appl. Microbiol.*, **15**, 819–821.
Newmark, H.L. (1992) Plant phenolic compounds as inhibitors of mutagenesis and carcinogenesis. In *Phenolic Compounds in Food and their Effect on Health II, Antioxidants and Cancer Prevention* (eds Mou-Than Huang, Chi-Tang Ho, Chang Y. Lee), pp. 48–52. ACS symposium Series No. 507, Washington, DC.
Nychas, G.J.E., Tassou, C.C. and Board, R.G. (1990) Phenolic extract from olives: inhibition of *Staphylococcus aureus*. *Ltr Appl. Microbiol.*, **10**, 217–220.
Okuda, Y., Yoshida, T. and Hatano, T. (1992) Polyphenols from Asian plants. In *Phenolic Compounds in Food and their Effect on Health II, Antioxidants and Cancer Prevention* (eds Mou-Than Huang, Chi-Tang Ho, Chang Y. Lee), pp. 160–183. ACS symposium Series No. 507, Washington, DC.
Omar, M.M. (1992) Phenolic compounds in botanical extracts used in foods, flavors, cosmetics and pharmaceuticals. In *Phenolic Compounds in Food and their Effect on Health I, Analysis, Occurrence, and Chemistry* (eds Chi-Tang Ho, Chang Y. Lee, Mou-Tuan Huang). ACS symposium Series, Washington, DC.
Panizzi, L.M., Scarpati, J.M. and Oriente, E.G (1960) Constituzione della oleuropeina, glucoside amaro e ad azione ipotensiva dell'olivo. *Nota II Gazz. Chim. Ital.*, **90**, 1449–1485.
Paredes, M.J., Moreno, E., Ramos-Cormezana, A. and Martinez, J. (1987) Characteristics of soil after pollution with waste waters from olive oil extraction plants. *Chemosphere*, **16**, 1557–1564.
Paredes, M.J., Monteoliva Sanchez, M., Moreno, E., Perez, J., Ramos Cormezana, A. and Martinez, J. (1986) Effect of wastewaters from olive oil extraction plants on the bacterial population of soil. *Chemosphere*, **15**, 659–664.
Parker, M.S. and Bradley, T.J. (1968) A reversible inhibition of the germination of bacterial spores. *Can. J. Microbiol.*, **14**, 745–746.
Paster, N., Juven, B.J. and Harshemesh, H. (1988) Antimicrobial activity and inhibition of aflatoxin B1 formation by olive plant tissue constituents. *J. Appl. Bacteriol.*, **64**, 293–297.
Paster, N., Juven, B.J., Shaaya, E., Menasherov, M., Nitzan, R., Weisslowicz, H. and Ravid, U. (1990) Inhibitory effect of oregano and thyme essential oils on moulds and foodborne bacteria. *Ltr Appl. Microbiol.*, **11**, 33–37.
Payne, K., Rico Munoz, E. and Davidson, P.M. (1989) The antimicrobial activity of phenolic compounds against *Listeria monocytogenes* and their effectiveness in a model milk system. *J. Food Protect.*, **52(3)**, 151–153.
Perez, J., De la Rubia, T., Moreno, J. and Martinez, J. (1992) Phenolic content and antibacterial activity of olive oil waste waters. *Envir. Toxicol. Chem.*, **11**, 489–495.
Pierson, M.D., Smoot, L.A. and Vantassell, K.R. (1980) Inhibition of *Salmonella typhimurium* and *Staphylococcus aureus* by butylated hydroxyanisole and the propyl ester of *p*-hydroxybenzoic acid. *J. Food Protect.*, **43(3)**, 191–194.
Prindle, R.F. and Wright, E.S. (1977) Phenolic compounds. In *Disinfection, Sterilisation and Preservation*, ed. Block, S.S., pp. 219–251. Philadelphia: Lea and Febiger.
Prochaska, H.J. and Talalay, P. (1992) Phenolic antioxidants as inducers of anticarcinogenic enzymes. In *Phenolic Compounds in Food and their Effect on Health I, Analysis, Occurrence, and Chemistry* (eds Chi-Tang Ho, Chang Y. Lee, Mou-Tuan Huang). ACS symposium Series, Washington, DC.
Radford, S.A., Tassou, C.C., Nychas, G.J.E. and Board, R.G. (1991) The effect of different

oils on the death rate of *Salmonella enteritidis* in homemade mayonnaise. *Ltr Appl. Microbiol.*, **12**, 125–128.
Ragazzi, E. and Veronese, G. (1967) Richerche sui constituenti idrosolubili delle olive. Nota I. Zuccheri e fenoli. *Ann. Chim.*, **57**, 1396–1397.
Rico Munoz, E. and Davidson, P.M. (1983) Effect of corn oil and casein on the antimicrobial activity of phenolic antioxidants. *J. Food Sci.*, **48**, 1284–1288.
Rico Munoz, E., Bargiota, E. and Davidson, P.M. (1987) Effect of selected phenolic compounds on the membrane-bound adenosine triphosphate of *Staphylococcus aureus*. *Food Microbiol.*, **4**, 239–249.
Robach, M.C., Smoot, L.A. and Pierson, M.D. (1977) Inhibition of *Vibrio parahaemoliticus* O4:K11 by BHA. *J. Food Protect.*, **40**, 549–551.
Robach, M.C. and Pierson, M.D. (1979) Inhibition of *Clostrodium botulinum* Types A and B by phenolic antioxidants. *J. Food Protect.*, **42**, 858–861.
Robach, M.C. and Stateler, C.L. (1980) Inhibition of *Staphylococcus aureus* by potassium sorbate in combination with sodium chloride, tertiary butylhydroquinone, butylated hydroxyanisole or EDTA. *J. Food Protect.*, **43**, 208–211.
Robishaud, J.L. and Noble, A.C. (1990) Astrigency and bitterness of selected phenolics in wine. *J. Sci. Food Agric.*, **53**, 343–353.
Rodriguez, M.M., Perez, J., Ramos Cormenzana, A. and Martinez, J. (1988) Effect of extracts obtained from olive oil mill wastewaters on *Bacillus megaterium* ATCC 33085. *J. Appl. Bacteriol.*, **64**, 219–226.
Ruiz-Barba, J.L. and Jimenez-Diaz, R. (1989) Actividad antibacteriana de las salmueras de aceitunas verdes sobre *Lactobacillus plantarum*. *Grasas y Aceites*, **40**, 102–108.
Ruiz-Barba, J.L., Garrido-Fernandez, A.G. and Jimenez-Diaz, R. (1991) Bactericidal action of oleuropein extracted from green olives against *Lactobacillus plantarum*. *Ltr Appl. Microbiol.*, **12**, 65–68.
Ruiz Barba, J.L., Brenes Balbuena, R., Jimenez Diaz, R., Garcia Garcia, P. and Garrido Fernandez, A. (1993) Inhibition of *Lactobacillus plantarum* by polyphenols extracted from two different kinds of olive brine. *J. Appl. Bacteriol.*, **74**, 15–19.
Ruiz-Barba, J.L., Rios-Sanchez, R.M., Fedriani-Triso, C., Olias, J.M., Rios, J.L. and Jimenez-Diaz, R. (1990) Bactericidal effect of phenolic compounds from green olives against *L. plantarum*. *System. Appl. Microbiol.*, **13**, 199–205.
Sakanaka, S., Kim, M., Taniguchi, M. and Yamamoto, T. (1989) Antibacterial substances in Japanese green tea extract against *Streptococcus mutans*, a Cariogenic Bacterium. *Agric. Biol. Chem.*, **53**, 2307–2311.
Sakanaka, S., Sato, T., Kim, M. and Yamamoto, T. (1990) Inhibitory effects of green tea polyphenols on glucan synthesis and cellular adherence of cariogenic Streptococci. *Agric. Biol. Chem.*, **54**, 2925–2929.
Saleem, Z.M. and Al-Delaimy, K.S. (1982) Inhibition of Bacillus ceresu by garlic extracts. *J. Food Protect.*, **45**, 1007–1010.
Salmeron, J., Jordano, R. and Pozo, R. (1990) Antimycotic and antiaflatoxigenic activity of oregano (*Origanum vulgare*, L.) and Thyme (*Thymus vulgaris*, L.). *J. Food Protect.*, **53**, 697–700.
Sankaran, R. (1976) Comparative antimicrobial action of certain antioxidants and preservatives. *J. Food Sci. Technol.*, **13**, 203–204.
Shahidi, F. (1992) Phenolic compounds of Brassica oilseeds. In *Phenolic Compounds in Food and their Effect on Health I, Analysis, Occurrence, and Chemistry*, pp. 130–142 (eds Chi-Tang Ho, Chang Y. Lee, Mou-Tuan Huang). ACS symposium Series No. 506, Washington, DC.
Sharma, A.A., Tewari, G.M., Shrikhande, A.J., Padwal-Desai, S.R. and Bandyopadhyay, C. (1979) Inhibition of aflatoxin-producing fungi by onion extracts. *J. Food Sci.*, **44**, 1545–1547.
Shelef, L.A. (1983) Antimicrobial effects of spices. *J. Food Safety*, **6**, 29–44.
Shelef, L.A., Naglik, O.A., Bogen, D.W. (1980) Sensitivity of some common food-borne bacteria to the spices sage, rosemary and allspice. *J. Food Sci.*, **45**, 1042–1044.
Sierra, G. (1970) Inhibition of the amino acid induced initiation of germination of bacterial spores by chlorocresol. *Can. J. Microbiol.*, **16**, 51–52.
Spanos, G.A. and Wrolstad, R.E. (1992) Phenolics of apple, pear, and white grape juices and their changes with processing and storage. *J. Agric. Food Chem.*, **40**, 1478–1487.

Spanos, G.A., Wrolstad, R.E. and Heatherbell, D.A. (1990) Influence of processing and storage on the phenolic composition of apple juice. *J. Agric. Food Chem.*, **38**, 1572–1579.
Stecchini, M.L., Sarais, I. and Giavedoni, P. (1993) Effect of essential oils on *Aeromonas hydrophila* in a culture medium and in cooked pork. *J. Food Protect.*, **56**, 406–409.
Stern, N.J., Smoot, L.A. and Pierson, M.D. (1979) Inhibition of *Staphylococcus aureus* growth by combinations of butylated hydroxyanisole, sodium chloride and pH. *J. Food Sci.*, **44**, 710–712.
Tassou, C.C. (1993) *Microbiology of olives with emphasis on the antimicrobial activity of phenolic compounds*. Ph.D Thesis, Univ. of Bath, Bath, UK.
Tassou, C.C. and Nychas, G.J.E. (1994) Inhibition of *Staphylococcus aureus* by olive phenolics in broth and in a model food system. *J. Food Protect.*, **56**.
Tassou, C.C., Nychas, G.J.E. and Board, R.G. (1991) Effect of phenolic compounds and oleuropein on germination of *Bacillus cereus* T spores. *Biotech. Appl. Biochem.*, **13**, 231–237.
Terazawa, M. (1986) Phenolic compounds in living tissues of woods. VI. Ligstroside and oleuropein in *Fraxinus mandshurica var. japonica and Syringa vulgaris* L. and their seasonal variation in them. *Res. Bull. of Coll. Exper. Forests*, **43**, 109–126.
Thompson, D.P. (1991) Effect of butylated hydroxyanisole on conidial germination of toxigenic species of *Aspergillus flavus* and *Aspergillus parasiticus*. *J. Food Protect.*, **54**, 375 –377.
Ting, E.W.T. and Deibel, K.E. (1992) Sensitivity of *Listeria monocytogenes* to spices at two temperatures. *J. Food Safety*, **12**, 129–137.
Toda, M., Okubo, S., Hiyoshi, R. and Shimamura, T. (1989) The bactericidal activity of tea and coffee. *Ltr Appl. Microbiol.*, **8**, 123–125.
Toda, M., Okubo, S., Ikigai, H., Suzuki, T., Suzuki, Y. and Shimamura, T. (1991) The protective activity of tea against infection by *Vibrio cholerae* O1. *J. Appl. Bacteriol.*, **70**, 109–112.
Tokutake, N., Miyoshi, H. and Iwamura, H. (1992) Effects of phenolic respiration inhibitors on cytochrome bc1 complex of rat-liver mitochondria. *Biosc. Biotech. Biochem.*, **56**, 919–923.
Torres, A.M., Mau-Lastovicka, T. and Rezaaiyan, R. (1987) Total phenolics and HPLC of phenolic acids of avocado. *J. Agric. Food Chem.*, **35**, 921–925.
Tranter, H.S., Tassou, C.C. and Nychas, G.J. (1993) The effect of the olive phenolic compound, oleuropein, on growth and enterotoxin B production by *Staphylococcus aureus*. *J. Appl. Bacteriol.*, **74**, 253–260.
Tsukamoto, H., Hisada, S. and Nishibe, S. (1985) Coumarin and secoiridoid glucosides from bark of *Olea africana* and *Olea capensis*. *Chem. Pharm. Bull.*, **33**, 396–399.
Vas, K. (1953) Mechanism of antimicrobial action. Interference with the cytoplasmic membrane. *Agrokemia es Talajtan*, **2**, 1–16 (*Chem. Abstr.*, 1954, **48**, 794d).
Vazquez Roncero, A., Graciani-Constante, E. and Maestro-Duran, R. (1974) Phenolic compounds in olive fruits. I. Poliphenols in pulpe. *Grasas y Aceites*, **25**, 269–279.
Vinter, V. (1970) Germination and outgrowth: Effect of inhibitors. *J. Appl. Bacteriol.*, **33**, 50–59.
Walter, W.M., Fleming, H.P. and Etchells, J.L. (1973) Preparation of antimicrobial compounds by hydrolysis of oleuropein from green olives. *Appl. Microbiol.*, **26**, 773–776.
Wanda, P., Cupp, J., Snipes, W., Keith, A., Rucinsky, T., Polish, L. and Sands, J. (1976) Inactivation of the enveloped bacteriophage o6 by butylated hydro-xytoluene and butylated hydroxyanisole. *Antimicrob. Agents Chemother.*, **10**, 96.
Weinstein, L.I. and Albersheirm, P. (1983) Host pathogen interactions. XXIII. The mechanism of antimicrobial action of glycinol, a plerocarpan phytoalexin synthesized by soybeans. *Plant Physiol.*, **72**, 557–560.
Weisburger, J.H. (1992) Mutagenic, Carcinogenic and Chemopreventine effects of phenols and catechols. In *Phenolic Compounds in Food and their Effect on Health II, Antioxidants and Cancer Prevention* (eds Mou-Than Huang, Chi-Tang Ho, Chang Y. Lee), pp. 35–48. ACS symposium Series No. 507, Washington, DC.
Wilkins, K.M. and Board, R.G. (1989) Natural antimicrobial systems. In *Mechanisms of Action and Food Preservation Procedures*, ed. G.W. Gould, pp. 285–362. Elsevier, London.

5 Food irradiation: current status and future prospects

P. LOAHARANU

5.1 Introduction

After five decades of research and development, proof of wholesomeness, political debates and public scrutiny, food irradiation is established as a safe and effective method of food processing and preservation. Similar to other food preservation methods, food irradiation offers advantages and disadvantages relative to the types of food to be treated. It is increasingly recognized as an alternative method to fumigation of food to control insect infestation and microbial contamination, as well as facilitating trade in fresh and dried foods. Its role as a method to ensure hygienic quality of the more solid foods, e.g. meat, poultry and seafood, is increasingly advocated and used in some countries. Its introduction into commercial practice has been somewhat slow because many governments required comprehensive data to support the wholesomeness of irradiated food, lengthy regulatory procedures in granting approval for the food or the process, often with the involvement of the public. In some cases, the public involvement led to debate all the way to the parliament. This review will attempt to examine what has happened to this well studied technology, the status of its application and, more importantly, its future in this constantly changing world.

5.2 Development of national regulations

After the establishment of the wholesomeness of irradiated food through the recommendations of the Joint FAO/IAEA/WHO Expert Committee on the Wholesomeness of Irradiated Food (JECFI), which was convened from time to time since the early 1960s, it is widely accepted that food subjected to ionizing radiation treatment under properly controlled conditions is safe for consumption. The last JECFI convened in 1980 made a fundamental recommendation on this aspect by stating that "Irradiation of any food commodity up to an overall average dose of 10 kGy causes no toxicological hazard; hence, toxicological testing of food so treated is no longer required" (WHO, 1981). The JECFI also stated that irradiation of food up to a dose of 10 kGy introduced no special microbiological and nutritional problems. In subsequent years, a number of national and regional authorities convened their own expert committees independently to evaluate these and other data on wholesomeness of irradiated food. All these committees arrived at similar conclusions as the JECFI. The most recent Expert Consultation convened by the WHO in 1992 to review all data on wholesomeness of food irradiated with doses up to 10 kGy also reaffirmed this conclusion (WHO, 1994).

The conclusions and recommendations of the 1980 JECFI were elaborated into an international standard under the procedures of the Codex Alimentarius Commission (CAC) of the Joint FAO/WHO Food Standard Programme. The CAC has its primary objectives in protecting consumer health and ensuring fair practice in the food trade. The procedures of the CAC require several rounds of governments' comments. In 1983, the CAC represented by some 130 governments, decided to adopt a Codex General Standard for Irradiated Foods and a Recommended International Code of Practice for Operation of Radiation Facilities Used for Treatment of Food (FAO, 1992). The Codex Standard and its associated Code of Practice provide important principles for proper irradiation of food (up to an overall average dose of 10 kGy) and essential control procedures. The Codex Standard was recommended by the CAC to all its member countries for acceptance in 1984.

The Codex Standard provided an important incentive for national authorities in introducing regulations on food irradiation. In fact, starting from the early 1980s several advanced countries including Canada, Denmark, France, the Netherlands, UK, USA, etc. and many developing countries including Bangladesh, Brazil, Chile, China, India, Republic of Korea, Mexico, Thailand, etc. have introduced regulations on food irradiation following the principles of the Codex Standard. In doing so, several countries invited public comments prior to finalizing their national regulations. Several other countries, while they have not yet introduced regulations on food irradiation, have approved the production and sale of several irradiated food items. Collectively, 37 countries have approved more than 100 irradiated food items or groups of food for consumption, either on an unconditional or restricted basis.

5.3 Technical advantages and limitations of food irradiation

Any food processing technology will be utilized only when it shows advantages in terms of effectiveness, public health, cost, convenience, etc. over competing technologies. Irradiation has clear advantages as well as limitations in relation to other technologies, according to the following:

5.3.1 *Techno-economic advantages*

(a) An alternative to fumigation of food. Fumigation of food and food ingredients with various chemicals such as ethylene dibromide (EDB), methyl bromide (MB), ethylene oxide (ETO), has been either prohibited or is being increasingly restricted in most advanced countries for health, environmental or occupational safety reasons. EDB was banned by the US Environmental Protection Agency in 1984 (EPA, 1983). The import of any food treated with EDB from other countries is also prohibited for sale in the USA. Most countries have followed the USA in banning the use of EDB for fumigating food and food ingredients for insect disinfestation purposes.

Methyl bromide (MB), the most widely used fumigant for food and agricultural products against pests such as insects and nematodes, appears to suffer the same fate as EDB. Based on data evaluated by the US National Aeronautics and Space Administration, methyl bromide has been listed under the Montreal Protocol (an international treaty for the regulation of ozone depleting substances worldwide and under the auspices of the United Nations Environmental Programme) as one of the substances which causes depletion of the ozone layer. Under the terms of the Montreal Protocol, chemical substances which cause depletion of the ozone layer will have to be phased out by the year 2000. In addition, the USA has a Clean Air Act which requires that any chemical substance which has an Ozone Depleting Potential (ODP) of 0.2 and above must be deregistered for use. MB is reported to have ODP of 0.7. The US Environmental Protection Agency has recently published the final rule for MB on 30 November 1993 which will terminate production and consumption of MB on 31 December 2000. Thus, it appears that MB will not be available for controlling pests of food and agricultural products in the near future.

The ban on EDB and the impending ban on MB will deprive the food industry of broad spectrum fumigants used for controlling insect infestation of food. Countries which depend on trade in fresh and dried food which require fumigation, for their revenue, will have to search urgently for alternative treatments to maintain their trade.

Ethylene oxide (ETO), another widely used fumigant for decontaminating insects and microorganisms in dry food ingredients, particularly spices and vegetable seasonings, was banned by the European Union starting from 1 January 1991 (Dickman, 1991). Other advanced countries such as USA and Canada are under a strong pressure to ban the use of ETO for health and occupational safety reasons.

Irradiation has been demonstrated as an effective alternative to the fumigants mentioned above. Low-dose irradiation between 0.2 and 0.7 kGy can control insect infestation of grain and other stored products

(IAEA, 1991a). Unlike fumigation, irradiation does not leave any residues in the products. To prevent reinfestation, grain or any stored product must be properly packaged in insect-proof containers. Low-dose irradiation is also being considered as an alternative to EDB and MB fumigation of fresh fruits and vegetables to overcome quarantine barriers. A minimum dose of 0.15 kGy is effective as a quarantine treatment of fresh horticultural produce against fruit-fly of the Tephritidae family, and a minimum dose of 0.3 kGy effective against other insect species (IAEA, 1991b). Unlike other competing techniques, irradiation is considered a broad spectrum quarantine treatment, not specific to insect species or host commodities. Its application as a quarantine treatment is endorsed by regional plant protection organizations operated within the International Plant Protection Convention, i.e. North American Plant Protection Organization, European Plant Protection Organization, Asia and the Pacific Plant Protection Commission, etc.

The use of irradiation instead of ETO to ensure hygienic quality of spices and vegetables seasonings has increased significantly following the ban on ETO by the EU in 1991 (Figure 5.1). Irradiation offers a broad spectrum treatment for sanitizing spices and vegetable seasonings, often at a more competitive cost than other processes. It is possible that irradiation will be

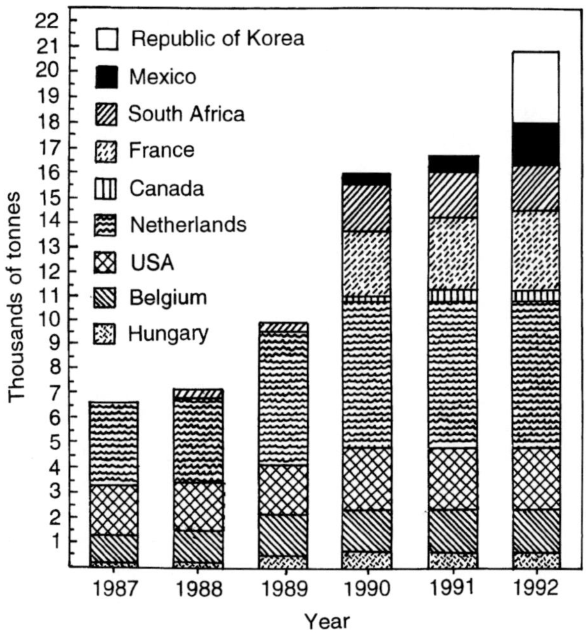

Figure 5.1 Commercial-scale irradiation of spices and vegetable seasonings.

a routine process for sanitizing spices and other dried food ingredients in the near future.

(b) A method to ensure hygienic quality of solid food. While thermal pasteurization of liquid food is a well established and widely accepted means of terminal treatment, it is not suited for solid food such as meat, poultry, seafood, etc., and dried ingredients or fresh foods whose raw characteristics must be maintained in order to meet market requirements. Such foods of animal origin, especially poultry, have been implicated in many outbreaks of food poisoning and, in some countries, investigations have shown that up to 100% of poultry may be contaminated with food-borne pathogens such as *Salmonella* and/or *Campylobacter* (WHO, 1992). Alternative chemical sanitizing procedures for such food have regulatory limitations and/or inherent public health problems due to toxic residues and environmental pollution.

Many research data and commercial experience demonstrate that irradiation can play an important role in reducing some of the food-borne diseases. At the Task Force Meeting on the Use of Irradiation to Ensure Hygienic Quality of Food held in Vienna in 1986 (WHO), it was concluded that, at that time and for the foreseeable future, no technology was available to produce raw foods of animal origin, particularly poultry and pork in which the absence of certain pathogenic microorganisms and parasites, such as *Salmonella*, *Campylobacter*, *Trichinella*, *Toxoplasma*, etc., can be guaranteed. These foods, therefore, pose significant threats to public health. Thus, the Task Force believed that, where such foods are important in the epidemiology of food-borne diseases, irradiation decontamination/disinfection must be seriously considered. It also pointed out that irradiation has a particular value in being adaptable to the terminal treatment of hermatically packaged food. Other new techniques, or improvement of current techniques, cannot be expected to achieve the same result except at exceptionally high cost.

In addition to the pathogenic microorganisms mentioned above, *Vibrio* spp., especially *V. cholerae*, *V. parahaemolyticus*, and *V. vulnificus* have caused a number of disease outbreaks and deaths in recent years. In the past few years, 15 countries in Latin America had reported some 700 000 cases of cholera and over 5000 deaths (PAHO). The most important cause of transmission of this disease was the consumption of contaminated water and food. *V. vulnificus* has caused some 50 deaths in the USA alone, especially among those with liver diseases and the immuno-compromised (Kilgen, 1993). Eating raw shellfish, especially oysters and clams, was the mode of transmission of this disease. *V. parahaemolyticus*, which is ubiquitous in the marine environment, is an important cause of food-borne infection resulting from the consumption of raw or improperly cooked seafood in several countries. *Vibrio* spp. have little resistance to irradiation. The D-values of *V. vulnificus*, *V. cholerae* and *V. parahaemolyticus* are in the range of 0.1–0.2 kGy (Rashid *et al.*, 1992; Loaharanu, 1973; Rodrick, 1993). It is therefore possible to sufficiently inactivate *Vibrio* spp. in food by a dose as low as 1 kGy. The use of irradiation to control *V. cholerae* in food has been recommended by the World Health Organization (WHO, 1991).

Other food-borne pathogenic microorganisms such as *Listeria monocytogenes* and *Escherichia coli* 0157:H7 have played an important role in the epidemiology of food-borne diseases in recent years. The US Center for Disease Control estimates that 1850 cases of human listeriosis occur each year and 425 deaths are related to *L. monocytogenes* (Anon., 1992). A number of ready-to-eat foods (cheeses, processed meats, pates, salad spreads, etc.) have been found to be contaminated by *L. monocytogenes*. High risk individuals including pregnant women, the elderly, people with conditions that suppress immunity and others who may be highly susceptible to listeriosis, are vulnerable. Some strains of *L. monocytogenes* have developed resistance to antibiotics such as tetracycline, erghromycin, cotrimoxazole and clindamycin (Barbuti *et al.*, 1992). The US-FDA and USDA have introduced zero tolerance for *L. monocytogenes* in ready-to-eat food since 1989 (Tompkin *et al.*, 1992). *L. monocytogenes* is of particular concern to the food industry as it can resist heat, salt, nitrite and acidity much better than many other pathogenic microorganisms as well as being able to multiply slowly at temperatures as low as 1°C. In early 1993, an outbreak of *E. coli* 0157:H7 in the west coast of the USA resulted in deaths of several children and hospitalization of hundreds of people who ate undercooked hamburgers contaminated by this bacterium.

Both *L. monocytogenes* and *E. coli* 0157:H7 are relatively sensitive to low dose irradiation. The D-values of the former have been reported to vary from 0.27 to 1.0 kGy depending on strains, growing media and irradiation conditions (Patterson, 1989; Huhtanen *et al.*, 1989; El-Shenawy *et al.*, 1989). The D-value of *E. coli* 0157:H7 varies from 0.16 to 0.44 kGy in ground beef and mechanically deboned chicken meat (Thayer and Boyd, 1993). The D-values of these two pathogens appear to be of the same order of magnitude as those of *Salmonella*. These pathogens in food could, therefore, be controlled by irradiation within the range of 1.5 to 3.0 kGy.

A number of food-borne parasitic disease such as trichinosis, toxoplasmosis, taeniasis, opithorchiasis, etc. which are prevalent in a number of developing countries, can also be controlled by low-dose irradiation. In most cases, a dose as low as 1 kGy can render these parasites non-infective without causing significant change in the physico-chemical or sensory properties of the treated food (Loaharanu and Murrell, 1994).

(c) An energy saving process. Compared with the energy required for canning, refrigeration and/or frozen storage, the energy used for irradia-

tion preservation of food is small. According to Brynjolfsson (1978), the total energy used for refrigerated raw, cut-up chicken is 17 760 kJ/kg, for frozen raw cut-up chicken including 3–5 weeks of frozen storage is 46 600 kJ/kg (further storage requires additional energy expenditure) and for canned chicken meat is 20 180 kJ/kg. In comparison, refrigerated and irradiated raw cut-up chicken requires a total energy of 17 860 kJ/kg. Thus, irradiation together with refrigeration could be used as a substitute for frozen storage for distribution of raw food, e.g. chicken, where feasible, e.g. for the domestic market, with a substantial reduction in energy and cost.

Because of the Montreal Protocol mentioned earlier, CFC refrigerants which are widely used by the refrigeration and food industries, will no longer be available after the year 2000. The ban on CFCs could result in higher cost of refrigeration of food which developing countries could find difficult to afford. In many developing countries, the cold chain (from harvesting, processing, distribution and refrigeration storage at home) simply does not exist or is too costly for the majority of the population. Alternative technologies which can reduce dependence on refrigeration of food should, therefore, be considered. Irradiation may reduce dependence on refrigeration of food for the following applications.

Replacing frozen distribution of food by irradiation in combination with chilling. A number of food products, especially those of animal origin, are distributed under frozen condition. In a number of instances, such distribution may be replaced by using low-dose irradiation (2–3 kGy) together with chilling at 0°–3°C to obtain sufficient shelf-life for marketing. A study conducted by a large poultry processor in Brazil showed that using such a combined treatment instead of freezing, could result in a benefit to the company of $0.18/kg of poultry with sufficient shelf-life for marketing requirements (Moretti, 1993).

High quality, shelf-stable food of animal origin. Food products, especially meat, seafood, sausages, etc., often marketed under frozen condition which required a tremendous amount of energy, may be sterilized by a combined treatment of blanching to inactivate proteolytic enzymes, freezing and irradiation with doses of 30–50 kGy to render them shelf-stable, with modest energy requirements. Such products (beef steak, corned beef, turkey slices, ham, pork sausages) have been used in many manned space flights (Josephson, 1983). Other radiation-sterilized products which have been developed by the US Army Natick Laboratories and South African Atomic Energy Authority included cooked salami, shrimp, codfish cake, bacon, pork shop, meat loaf, smoked Viennas, roast chicken, country sausages, beef curry, roast beef, etc. (Josephson, 1983; McGill, 1993).

(d) Cost-effectiveness. Irradiation is a cost-effective method of food processing for reducing the incidences of food-borne diseases, reducing post-

harvest food losses and facilitating trade in a number of food commodities. The increasing incidences of food-borne diseases have not only affected health but reduced the economic productivity of populations in both advanced and developing countries (WHO, 1984). For example, economic loss from trichinosis, toxoplasmosis, salmonellosis, campylobacteriosis and beef tapeworm in the USA in 1985 was estimated to be over US$ 1.5 billion (Morrison and Roberts, 1985). The cost of condemnation of pork infected with the tapeworm, *C. cellulosae*, in Mexico was estimated to be as high as $43 million in 1980 (Acevedo, 1982). The US Center for Disease Control and Prevention estimate that 6.5 to 33 million cases of food-borne diseases occur each year (Kvenberg and Archer, 1987). Salmonellosis alone accounts for about 2 million of these cases and is estimated to cost the US economy $2,540 million annually (Roberts, 1988). The recent food-borne outbreaks in the west coast of the USA due to *Escherichia coli* 0157:H7-contaminated hamburgers resulted in numerous illnesses and deaths and will undoubtedly prove extremely costly to the responsible retailer and wholesaler. The cost of food-borne toxoplasmosis in the USA is estimated to be up to $4,400 million a year (Roberts and Murrell, 1993). In Thailand, the annual cost of human infection with liver fluke *Opisthorchis viverrini* is estimated at about $99 million (Loaharanu and Sornmani, 1991).

Low-dose irradiation is effective in inactivating non-spore forming pathogenic bacteria and food-borne parasites mentioned above. Most food-borne parasitic diseases may be inactivated or destroyed by a dose as low as 1 kGy. A dose in the range of 3 to 5 kGy can inactivate non-spore forming pathogenic bacteria in food without causing significant change in physico-chemical and sensory properties of the product.

Food irradiation offers strong potentials to counter insect infestation, microbial spoilage, premature ripening and physiological changes of food crops which resulted in high post-harvest food losses in developing countries at extremely high cost. Often, developing countries depend on the use of fumigation (from imported fumigants) to combat insect infestation of their food crops. Irradiation can be used as an economical treatment for commodities such as cocoa beans which are subject to insect infestation; potatoes, onions, yams, and garlic, which are subject to sprouting; mushrooms and asparagus which are subject to rapid physiological changes. In Pakistan, the cost of irradiating potatoes and onions is estimated to be $4–5/metric ton with a net benefit of 40% (Khan, 1993). In Syria, it was estimated that irradiating potatoes and onions would yield a net benefit of $37.50/metric ton (Al-Bachir, 1993). The cost of irradiation, however, may vary depending upon many factors, including the capital investment required (type of facility, strength of cobalt-60 source, throughput of products and transport and storage cost).

Irradiation is facilitating trade by replacing the use of fumigants such as ethylene dibromide (EDB), methyl bromide (MB) and ethylene oxide

(ETO) which are/were widely used to overcome trade barriers with regard to insect infestation and microbial contamination. As mentioned earlier, the use of these fumigants have either been prohibited or increasingly restricted because of health, environmental or occupational safety reasons. Irradiation has emerged as a cost-effective alternative to fumigation by these chemicals. For example, the estimated benefit from using irradiation instead of methyl bromide fumigation in the USA for imported grapes, nectarines, peaches, okra and plums showed a net gain of $650 to 1,100 million for the five year period analysed (Forsythe, 1993). The estimated costs of irradiation for these commodities ranged from $0.03–0.07/kg. Irradiation is replacing ETO fumigation of spices and dried vegetable seasonings, especially in the European Union where the use of this fumigant is prohibited. Irradiation is not only more cost effective, but more versatile with respect to the types of products treated than other alternative processes. Because of its cost-effectiveness, irradiation is likely to become standard treatment for spices and other food ingredients to meet microbiological standard required by the trade and food industry.

Because of the marked consumer preference and price incentive for fresh (rather than frozen) poultry, a study in Brazil showed that irradiation together with refrigeration (above freezing) would not only be more economical than freezing but would satisfy the market life required for distribution of fresh poultry in the country. The estimated cost of irradiation ($0.02/kg) is greatly offset by the increased market price differential ($0.20) of fresh over frozen poultry and results in an additional margin of $0.18/kg for all poultry distributed in markets which were previously inaccessible to fresh product (Moretti, 1993).

5.4 Limitations of food irradiation

5.4.1 Technical

Food irradiation can reduce specific food loss problems and can complement other food processes (e.g. refrigeration) in maintaining the quality and wholesomeness of food. It can neither replace good manufacturing practices nor is it applicable to all food. For example, dairy products such as milk and butter can develop an off-odour when treated by irradiation. Many food products (e.g. meat, chicken, fish, etc.) have threshold doses above which organoleptic changes occur. Some of the changes can be offset, however, if the food is treated in the frozen state.

At the doses recommended for treating food at present, irradiation of certain foods, similarly to other sub-sterilizing food processes, will not eliminate all microorganisms or their toxins. Low-dose irradiation will not destroy all bacterial spores. Thus, irradiated meat, chicken, seafood will require appropriate temperature control similar to heat pasteurized or controlled atmosphere storage of these products, e.g. to prevent germination and toxin formation by *Clostridium botulinum*. Microbial toxins such as staphylococcus enterotoxin and aflatoxin, cannot be easily inactivated by irradiation. Therefore, food prone to be contaminated by microorganisms which produce these toxins must be handled in accordance with good manufacturing practices (e.g. chilling, low moisture content, proper packaging and storage, etc.) prior to and after processing by any substerilization method including irradiation. Viruses also cannot be easily destroyed by low-dose irradiation.

5.4.2 Infrastructure and economics

Successful implementation of a new technology depends upon the availability of a proper infrastructure within a given country. In general, food irradiation processing requires the same infrastructure as other physical processes such as canning, freezing, drying, etc. For example, a factory using any of these processes must be located at a central point where sufficient amounts of food are produced and transported to the plant for treatment and storage before sending it to the market.

Understandably, any food treatment adds costs to the product. Like other physical food processes, irradiation has high capital costs and requires a critical minimum capacity and product volume(s) for economic operation (Urbain, 1982). But unlike other physical processes, irradiation has a low operating cost, especially with regard to energy requirements. In technologically advanced countries the infrastructure required for setting up processing plants already exists. Similar infrastructure also exists in many developing countries which process food by canning or freezing on a large scale, many of these foods being for export.

5.4.3 Consumer concerns

As food irradiation is perceived to be associated with nuclear technology, any introduction of irradiated food can be erroneously connected with radioactive materials. Thus, it comes as no surprise that there appears to be a widely held opinion among some national authorities and the food industry that consumers would be apprehensive about foods treated by irradiation because of the perceived association with radioactivity. Most consumers do not understand why food should be irradiated while the same foods (but not irradiated) are available in the market. Although consumers are more aware of the risk of chemicals in the food chain, they are not as well aware of pathogenic microorganisms which are prevalent in foods of animal origin such as poultry and red meat. Consumers are also concerned that irradiation may result in the generation of chemical compounds of unknown toxicity as well as the development of radiation resistant strains of microorganisms. Thus, the introduction of food irradiation on a commercial scale requires an education campaign on not only how and why foods are normally treated before they reach the market but on also how and why irradiation could provide consumers with a wider choice or a safer food supply.

5.5 Consumer acceptance of irradiated food

The controversy over food irradiation started in the early 1980s when several advanced countries, e.g. Australia, Canada, UK and USA sought public comments in the process of introducing national regulations of food irradiation. Numerous self-appointed 'consumer groups' opposing the introduction of food irradiation emerged from the 'middle of nowhere'. These groups, mostly aligned with the anti-nuclear movement or without technical background or represented by scientists specializing in disciplines unrelated to food irradiation, often claimed that they represented the consumer-at-large. Their tactics ranged from spreading inaccurate and unscientific propaganda against the technology or lobbying members of congresses/parliaments, issuing threats to food companies that expressed an interest in using food irradiation, picketing food stores which conducted (or wish to do so) market trials of irradiated food. Their sensational actions were often picked up by the media which caused further confusion for the public.

5.5.1 Consumer attitude surveys

Because of the controversy and public debate on food irradiation during the 1980s which, in some small ways, is spilling over into this decade, several consumer groups, academic institutions, food industries and companies specializing in measuring consumer responses, conducted a number of attitude surveys to determine whether consumers would be willing or prepared to accept irradiated food. Most of these surveys were conducted through telephone interviews, home interviews or mailed questionnaires mainly in advanced countries such as Canada, UK and USA at the time when irradiated foods were not available for retail sale. Early surveys indicated that consumers were unfamiliar with food irradiation (Anon., 1984, 1986; Titelbaum *et al.*, 1983; Weise, 1984; Bruhn *et al.*, 1986a). The outcome of these surveys led to further public debate whether consumers will accept irradiated food or not. An interesting outcome of these surveys showed that the public and health professionals gained a better understanding of the technology and often made positive statements about it. From these surveys, Bord and O'Connor (1989) concluded that the extent to which the public will accept or reject irradiated food depends on the presence or absence of information about the topic. Interviews with those who knew something about irradiation and responded correctly to information about the technology were significantly more willing to accept irradiated food.

Starting from the late 1980s, universities, professional societies and industry groups have included irradiation in their public information programmes and information is slowly reaching the public. Initially, coverage by the media, the primary source of information, has frequently focused on special interest groups who oppose irradiation (Bruhn, 1993). More recently, however, major newspapers especially in the USA have gained a better understanding of various issues related to irradiation and reported to the public. The public in the USA was accurately informed about this technology when several television networks, notably ABC's 20/20 program, began to air accurate information regarding this topic.

Following the outbreaks of *E. coli* 0157:H7 in the west coast of the USA in early 1993 and the increasing interest in the use of irradiation to ensure hygienic quality of food of animal origin, the American Meat Institute funded a nation-wide three-part study conducted by Gallup Organization, Abt Associates and the Center for Food Safety and Quality Enhancement, University of Georgia, to measure consumer attitudes to irradiation in relation to food safety (AMI Foundation, 1993). The Gallup survey found that while most consumers were aware of food irradiation, few were knowledgeable of the process. Seventy three percent of consumers had heard of irradiation. However, only five percent knew a lot about the process, while 19% knew something about it. According to the survey, after the benefits of irradiation are explained and endorsements by health organizations such as American Medical Association, Food and Drug Administration, World Health Organization are mentioned, 54% of those interviewed said they were more willing to purchase irradiated meat rather than non-irradiated meat. Sixty percent of the survey participants said they would be willing to pay a 5% premium for hamburgers with bacterial counts greatly reduced by irradiation. Consumers viewed irradiation as more necessary for meat, seafood, and poultry products than for fruits and vegetables.

In the simulated supermarket selling study conducted by the Center for Food Safety and Quality Enhancement, University of Georgia, 50% of consumers tested chose irradiated ground beef over non-irradiated ground beef. After the consumers tested learned more about the irradiation process and how it affects raw meats, those choosing irradiated beef increased to 70% of the sample size.

The conclusion from various consumer attitude surveys, conducted mainly in advanced countries, showed that consumers at large are still not knowledgeable about food irradiation. They need accurate information about the safety, benefits and limitations of food irradiation to be able to make an informed decision whether they will accept irradiated food or not. Similar conclusions could be made about consumers in developing countries.

5.5.2 Market testings and retail sales of irradiated food

The opinion of consumers about irradiated food is quite different when they are given the opportunity to select and purchase the food. During the past decade, a number of market trials of several irradiated food items with clear labelling indicating the treatment, were carried out in both advanced and developing countries. A variety of irradiated foods, e.g. onions, potatoes, garlic, mango, papaya, strawberries, dried fish, fermented pork sausages, etc. were put on sale, often alongside non-irradiated ones. Positive results, which showed that consumers either preferred irradiated food or were willing to accept irradiated food at the same level as non-irradiated ones, were registered in all countries, i.e. irradiated mangoes and papaya in the USA in 1986 and 1987 (Giddings, 1986; Bruhn and Noel, 1987), irradiated strawberries and camembert cheese in France (Timsit, 1987; Bougle and Stahl, 1993); irradiated potatoes in Poland (Fiszer, 1988), irradiated apples, garlic, ginger, hot pepper and meat products in China (Zhu, 1994), irradiated fermented pork sausages locally known as Nham in Thailand, which is almost always consumed raw (Prachasitthisak, 1989), irradiated onions in the Philippines (Lustre *et al.*, 1985), irradiated dried fish in Bangladesh (Matin *et al.*, 1988), irradiated onions and garlic in Argentina (Curzio *et al.*, 1986), etc. In none of these trials, actually carried out in market places where consumers could make their own choice whether to buy irradiated food or not, was there any evidence to indicate that informed consumers will not accept irradiated food.

The positive consumer response to irradiated food in the market places has motivated retail stores to start marketing irradiated food. Starting from early 1992 after the Vindicator, Co., the first commercial food irradiator in the USA, started operation, Laurenzo Supermarket in North Miami, Florida and Carrot Top, a small retail grocery in Northbrook, Illinois started marketing irradiated fruits and vegetables including strawberries, onions, mushrooms and oranges, with success. Often, the demand for irradiated food (because of its superior quality) was overwhelming. Carrot Top reported that the sales of irradiated strawberries over non-irradiated ones in 1992 was 11 to 1 and in 1993 was 20 to 1. In Thailand, the consumers were so convinced that irradiation has removed the risk from microbial and parasitic infection, often associated with consumption of raw fermented pork sausages (Nham), that several manufacturers of the products have been marketing irradiated Nham at retail level in Bangkok with great success. Similarly, consumers in Bangladesh were pleased with irradiated dried fish instead of products which were previously fumigated with various pesticides including DDT. Irradiated dried fish has been marketed with success since a semi-commercial scale food irradiator came into operation at Chittagong in early 1994. Because of strict microbiological specifications in France, irradiated frog legs have been marketed routinely at retail level in the past ten years.

5.6 Commercial application of food irradiation

The number of countries which use irradiation for processing food for commercial purposes increased steadily from 19 in 1987 to 28 in early 1994. The majority of the increase in recent years has been in developing countries which either need the irradiated products for their domestic market or see an opportunity to develop markets overseas. The current situation with regard to the commercial use of food irradiation is summarized in Table 5.1. Spices and vegetable seasonings are the most common products which have been irradiated, in some 20 countries. Since 1990 when the European Union proposed a ban on the use of ethylene oxide for fumigating food and food ingredients, the use of irradiation to ensure hygienic quality of spices and vegetable seasoning has increased significantly, to over 20 000 metric tonnes in 1993 (Figure 5.1). The ban became effective starting 1 January 1991.

The most significant event which created an awareness among the governments, food industry, and the media was the opening of the first commercial food irradiator in the USA which is located at Malburry, near Tampa, Florida. The facility initially treated strawberries for spoilage control and started marketing the product in the Miami and Chicago areas in early 1992 with great success (Marcotte, 1992; Pszczola, 1992). Apparently, consumers were attracted by the premium 'natural field ripe' quality of irradiated products as compared to the half- to three-quarter-ripe non-irradiated strawberries normally available to them. The success of sale of other irradiated produce such as onions, mushrooms, citrus fruits, etc. was similar to that of strawberries. The retail stores which carried out the sales also reported significant savings by reducing spoilage losses of about 10% for non-irradiated strawberries to about 2% for irradiated ones. This reduction not only provided them with additional profit but enabled them to compete with larger retailers by offering better quality products at the same price charged by other stores for lower quality products (Corrigan, 1993). The success of the sale of irradiated produce has led to the sale of irradiated chicken at the retail level in the USA in 1993, which was also successful (Pszczola, 1993).

Table 5.1 Countries with irradiation facilities available for food processing (January 1994). Facilities in italics are under construction or planned; countries in italics are irradiating food for commercial use

Country	Location (starting date for food irradiation)	Products
Algeria	*Mascara*	Potatoes
Argentina	Buenos Aires (1986)	Spices, spinach, cocoa powder
Bangladesh	Chittagong (1993)	Potatoes, onions, dried fish
Belgium	Fleurus (1981)	Spices, dehydrated vegetables, deep frozen foods
Brazil	Sào Paulo (1985)	Spices, dehydrated vegetables
Canada	Laval (1989)	Spices
Chile	Santiago (1983)	Spices, dehydrated vegetables, onions, potatoes, poultry meat
China	Chengdu (1978)	Spices and vegetable seasonings, Chinese sausage, garlic
	Shanghai (1986)	Apple, potatoes, onions, garlic, dehydrated vegetables
	Zhengzhou (1986)	Garlic, seasonings, sauces
	Nanjing (1987)	Tomatoes
	Jinan (1987)	Not specified
	Lanzhou (1988)	Not specified
	Beijing (1988)	Not specified
	Tienjin (1988)	Not specified
	Daqing (1988)	Not specified
	Jianou (1991)	Not specified
Côte d'Ivoire	*Abijan*	Yams, cocoa beans
Croatia	Zagreb (1985)	Spices, rich, food ingredients
Czech Republic	Prague (1993)	Species, dry food ingredients
Cuba	Havana (1987)	Potatoes, onions, beans
Denmark	Riso (1986)	Spices
Finland	Ilomantsi (1986)	Spices
France	Lyon (1982)	Spices
	Paris (1986)	Spices, vegetable seasonings
	Nice (1986)	Spices
	Vannes (1987)	Poultry (frozen deboned chicken)
	Marseille (1989)	Spices, vegetable seasonings, dried fruit, frozen frog legs, shrimp
	Pouzauges (1991)	Not specified
	Osmanville	Not specified
	Sablé-sur-Sarthe (1992)	Camembert
Hungary	Budapest (1982)	Spices, onions, wine cork, enzyme
India	*Bombay*	Spices
	Nasik	Onions
Indonesia	Pasr Jumat (1988)	Spices
	Cibitung (1992)	
Iran	Tehran (1991)	Spices

Table 5.1 (*continued*)

Country	Location (starting date for food irradiation)	Products
Israel	Yavne (1986)	Spices, condiments, dry ingredients
Japan	Hokkaido (1973)	Potatoes
Korea, Rep.	Seoul (1986)	Garlic powder, spices and condiments
Mexico	Mexico City (1988)	Spices and dry food ingredients
Netherlands	Ede (1981)	Spices, frozen products, poultry, dehydrated vegetables, rice, egg powder, packaging material
Norway	Kjeller (1982)	Spices
Philippines	Quezon City (1989)	Not specified
Poland	*Warsaw* (1984) *Wlochy* (1991) *Lodz* (1984)	
South Africa	Pretoria (1968) Pretoria (1971) Pretoria (1980) Tzaneen (1981) Kempton Park (1981) Mulnerton (1986)	Potatoes, onions Fruits Spice, meat, fish, chicken, fruits, spices Onions, potatoes, processed products Fruits, spices, potatoes Fruits, spices
Thailand	Bangkok (1971) Patumthani (1989)	Onions Fermented pork sausages, enzymes, spices
Ukraine	Odessa (1983)	Grain
UK	Swindon (1991)	Spices
USA	Rockaway, NJ (1984) Whippany, NJ (1984) Irvine, CA (1984) Gainsville, FL (1993) *Ames*, IA (1993) Mulberry, FL (1992	Spices Spices Spices Strawberries, poultry Not specified Fruits, vegetables, poultry
Viet Nam	*Hanoi* (1991)	Onions, potatoes, seafood, spices, rice
Yugoslavia	Belgrade (1986)	Spices

5.7 International co-operation in the field of food irradiation

International co-operation is essential for a complex issue such as food irradiation which encompasses research and development, proof of wholesomeness, regulations, consumer acceptance, and trade. Without excellent international co-operation from its early stage, it is highly possible that food irradiation would not have made such significant progress and achievements to date. Although there has been much co-operation at national and regional levels, this review will address only international co-operations especially those which involved United Nations Agencies in the past four decades.

5.7.1 Co-operation between FAO, IAEA and WHO

FAO, IAEA and WHO have recognized the potential role of food irradiation as a method to increase food availability and improve food safety since the late 1950s. Co-operation first started in 1961 when the three Organizations convened an Expert Panel on Wholesomeness of Irradiated Foods, Brussels (FAO, 1962). The Panel recommended the evaluation of the wholesomeness of irradiated food prior to its approval for consumption. This recommendation led to the establishment of a Joint FAO/IAEA/WHO Expert Committee on the Wholesomeness of Irradiated Foods (JECFI). JECFI was subsequently convened in 1964, 1969, 1976 and 1980. As mentioned earlier, the 1980 JECFI made the most fundamental recommendation on food irradiation by stating that "irradiation of any food commodity up to an overall average dose of 10 kGy causes no toxicological hazard; hence, toxicological testing of food so treated is no longer required". The JECFI also recognized that irradiation of food up to an overall average dose of 10 kGy introduces no special microbiological and nutritional problems.

The co-operation of the three UN agencies was strengthened when the Joint FAO/IAEA Division of Nuclear Techniques in Food and Agriculture was created in Vienna in 1964. In addition to the JECFI, the three Organizations co-operated closely through the following.

(a) International Project in the Field of Food Irradiation (IFIP). FAO, IAEA, OECD(NEA) and a total of 24 countries cooperated between 1970 and 1981 to prove the wholesomeness of irradiated food and to supply data to be evaluated by the JECFI. WHO participated in IFIP on an advisory basis. IFIP was based at the Federal Research Institute for Nutrition, Karlsruhe, Germany. IFIP played a crucial role in developing, co-ordinating and compiling all data related to the wholesomeness of irradiated food, which included toxicological, nutritional, microbiological and radiation chemistry aspects of food irradiation. Largely through the effort of IFIP, the 1980 JECFI was able to reach such an important conclusion stated above.

(b) International Facility for Food Irradiation Technology (IFFIT). Recognizing the potential role of food irradiation in overcoming problems related to food losses and food safety in developing countries, FAO, IAEA and the Ministry of Agriculture and Fisheries of the Netherlands co-operated in establishing IFFIT at the Institute of Atomic Research in Agriculture, Wageningen, the Netherlands between 1978 and 1990. The

objectives of IFFIT were to conduct training, research and feasibility studies for the benefit of scientists from developing countries. Over the twelve year period, IFFIT trained over 400 scientists/officials from more than 50 countries in various aspects of food irradiation. Trained manpower on food irradiation for developing countries therefore has been established mainly through IFFIT.

(c) International Consultative Group on Food Irradiation (ICGFI). The success of international co-operation in solving the wholesomeness of irradiated food up to a maximum dose of 10 kGy which covers most applications of food irradiation, led the Board of Management of IFIP to recommend to FAO, IAEA and WHO the consideration of another form of international co-operation to harmonize regulations in different countries and to facilitate commercial application of the technology for the benefit of mankind. At the invitation of the Directors General of the three Organizations, representatives of 19 governments, most of which were former members of IFIP, drafted a Declaration to establish the ICGFI in Vienna in early 1983. The Declaration provides the legal framework for governments which are members of the three Organizations to co-operate to further the work on food irradiation with emphasis on the development and harmonization of regulation, trade promotion, public information and consumer acceptance, control of irradiation facilities and the process, etc. Specifically, the ICGFI has the following functions:

(i) To evaluate global developments in the field of food irradiation.
(ii) To provide a focal point of advice on the application of food irradiation to FAO, IAEA, WHO and their Member States.
(iii) To furnish information as required, through the Organizations, to the Joint FAO/IAEA/WHO Expert Committee on the Wholesomeness of Irradiated Food, and the Codex Alimentarius Commission.

The membership of ICGFI has more than doubled from 15 governments in 1984 to 40 in 1994, the majority of which are from developing countries. These governments provide cash or in-kind support to the activities of ICGFI. The mandate of the ICGFI, initially planned for 5 years, has been extended twice, in 1989 and 1994, each for a further 5 years duration. The achievements of the ICGFI to date may be summarized as follows.

5.7.2 Co-operation with the Codex Alimentarius Commission (CAC)

Recognizing the important role which the CAC has played in protecting consumer health and ensuring fair practice in international food trade close co-operation with this international body in the field of food irradiation started in the late 1970s after the 1976 JECFI made a number of positive recommendations on the wholesomeness of several food items. Such co-operation led to an initial adoption of a Recommended International Standard for Irradiation of Food covering the food items recommended by the JECFI, by the CAC in 1979. After the JECFI made another recommendation in 1980 for the wholesomeness of irradiated food treated up to an overall average dose of 10 kGy, the CAC elaborated such recommendation under its procedures and reached a final adoption of the Codex General Standard for Irradiated Foods and Recommended International Code of Practice for Operation of Radiation Facilities for Treatment of Food, in 1983. Co-operation with the CAC is continuing through the activities of the ICGFI.

5.7.3 Co-operation leading to international trade in irradiated food

The role of irradiation in overcoming problems in food trade has long been recognized. Although national regulations on food irradiation, despite the adoption of the Codex General Standard for Irradiated Foods, remain largely unharmonized, efforts have been made to evaluate the role of irradiation in overcoming certain food trade problems in the past decade. In 1988, an International Conference on the Acceptance, Control of and Trade in Irradiated Foods was organized by FAO, IAEA, WHO and the International Trade Center (ITC)/UNCTAD/GATT, in Geneva, Switzerland. The Conference addressed issues related to trade in irradiated food, especially the lack of harmonization of national regulations and the lack of acceptance by some governments which do not see a need for this technology in their own countries, but which may hamper developments in other countries where its use could significantly improve consumer health and nutrition, as well as national economic and trading potential. The Conference was attended by 220 experts designated by 57 Member States including those who were authorized to speak on behalf of their governments, and representatives of 14 observer organizations. The Conference adopted by concensus an International Document on Food Irradiation which highlights the major issues related to the acceptance of irradiated food by consumers, governmental and intergovernmental activities, the control of the process, and trade (IAEA).

The positive conclusion of the Uruguay Round of the GATT Multilateral Trade Negotiations in December 1993, especially the Agreement on Sanitary and Phytosanitary Measures, will add further incentive to international trade in irradiated food. Under the Agreement, no country which is a signatory of GATT or a member of the World Trade Organization can deny entry of foods (including irradiated food) which have been processed according to international standards, unless it can demonstrate that such foods would endanger health of their citizens, plants or animals. The Agreement recognizes competent international expert bodies such as the Codex Alimentarius Commission, the International Plant Protection Convention, and the International Organization of Epizootics, to advise the GATT/World Trade Organization to settle disputes in food trade. With regard to irradiated food, there is already a worldwide standard, i.e. Codex General Standard for Irradiated Foods which recognizes the safety and effectiveness of food treated up to an overall average dose of 10 kGy. Thus, one can be optimistic of the future of international trade in irradiated foods based on the developments mentioned above.

5.8 Conclusions

After decades of research, development and public debate, food irradiation is moving to centre stage along with other technologies in its quest to feed the world's increasing population and to overcome barriers to food trade. At the top of the list of applications of this technology is its use to ensure the hygienic quality of food of animal origin and spices; as a quarantine treatment to overcome barriers in trade; to protect grain and other stored products from insect infestation; and as an innovative method, in combination with other food processing procedures, as an energy saving process to maintain food quality. The positive conclusion in 1993 of the Uruguay Round of the GATT Multilateral Trade Negotiations, especially the adoption of the Agreement on the Application of Sanitary and Phytosanitary Measures, in 1993, has added further incentive to international trade in irradiated foods.

References

Acevedo, H. (1982) Economic impact of porcine cysticercosis. In *Cysticercosis, Present State of Knowledge and Perspectives*. A. Flisser *et al.* (ed.). Academic Press, New York.

Al-Bachir, M., El-Sharabi, N. and Ayman Midani, M. Economic feasibility study of potato and onion irradiation in the Syrian Arab Republic. In *Cost–Benefit Aspects of Food Irradiation Processing*, STI/PUB/905. IAEA, Vienna.

AMI Foundation (1993) *Consumer Awareness, Knowledge, and Acceptance of Food Irradiation*. The American Meat Institute Foundation, Arlington, Virginia.

Anon. (1992) Deli items may dish up dose of *Listeria*. FDA Consumer. July–August 1992. FDA, Washington, DC.

Anon. (1984) Marketability testing of irradiated fish and seafood (for the Department of Fisheries and Oceans, Canada). Canadian Gallop Poll Ltd.

Anon. (1986) Food irradiation – tomorrow's technology. *Which?* Consumers' Association, London.

Barbuti, S., Maggi, A. and Casoli, C. (1992) Antibiotic resistance in strains of *Listeria* spp. from meat products. *Lett. Appl. Microbiol.*, **15**, 56–58.

Bord, R.J. and O'Connor, R.E. (1989) Who wants irradiated food? Untangling complex public opinion. *Food Technol.*, **43(10)**, 87–90.

Bougle, D. and Stahl, V. (1993) Eradication de bacteries pathogens (*Listeria monocytogene et Salmonella*) de camemberts au laid cru par rayonnements ionisants. In *Cost–Benefit Aspects of Food Irradiation Processing*, STI/PUB/905. IAEA, Vienna.

Bruhn, C.M. and Noel, J.W. (1987) Consumer in-store response to irradiated papaya. *Food Technol.*, **41(9)**, 83.

Bruhn, C.M., Schutz, H.G. and Sommer, R. (1986) Attitude change toward food irradiation among conventional and alternative consumers. *Food Technol.*, **40(1)**, 86–91.

Bruhn, C.M. (1993) Consumer attitude and market response to irradiated food. Paper presented at an ICGFI/IOCU Seminar on Food Irradiation and Consumers, Hoofdorp, The Netherlands, September 1993. IAEA, Vienna.

Brynjolfsson, A. (1978) Energy and food irradiation. In *Food Preservation by Irradiation*, Vol. II, STI/PUB/470. IAEA, Vienna.

Corrigan, J.P. (1993) Experiences in selling irradiated food at the retail level. In *Cost–Benefit Aspects of Food Irradiation Processing*, STI/PUB/905. IAEA, Vienna.

Curzio, O.A., Croci, C.A. and Urioste, A.M. (1986) Marketing trials with irradiated onions and garlic in Argentina. *Food Irradiation Newsletter*, **10(1)**, 44–45.

Dickman, S. (1991) Compromise eludes EC. *Nature*, **349**, 273.

El-Shenawy, M.A., Yousef, A.E. and Marth, E.M. (1989) Radiation sensitivity of *Listeria monocytogenes* in broth or in raw ground beef. *Lebensm.-Wiss. u. Technol.*, **22**, 387–390.

Environmental Protection Agency (EPA) (1993) EPA Acts to ban EDB pesticides. *Environmental News*. US Environmental Protection Agency, Washington, DC.

FAO (1962) *Report of the Meeting on the Wholesomeness of Irradiated Foods*. Brussels, Oct. 1961. FAO, Rome.

FAO (1992) Codex general standard for irradiated foods. CODEX STAN 106-1983. In *CODEX ALIMENTARIUS – General Requirements*, 2nd edn. FAO, Rome.

Fiszer, W. (1988) Status report on food irradiation in Poland. Submitted to the Fifth Meeting of the International Consultative Group on Food Irradiation. IAEA, Vienna.

Forsythe, K.W. and Evangelou, P. (1993) Cost and benefit of irradiation and other selected quarantine treatments for fruits and vegetables imported to the United States of America. In *Cost–Benefit Aspects of Food Irradiation Processing*, STI/PUB/905. IAEA, Vienna.

Giddings, G.G. (1986) Summary of the Puerto Rico mango consumer test market. *Food Irradiation Newsletter*, Vol. 10(2). IAEA, Vienna.

Huhtanen, C.N., Jenkins, R.K. and Thayer, D.W. (1989) Gamma radiation sensitivity of *Listeria monocytogenes*. *J. Food Protect.*, **52(9)**, 610–613.

IAEA (1991a) *Insect Disinfection of Food and Agricultural Products by Irradiation*. STI/PUB/ 895. IAEA, Vienna.

IAEA (1991b) Report of a Task Force Meeting on Irradiation as a Quarantine Treatment of Fresh Fruits and Vegetables, convened by the International Consultative Group on Food Irradiation (ICGFI), ICGFI Document 13, IAEA, Vienna.

Josephson, E.S. (1983) Radappertization of meat, poultry, finfish, shellfish and special diets. In *Preservation of Food by Ionizing Radiation*, Vol. III (eds E.S. Josephson and M.S. Peterson). CRC Press, Boca Raton, Florida.

Khan, I. (1993) Techno-economic evaluation of food irradiation in Pakistan. In *Cost–Benefit Aspects of Food Irradiation Processing*, STI/PUB/905. IAEA, Vienna.

Kilgen, M. (1993) Economic benefits of irradiation of molluscan shellfish in Louisiana. In *Cost–Benefit Aspects of Food Irradiation Processing*, STI/PUB/905. IAEA, Vienna.

Kvenberg, J.E. and Archer, D.L. (1987) Economic impact of colonization control on food-borne disease. *Food Technol.*, **41**, 77, 80, 81, 98.

Loaharanu, P. and Murrell, K.D. (1994) A role of irradiation in the control of food-borne parasites. *Trends Food Sci. Technol.*, **5(6)**, 190–195.

Loaharanu, P. and Sornmani, S. (1991) Preliminary estimates of economic impact of liver fluke infection in Thailand and the feasibility of irradiation as a control measure. *Southeast Asian J. Trop. Med. Publ. Hlth*, **22**, 384–390.

Loaharanu, P. (1973) Gamma irradiation of fishery products in Thailand with special reference to their microbiological and entomological aspects. In *Aspects of the Introduction of Food Irradiation in Developing Countries*, STI/PUB/362. IAEA, Vienna.

Lustre, A.O., Ang, L., Dianco, A., Cabalfin, E.F. and Navarro, Q.O. (1985) Philippine's experience in marketing irradiated foods. Paper presented at the ASEAN Workshop of Food Irradiation, Bangkok, Thailand, 1985. ASEAN Food Handling Bureau, Kuala Lumpur.

Marcotte, M. (1992) Irradiated strawberries enter the US market. *Food Technol.*, **46(5)**, 80–86.

Matin, M.A., Bhuiya, A.D., Ahmed, M., *et al.* (1992) Commercialization, storage, and transportation studies of irradiated dried fish and onions. In *Asian Regional Co-operative Project on Food Irradiation: Technology Transfer*, STI/PUB/883. IAEA, Vienna.

McGill, A. (1993) Shelf-stable meal components – the South African Experience (unpublished observations).

Moretti, R.H. (1993) Cost–benefit of poultry irradiation in Brazil. In *Cost–Benefit Aspects of Food Irradiation Processing*, STI/PUB/905. IAEA, Vienna.

Morrison, R.M. and Roberts, T. (1985) *Food Irradiation: New Perspectives on a Controversial Technology*. Office of Technology Assessment, Washington, DC.

Patterson, M. (1989) Sensitivity of *Listeria monocytogenes* to irradiation on poultry meat and in phosphate-buffered saline. *Lett. Appl. Microbiol.*, **8**, 181–184.

Prachasitthisak, Y., Pringsulaka, V. and Chareon, S. (1989) Consumer acceptance of irradiated Nham (fermented pork sausages). *Food Irradiation Newsletter*, **13(1)**, 65–67. IAEA, Vienna.

Pszczola, D. (1993) Irradiated poultry makes US debut in Midwest and Florida markets. *Food Technol.*, **47(11)**, 89–96.

Pszczola, D. (1992) Irradiated produce reaches Midwest market. *Food Technol.*, **46(5)**, 89–92.

Rashid, H.O., Ito, H. and Ishigaki, I. (1992) Distribution of pathogenic vibrios and other bacteria in imported frozen shrimps and their decontamination by gamma-irradiation. *World J. Microbiol. Biotechnol.*, **8**, 494–499.

Roberts, T. and Murrell, K.D. (1993) Economic losses caused by food-borne parasitic diseases. In *Cost–Benefit Aspects of Food Irradiation Processing*, STI/PUB/905. IAEA, Vienna.

Roberts, T. (1988) Salmonellosis control: estimated economic costs. *Poultry Sci.*, **67**, 936–943.

Rodrick, G.E. and Dixon, D.W. (1993) The effects of gamma radiation upon shellshock oysters with respect to shelf-life, and *Vibrio vulnificus* levels. Paper presented at 2nd Research Co-ordination Meeting on Irradiation in Combination with other Process to Improve Food Quality, St. Hyacinthe, Canada, June 1993. IAEA, Vienna.

Thayer, D.W. and Boyd, G. (1993) Elimination of *Escherichia coli* 0157:H7 in meats by gamma irradiation. *Appl. Environ. Microbiol.*, **59(4)**, 1030–1034.

Timsit, M. (1987) *Marche'-test de la fraise fraiche*. ADIA, Bordeaux, France.

Titlebaum, L.F., Dubin, E.Z. and Doyle, M. (1983) Will consumers accept irradiated foods? *J. Food Safety*, **5**, 219–228.

Tompkin, R.B., Christiansen, L.N., Shaparis, A.B., Baker, R.L., and Schroeder, J.M. (1992) Control of *Listeria monocytogenes* in processed meats. *Food Australia*, **44(8)**, 370–376.

Urbain, W.M. (1982) *Food Irradiation*. Academic Press, New York.

WHO (1986) *Report of a Task Force on the Use of Irradiation to Ensure Hygienic Quality of Food*, convened by the International Consultative Group on Food Irradiation (ICGFI), WHO/EHE/FOS/87.2. WHO, Geneva.

WHO (1991) *A statement by the Global Task Force on Cholera Control to the Forty-fourth World Health Assembly* (10 May 1991).

WHO (1984) *The Role of Food Safety in Health and Development*. Technical Report Series No. 705. WHO, Geneva.

WHO (1981) *Wholesomeness of Irradiated Food*. Report of a Joint FAO/IAEA/WHO Expert Committee. Technical Report 651, WHO, Geneva.

WHO (1994) *Review of the Safety and Nutritional Adequacy of Irradiated Food*, Report of a WHO Consultation, Geneva, 20–22 May 1992. WHO, Geneva (in press).

WHO (1992) *Global Health Situation and Projections Estimates*. WHO, Geneva.

Wiese Research Associates (1984) *Consumer Reaction to the Irradiation Concept. A Summary Report*. Wiese Research Associates, Inc. US Department of Energy By-Product Utilization Program. March.

Zhu, J. (1994) *Food Irradiation in China – with emphasis on market testing and public acceptance*. Paper presented at the 21st Japan Conference on Radiation and Radioisotopes, Tokyo, Japan, 2–4 Feb. 1994.

6 Microwave processing

J. MULLIN

6.1 Introduction

It hardly needs to be said to the readership of a book on *New Methods of Food Preservation*, but there is a well-established hypothesis that UHT (ultra-high heat treatment) enhances the quality of many food products, whether for chill or ambient distribution. Add to that the very clear consumer demand for the same higher quality and the simple conclusion is that one can expect a market pull for technologies which enhance the current rates of heating used in food processing. The results of this pull have led to many of the innovations discussed elsewhere in this book. This chapter concentrates on the long recognised potential of microwave heating to lead to a step change in the ability of the food processor to achieve the heating rates necessary to deliver profiles which could improve current UHT process routes. In the wider field of preservation microwaves have been used in drying, blanching and vacuum drying. Typical product areas where microwaves have been used commercially include blanching of vegetables, where it is claimed that there is less need for mechanical handling with consequent better product. Also, microwaves in combination with hot air have been shown to be a positive route to drying of foodstuffs, if only in selective product areas. Finally, microwave vacuum drying has found some outlets in producing fruit juices and meat extracts.

The principle behind this potential of microwaves to enhance heating rates is simple: in conventional heating one relies on the processes of conduction and convection to transport heat from the heating source (a hot surface or fluid) to the bulk of the product. These mechanisms require a finite time. Research on devices such as scraped surface heat exchangers has greatly enhanced many such processes. However, there are limitations of such technologies, e.g. in the processing of products which contain particulates or products which one wishes to post-pack preserve. In this respect it is clear that a volumetric heating mechanism has the potential to deliver heat instantly throughout the product and in so doing deliver the mildest heat treatment consistent with the required microbial lethality for the product. One such volumetric heating process is microwave heating. Unfortunately the full potential of microwave heating, recognised as long ago as 1950, has not yet been achieved. The reasons for this are a mixture of commercial and technical barriers. For the former the reader is referred to a paper by Gerling (1986) where he explores the commercial relationship between the microwave equipment supplier and the food manufacturer.

The work presented here concentrates on the technical aspects of the technology where the biggest barrier to success has been to harness the volumetric heating benefit without the deleterious effects of uneven heating observed in most aspects of microwave heating. The nature of microwave heating is discussed both from the materials point of view but also from the technology perspective and chronicles the gradual improvements which have been achieved in that area over the last ten years.

6.2 Introduction to microwaves and their interaction with food materials

6.2.1 Basics

Microwaves cover the broad range of radio frequencies from 300 MHz (million cycles per second) to 300 GHz (billion cycles per second), corresponding to wavelengths between 1 metre and 1 millimetre according to the equation:

$$\text{Wavelength (cm)} = \frac{\text{Speed of light (cm/s)}}{\text{Frequency (Hz)}} \tag{6.1}$$

The International Telecommunication Unions (ITU), to which most nations subscribe, is responsible for the allocation of the various frequency bands of the electromagnetic spectrum. The ITU has set aside a number of frequency bands for Industrial, Scientific and Medical (ISM) use. They range from 13 MHz to 22125 MHz and are listed below:

13.56 Mhz
27.12 MHz
40.68 MHz
433.92 MHz
896,915 MHz*
2450,2375 MHz
5800 MHz
22125 MHz
(*US and UK only.)

For all practical purposes industrial applications are carried out at 915 MHz in the USA, 896 MHz in the UK and 2450 MHz worldwide. All commercial and consumer microwave ovens operate at 2450 MHz (or 2375MHz in the former Soviet block).

The principle reasons these frequencies are used in the food industry are:

(1) The availability of microwave generators at these frequencies at useful power levels and at reasonable cost;
(2) The fact that the wavelength at these frequencies is consistent with food product dimensions, which results in somewhat more efficient heating than would obtain at higher or lower frequencies.

Microwaves are produced by vacuum tube devices called magnetrons and klystrons that were originally developed for radar applications. Magnetrons are currently available at power levels from a few hundred watts to 50 kW, although not limited to this range. They are generally approximately 50% efficient. Although klystrons have been used in some microwave heating applications they are not used in current practice and will not be mentioned further.

6.2.2 How microwaves heat

Microwaves are reflected by metals, transmitted by electrically neutral materials such as glass, most plastics, ceramics and paper, and absorbed by electrically charged materials. Microwaves penetrate deeply into food materials and as they penetrate, the energy they carry is converted to heat, actually by the food material itself, mainly by the mechanisms of polar and ionic orientation (Figure 6.1).

Water molecules, which are the most common, polar molecules in food, are present in random orientation. When an electric field is applied, as for example when food is placed between two electrodes, the polar water molecules attempt to line up with the electric field. When the field is removed the molecules tend to return to random orientation. When the electric field is alternated rapidly, millions of times per second, as for example at 2450 MHz in a microwave oven, kinetic energy is expended as the polar molecules attempt to follow the field and this is evident as heat (Figure 6.1).

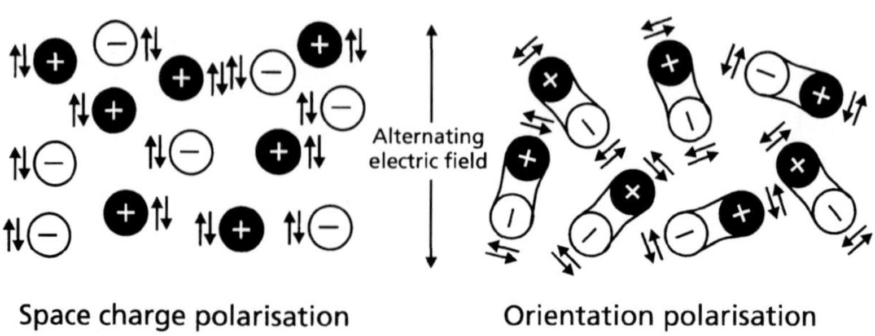

Figure 6.1 Ionic and polar orientation.

Ionic orientation occurs when ions in solution move as an electric field is applied. Kinetic energy is given up to the ions by the electric field so that when the field is alternating rapidly heat is generated.

6.2.3 *Power absorption*

The conversion of microwave energy to heat is expressed by the following equation:

$$P = 2\pi E^2 f \varepsilon_0 \varepsilon'' \tag{6.2}$$

Where P = power in Watts/m
E = the electric field strength in volts/m
f = the frequency in Hertz
ε_0 = the permittivity of free space in farad per metre
ε'' = the dielectric loss

There is a substantial body of dielectric data on foods and other materials in the literature. Such information is useful in calculating the penetration depth of high frequency energy into food and also in building mathematical models of microwave heat transfer.

However, the variation of ε' (dielectric constant) and ε'' with salt level, temperature and particularly with ice transition to water, etc. is not well documented and for truly predictive modelling these data must be available or acquired.

6.2.4 *Uniformity of heating*

The aim of using microwave processing in preservation in general and pasteurisation or sterilisation in particular is to deliver a more homogeneous heat treatment at a faster rate than conventional heating. Therefore, it is essential to understand the limitations of potential systems in this regard. It can be seen from the equation (6.2) that there are basic physical restrictions imposed on the heating uniformity for a given combination of frequency and material. As frequency increases so too does power absorbed until in some cases there can be severe differential heating between the centre and surface of a product. At the most widely available frequency of 2450 MHz this can be a major concern and certainly limits the maximum thickness of products to be treated to a few centimetres. One might reasonably expect that a lower frequency system may lead to more uniform heating and this is certainly the case in microwave tempering of frozen meat blocks where 896 or 915 MHz has been demonstrated superior to the higher frequency. However, this has not been the case in the area of pasteurisation and sterilisation systems where all the commercial systems use 2450 MHz. This may be because packaging formats introduce another level of complexity which actually dominates the lack of heating uniformity, e.g. pack geometry can lead to increased field strengths at corners and edges and as power absorbed depends on the square of electric field then this can be the dominant effect. At this stage the reasons for the apparent superiority of 2450 MHz in this respect are not established.

6.2.5 Material properties

If a material is placed in an electric field the individual molecules experience a force and so reorientate in the direction of the field giving rise to a nett electric field. If the applied field now alternates then as expected

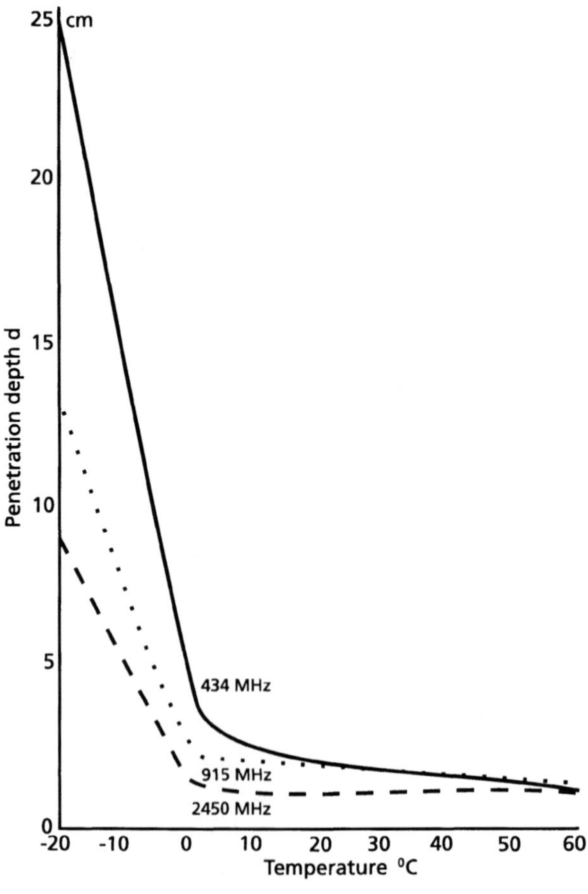

Figure 6.2 Penetration depth of the electromagnetic waves into raw beef as a function of temperature and for three different frequencies.

the individual molecules follow the field. Of course, there is a time lag in following the field and so there is work done against the field. The ability of the molecules to follow the field is characterised by the *permittivity* (ε), while their receptiveness to heating is known as the *loss factor* (ε''). In addition to the loss factor in food materials there will always be some ions present and this will result in conductive losses proportional to the *conductivity* (σ). The conductivity and loss factor are indistinguishable in that both lead to volumetric heating. However, they do complicate the relationship between frequency and uniformity of heating as many food products have increasing conductive losses as the frequency decreases giving rise to less penetration at lower frequencies for products such as ham with its high salt content.

The conventional method of characterising the relationship between dielectric properties and uniformity of heating is through the concept of penetration depth which is defined as 'the distance from the surface of the product before the intensity has decayed to 1/*e* (approx. 37%) of its original value':

$$d = \frac{\lambda}{2\pi\sqrt{\varepsilon_r}\ \tan\delta} \tag{6.3}$$

d = penetration depth in cm
λ = wavelength in cm
ε_r = relative dielectric constant
$\tan\delta$ = loss tangent

Figure 6.2 demonstrates the effect of temperature and frequency on penetration depth for raw meat.

6.3 Microwaves and microorganisms

Over the past four decades there has been a lot of debate on the 'lethality of microwaves'. Therefore it seems obligatory in this chapter to summarise the views and work on the topic. But we should first note the conclusion that *there is no substantiated mechanism for the interaction of microwaves with microorganisms other than volumetric heating*.

The literature on microbiological interactions can be divided into two chronological sections. A great deal of work was performed in the 1940s to the mid 1950s when the consensus was that there were no athermal effects. With the increased use of microwaves and radio frequency in the 1960s more work was undertaken in the area of health exposure effects, but also, on microbiological effects.

6.3.1 *Early work (1940–55)*

Fleming (1944) performed some of the earliest work in the field. He used a range of frequencies from 11–350 MHz and exposed suspensions of *Escherichia coli* to fields where up to 10 W were absorbed. In all cases he claimed to avoid significant temperature rise and yet in every instance observed a lethal effect. He also claimed that there was a frequency dependence in his results which indicated that 60 MHz was the most effective.

The results of Fleming stimulated Brown *et al.* (1947) to investigate the existence of a lethal microbiological effect. The work of Brown and co-workers extends more than 10 years and we can see a development of the work towards his eventual conclusion that all of the reported efforts can be attributed to thermal phenomena. The early work of Brown *et al.* (1947) consisted of repeating Fleming's work at low power. They report no agreement with Fleming's work, in fact, they observed thermal effects only. In the same paper Brown extended the work to look at milk pasteurisation using milk preheated to various temperatures and then exposing it to 27 MHz for short periods of time (0.07 s) and high field 700 V cm^{-1}. Once again, thermal effects alone were observed.

At the same time Nyrop (1946) reported work which spanned a four year period. He was working at much lower frequencies (ca. 20 kHz) and claimed to kill 99.5% of *Bacillus coli* irrespective of temperature! His stated conditions were 230 V cm^{-1} + applied for 7 s. In addition, he claimed to have inactivated the Foot and Mouth virus at very low temperatures (under 36°C). Other work tried to capitalise on these claims (Zarotochemzeff, 1944) but nothing practical came from this work.

Brown *et al.* continued their work on *E. coli* and in 1947 appeared to demonstrate substantial effects due to both 30 MHz and 190 kHz fields. However, later work by Brown himself (1954) showed that the temperature control necessary to investigate the effect had not been possible. He went on to show that at frequencies up to 600 MHz there were only thermal effects on *E. coli*. Further work at 2450 MHz by Proctor and Goldblith (1951) and by Ingram and Page (1953) on 20 MHz systems confirmed that there were no athermal effects on *E. coli*.

6.3.2 *Renewal of interest in the 1960s*

With the arrival of domestic and industrial microwave processing equipment in the 1960s there was a revival of interest in the subject. Desel *et al.* (1960) compared the effectiveness of 'conventional heating' versus microwave power in killing several bacterial species. This was done by heating inoculated eggs and custard. The results showed no significant differences. In particular, they noted that both methods failed to destroy spores of *Bacillus cereus*.

There was then a period in which people published 'beneficial athermal' claims and filed patents to take advantage of the so-called effects. These included Lacey (1965) who claimed substantial benefits in treating hospital meals by a combination of infra-red and microwave, and Hamid (1969, 1970) who considered the use of microwaves in pasteurisation of milk and its use in treatment of contaminated waste water. Also, Robe (1966) published an article on the benefits. All of the experiments which claimed to support the existence of athermal effects had one thing in common. They all had difficulties in temperature monitoring and/or control of temperature.

During this period, three papers were published which provided quantitative evidence disproving the existence of athermal effects. At least, proving that in these particular systems all of the effects could be adequately explained as thermal. Goldblith (1967, 1968) produced papers on the effect of 2450 MHz on two bacterial strains (*E. coli* and *Bacillus subtilis*) and its effect on thiamine. In all these cases the effects recorded were shown to be thermal effects. Slightly later Lechowich *et al.* (1969) studied the effect of 2450 MHz on both *Streptococcus faecalis* and *Sacchoromyces cerevisiae*. This study also concluded that no athermal effect could be detected in these systems.

The debate continued into the seventies followed by the more general biological debate between the 'Eastern Bloc' and the 'West'. The former continued to make claims regarding the effect of electromagnetic fields at very low intensity. Western experts continued to contradict the findings of their colleagues in the East, but there was a growing level of agreement that there were some phenomena which were associated with strangely small fields (see reviews by Cleary, 1977; Lorenz, 1976; Fung, 1980).

The data against an athermal effect in microbiological systems continued to grow with papers by Fruin (1982), Correlli (1977) and Hamrick (1973), all concluding that there were no effects other than the thermal effect in the system under consideration. Each of these authors concentrated either on 2450 MHz or, in the case of Correlli, a band close to the domestic frequency, namely 2.6–4.0 GHz. In each case the power levels quoted were low enough to avoid significant thermal efforts and, in the case of Correlli, infra-red analytical work was performed to provide chemical data which showed that not only was there no significant microbiological effect of microwaves but also, that the low level of microwaves had not caused any specific chemical change.

In the 1980s (Lacey, 1988) produced data which he interpreted as indicating a non-specific athermal effect but this has been disputed by the work of Cole *et al.* (1989).

6.3.3 Conclusion

The debate has gone on through five decades. However, at all points when an 'athermal effect' has been reproducibly identified the measurement and/ or control of temperature has been in considerable doubt.

In contrast, there are many experiments which show on specific systems that no athermal effect exists. This does not mean that athermal effects do *not* occur in biological systems. Indeed, there is growing evidence in other areas which support the idea. However, in the specific area of the interaction of microwaves with microbiological systems there is very little evidence to support the existence of an athermal effect.

6.4 Microwave processing equipment

6.4.1 The benefits of microwave processing

The major benefits of microwave processing can be summarised as follows:

Dry continuous processing. In applications where microwaves replace the need for water or steam as a heating source it is possible to manufacture continuously (rather than batch-wise) and without the need to dry package material.

Rapid heating and drying. Because this form of heating occurs *within* the material, the limitations of surface heating are reduced and the process can be greatly accelerated. This is particularly true for materials with a low thermal conductivity and for drying processes where the high dielectric loss properties of water result in selective heating of moist areas and rapid diffusion of water vapour to the surface.

Energy savings. Energy is not expended in heating the walls of the apparatus or the environment. Heat is developed where it is needed – in the material being processed. In drying processes, microwave energy is applied as the conventional drying process approaches the critical point, thus the heat of vaporisation is supplied by cheaper forms of energy that in the early stages of drying are very efficient.

Space savings. The increase in processing rate makes it possible to design much more compact equipment. In many cases plant capacity can be increased several fold without additional building space. Bread baking can be accomplished in 50% less time when microwave energy is used.

Fast start-up and shut-down. There is practically no thermal mass with microwave processing so that the system can be operational almost immediately, adding in some cases extra hours of operation. Clean-up after shut-down can begin sooner in most cases because the equipment is cooler and can be worked on sooner. Where conventional heat is used simultaneously usually lower temperatures can be applied.

Precise process control. Heating adjustments can be made immediately and the operation can be fine tuned easily. Systems can be made to respond instantaneously to feedback from sensors.

In spite of the clear advantages that we have outlined a survey of actual microwave installation shows quite a low success rate.

6.4.2 *Current status of microwave processing in food industry applications*

The current status of microwave processing in the food industry can be judged by examining Table 6.1, which is from Oda (1990). The absolute figures will, of course, have increased but the overview will still remain. Microwaves have not yet broken into the food industry significantly. The reasons for this are complex and include commercial and technical reasons. However, it has often been postulated that a major factor in this has been a lack of understanding of the basic interaction between product and process. That has been fuelled by a lack of sufficient collaboration between the equipment manufacturer and the user. However, that barrier seems to be less imposing today with some well published interactions between the two industry segments. In addition, the level of research into the basic problems appears to be growing and has resulted in several papers on the modelling of microwave power (Barrett, 1998; Railston, 1993). Concurrently, the technology in the area is clearly growing if the number of patents is a measure.

Table 6.1 Processes in which microwave energy plays a significant role

Process	No. of installations	Power (kW)
Tempering frozen foods	250 (est)	25 000
Drying pasta	30	1800
Cooking bacon	22	6000
Cooking beef patties	6	300
Cooking potatoes	1	30
Pasteurisation of bread	4	200
Pasteurisation of pasta	4 (est)	120
Drying snack foods	4	600
Drying onions	1	25
Drying rice cakes	1	30
Drying seaweed	1	30
Vacuum drying orange juice	1	48

It is our judgement that a major factor in the relatively small number of installations is that the high technology of the equipment manufacturer has not been matched by an improvement in the understanding of the system/product interaction. However, several factors have combined to change the situation:

(1) New applicator and system design, e.g. Alfa Star, Magnetronics, Berstorff.
(2) Improvements in generator efficiency has resulted in a decrease in capital costs (generator efficiency of 915 MHz tube up from 60% to 80%).
(3) Increase in the microwave oven market has increased the market for microwaveable products.

6.4.3 *Microwave patents in preservation*

Another indicator of progress made in the area is the patent activity. Figure 6.3 shows the level of patent activity worldwide in the field of microwaves and preservation. This clearly shows a peak in the late 1980s and so one can expect to enter the mature phase of the technology in the 1990s.

Patents, of course, only tell part of the story. There have been other developments which are seen in the market place, e.g. Omac in Italy have been very active in the pasteurisation of pasta.

The market trends in microwaveable products and the improvements in packaging materials have combined to create a demand for quality sterilised and pasteurised products. This pressure on manufacturers has led

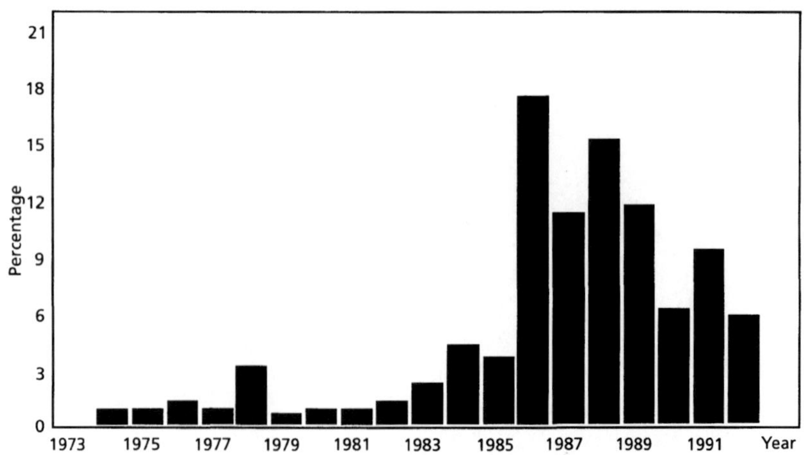

Figure 6.3 Patent activity for microwave sterilisation/pasteurisation applications.

Table 6.2 Microwave equipment

Company name	Form of preservation	Product
Alfastar	Steriliser	Salads/deli
Berstorff	Steriliser	Liquid food plus particulates
Le Houlme	Pasteuriser	Fruit and sugar mixtures
Omac	Steriliser	Pasta meals
Toppan	Steriliser	Liquid food plus particulates

to an opportunity for equipment suppliers to provide process systems. This challenge has been accepted by the equipment industry and has led to several innovations, outlined in Table 6.2.

Table 6.2 is not meant to be comprehensive. Rather it gives an indication of the breadth of product applications microwaves have found even in this narrow area. Equally, in a novel area one might expect a variety of approaches to the technology. This has certainly been the case in microwaves and preservation.

6.5 Case histories

The true test for any technology is its success in the market place. Therefore, I include some examples of microwaves in food (and related industries) preservation from an international perspective. The source of these case histories is the Union International Electroheat (UIE) who have many relevant publications in the area, e.g. *Dielectric Heating for Industrial Processes*. Each of these case histories follows the same format with the emphasis on the problem addressed and the quantifiable improvement offered by the microwave solution, rather than concentrating on the technology itself. The processes considered include blanching, drying pasteurisation and sterilisation.

Case History 1 Green tea drying/roasting system with microwave and far infra-red techniques

ITO-EN SHIZUKOA, SHIZUKOA PREF. JAPAN (AND OTHER USERS)

Background of installation. The manufacturing process for green tea can be broadly divided into two stages. One is a rough-processing stage for preparing half-finished tea from fresh leaves and the other is a finishing stage for making finished tea from the semi-manufactured goods. Half-finished tea has yet to go through the finishing stage which fully develops all desirable qualities of tea, including fine taste and fragrance. The principal aim of the finishing process is to further dry half-finished tea leaves and then give them a delightful flavour by roasting their surfaces.

However, a long time and good skill are required to fully remove the water content of tea leaves by external heating. The process also involves such adverse effects as the overheating of the surfaces of leaves, resulting in their discoloration, deformation and loss of fragrance.

After carefully studying various methods of improvement, the factory reported here decided to instal an integrated tea processing system using a microwave oven for drying and a long-wave infra-red heater for roasting (i.e. firing tea to give it good fragrance). The new installation has helped improve the product quality.

Features. In general, finished green tea is produced through a number of processes, including overall finishing, air stream grading, stem separation, drying and roasting.

The new system uses a microwave oven for the drying process to reduce the water content of half-finished tea from 8–10% to 1–2%. Dried tea leaves are then sent into a rotary drum with a long-wave infra-red heater, installed below the microwave oven, where they are roasted into tasty, fragrant, finished tea.

Advantages.

- Produces good tea which, when brewed, offers a delightfully coloured, fragrant beverage.
- Retards the fading or discoloration of tea during storage.
- Dries tea leaves satisfactorily and uniformly without scorching or discolouring their surface.
- Has good controllability to ensure stable product quality.
- Minimises heat radiation for environmental control purposes.

Outline of the system.

- Microwave power output: 7.2 kW
  (smaller power units are available)
- Far infra-red (FIR) heater: 14.0 kW
- Throughput: 120 kg/h
- Line power source: 200 V – 3 phase 50 Hz

Manufacturers.

- Microwave and FIR system: Shin Nihon Musen (New Japan Radio Co.)
- Full system: Ochiai Seisakusho.

Main users.

Ito-En Shizuoka, Shizuoka Pref.
Yamamotoyama Shimada, Shizuoka Pref.

Miake Co. Sendai, Miyagi Pref.
Fukuju-En Kyoto.
Kagoshima Chuo Seicha Co. Kagoshima, Kagoshima Pref.

Case history 2 Drying of pharmaceuticals

ICI PHARMACEUTICALS LIMITED – MACCLESFIELD (UK)

Process requirements. Drying of pharmaceutical granules.

Original plant. In the conventional granulation process, drying the granules creates the greatest problems. Granule attrition, biologically active dust emission, dust explosion hazards and the need to clean down between product changeovers are the worst of these problems.

Poor heat transfer and solids handling requirements in tray ovens led to the widespread use of fluidised bed dryers but many problems still remained. Operator protection relied on the use of air-fed suits, which are cumbersome and uncomfortable to work in. The protective clothing isolated the operator from the process and made it more difficult to obtain the very high quality consistent standards that are demanded by the pharmaceutical industry.

For these reasons, an alternative dryer that does not possess so many processing difficulties but which retains the high efficiency of the fluidised bed dryer should have attractions for the pharmaceutical industry.

New plant. The microwave/vacuum dryer has a cylindrical cavity with a dished top on which ten 5 kW magnetrons are mounted. The product is loaded into a polypropylene bowl which, in turn, is positioned on the dryer platen. This is also dished to withstand vacuum but has the advantage that it allows microwave energy to pass under the product. The platen lifts the bowl into the dryer and also forms the base of the cavity. The dryer shell and platen are fully jacketed to prevent condensation on the inner walls, which are mirror polished for ease of cleaning.

The dryer has a capacity of 80 kg (dry product) and typically 100 kg (wet weight) mixes are dried. The drying time for each mix is 35–45 minutes and the dry product moisture content is 1.5% to 2.5% water.

Power to the dryer is supplied by magnetrons. Each magnetron can be operated in the range from 500 W to 5 kW.

Control of the dryer can be either by residual electric field (E-field) in the cavity or by product temperature. The E-field in the cavity is monitored by diodes set into the dryer wall. As the load decreases i.e. whilst water is being evaporated, the E-field tends to rise but the automatic control turns down the output from the magnetrons and so keeps the E-field constant. When the magnetrons are at minimum power, the E-field is allowed to rise until it reaches a preset level at which point the power is switched off and the dryer vented to relieve vacuum. Similarly with temperature control, a fibre-optic temperature probe is forced into the product as the platen is lifted. The dryer controller maintains the product at a preset temperature by turning down magnetron output power as the product dries.

When the dryer is controlled on E-field, the output power of the magnetrons is continually adjusted and the energy matched to the load throughout the drying cycle. On the other hand, temperature control tends to be less at the boiling point of the water. However, this varies with vacuum level and that can vary during the drying cycle as the volume of vapour being removed increases or decreases. Some stability is achieved by condensing the vapour but vacuum variations cannot be eliminated completely.

The object of the drying end point control is to produce a mix of granules with the desired moisture content but with a uniform temperature and moisture across the bed. For this type of product, it has been found that it is better to operate at a low E-field set point such that the magnetrons are switched to the minimum power early in the drying cycle. This extends the drying time but does give a moisture content and temperature levelling effect across the bed of granules.

Technical data. The main characteristics of each unit are:

- capacity: typically 100 kg (moisture content 25% dwb)
- final moisture content: 1.5–2.5% (dwb)
- drying time per load: 45 minutes
- vacuum: 35 torr
- microwave output power: 10×5 kW
- operating frequency: 2450 MHz

Economics/advantages. The use of microwave energy in combination with vacuum for the drying of pharmaceutical granules enables the process to be carried out at lower temperatures, thus removing the risk of degradation for heat sensitive products. The technique has become well established at ICI's UK pharmaceutical plant and is in routine use for the drying of highly active products.

When the microwave dried granules are tabletted, they are as good as, if not better than, conventionally dried granules. In all cases, the microwave dried product is much less dusty than fluidised bed dried material.

Equipment supplier. Niro Fielder Limited, Eastleigh (UK).

Reference. This case history is based on a paper presented by Dr M.J. Cliff of ICI at the Heating and Process 1–3000 MHz Conference, September 1986.

Case History 3 Pasteurisation of fruit and sugar mixture

BRASFOUCARD INDUSTRIE – LE HOULME (FRANCE)

Process requirements. Mixtures of fruit and sugar syrup are incorporated in many dairy products such as yoghurts or ice creams. After mixing, this product has to be pasteurised and cooled, before being packed and dispatched. The pasteurisation temperature varies from 85°C to 125°C, depending on the mixture.

Conventional process. Pasteurisation of the fruit and sugar mixture was carried out in jacketed vats, heated with steam or hot water. This batch process caused local overheating near the tank walls, which may give an overcooked taste to the product. The long heating time led to a degradation of the aromatic components of the mixture.

Microwave process. The mixture is continuously pumped through a silica tube, transparent to microwaves. The tube goes through 3 successive microwave multi-mode cavities. Each cavity is fed by 15 microwave generators (800 W magnetrons). Microwave power is automatically controlled, using pressure and temperature measurements located at the input and output of each cavity.

The processing line is now fully automated. It includes syrup preparation, mixing of the components and, after microwave heating, cooling through a heat exchanger followed by aseptic packing in 850 litre containers.

Technical data. The main characteristics are:

- capacity 100–2000 kg/h.
- input temperature – depending on the product.
- output temperature (depending on the product) 85–125°C.
- range of dry solid rate (depending on the product) 10–90%.
- tube inner diameter 38 mm.
- tube length 12 m (3 × 4 m).
- product speed about 0.5 m/s.
- microwave power installed 36 kW.
- electric power installed 72 kW.
- frequency 2450 MHz.

Economics – advantages. Microwave heating allows continuous and automated processing. Warm-up time is shorter. The more even temperature distribution eliminates overcooking. A 75% reduction in the time for the high temperature treatment produces the same pasteurisation effect as conventional techniques. As a consequence, appearance and flavour of the processed product has been noticeably improved. The average energy consumption of the whole process has been reduced by 75%. The improvement of the productivity of the process is around 30%.

Manufacturer. SFAMO, France.

Case History 4 Sterilisation after packaging of pasta products

P & T FOODS – HULSHOUT (BELGIUM)

Process requirements. Over the last few years, there has been a growing market for retail packaged food. P & T Foods meets this demand with a variety of Italian prepared dishes: agnolotti, lasagna, tortellini, ravioli, tagliatelli, spaghetti. The special feature of their production installation is the use of microwave sterilisation for cooked fresh products after packaging. The main advantages of this process are the 9 months shelf-life and the fresh taste of the cooked food.

Conventional technologies. Classical sterilisation installations use steam for heating up the food: this results in a slow temperature rise inside the food due to its poor thermal conductivity. In order to get a given sterility, expressed in an *F*-value, a long time is needed such that the food loses organoleptic value and its fresh product taste.

New plant. The sterilisation installation of P & T Foods has been in operation since 1990: the quick temperature rise inside the dishes is ensured by microwave generators placed around the steriliser. *The expected F-value is obtained up to five times faster than with classical techniques*; the advantage is the good taste of freshly cooked products.

The individual dishes are first filled with pasta and sauce, and then sealed. They are then preheated by microwaves. The installed power amounts to 17 kW (14 computer-controlled magnetrons of 1200 W at 2450 MHz). The sterilisation process then takes place in a cylinder with a diameter of 2 m. A mechanical system enables the small packages to be introduced and removed from the pressure vessel in a continuous manner. The installed electric power of the computer-controlled steriliser is divided into:

- microwave: 72 kW (60 × 1200 W – 2450 MHz)
- resistances: 75 kW (hot air production).

The whole cylinder is maintained at a pressure of 2.2 bar in order to work at about 120°C without damage to the packages. The treatment requires four phases:

- quick heating with microwaves;
- even distribution of the temperature using microwaves and hot air;

- temperature holding using hot air;
- cooling with cold air.

The quality control is assumed at two levels:

- *Internal control*: Temperature control of the dishes in the process.
- *External control*: Samples are taken of each series and analysed in an independent laboratory after having been kept at ambient conditions for 7 days.

Benefits. The installation has been perfected and tuned during 14 months of close cooperation between P & T Foods and OMAC, the component manufacturer. The industrial line is now able to produce sterilised packaged dishes with only 3% of waste products.

The most important advantages proceed from the used technology:

- *Better quality*: Organoleptic characteristics and taste of freshly cooked products thanks to the quick heating-up.
- *Stable shelf life*: 9 months in ambient temperature instead of 2–3 weeks in a refrigerated case with pasteurised dishes.
- Complete automation of the sterilisation process.
- Continuous production line.
- Space saving.
- Less man-power.
- Better working conditions.

Today's capacity is a minimum of 6 million products per year (750 kg/h). Products with uniform good quality cannot be obtained with classical techniques.

Equipment supplier. OMAC Impianti Srl, Pratissolo di Scandiano, 49019 Reggio Emilia, Italy.

Case History 5 Pilot plant microwave steriliser

Despite the advances made in microwave processing the major hurdle of uniformity of heating in high temperature sterilisation (about 121°C) remains and so this last section concentrates on one approach to solving this dilemma. This section has been compiled with the help of Berstorff, the equipment supplier of the equipment described here.

The advantages ascribed to microwaves includes the rate of heating but in high temperature retorts the uniformity of heating and the degree of certainty of attained temperature are more important than speed of itself. Therefore the inherent non-uniformity of heating normally found in microwave heating is a major technical barrier to the exploitation of the technology in the preservation field. However, uneven heating is not

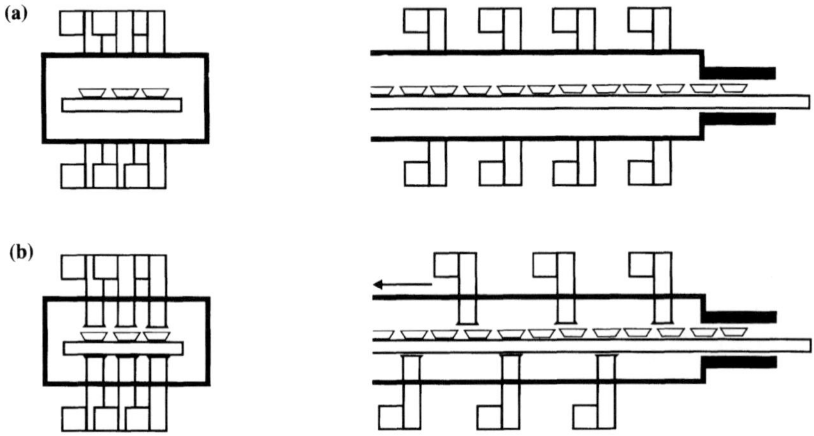

Figure 6.4 Microwave tunnels. (a) Multimode cavity. (b) Hybrid mode cavity.

always evident in microwave heating, e.g. microwaves can have a very reproducible heating effect in tempering frozen meat. Rather, it is *equipment design* which leads to this problem. Conventionally, microwave tunnels employ an applicator design known as *multimode* (Figure 6.4(a)). In this design the food product passes through a series of 'hot' and 'cold' spots which average out to deliver a roughly uniform heating pattern as the product travels along the conveyor belt. However, this is not sufficiently uniform to meet sterilisation requirements and so Berstorff designed a new approach in the mid-1980s which is described schematically in Figure 6.4(b), the *hybrid mode*. The principle is described in Figure 6.5. The heating potential offered by the electric field is restricted to be close to the 'horn'. When this is combined with control of individual power levels one can deliver customised heating profiles such as that seen in Figure 6.6.

This principle has been applied in a number of commercial pasteurisers and more recently has been applied to the more stringent area of sterilisation. Experiments have been performed in a pilot plant steriliser, shown diagrammatically in Figure 6.7. The result in product terms is a clear improvement in colour of all the components, most noticeably carrots which retain their colour while difficult vegetables like broccoli retain their colour and shape. It is assumed that the improvement in colour and texture is associated with the difference in heat treatment between microwave and conventional retort. This is indicated in the temperature profiles in Figure 6.8. The reduction in come-up time is most notable but there is also a reduction in the cooling time due to the smaller thermal mass for cooling. In conclusion, there has been progress made in this area, and the process may soon be on the market.

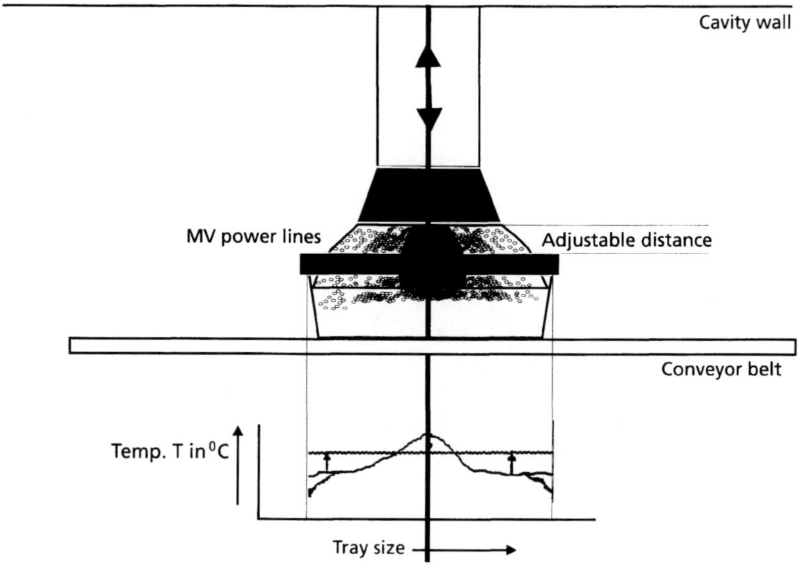

Figure 6.5 The principle of the hybrid mode.

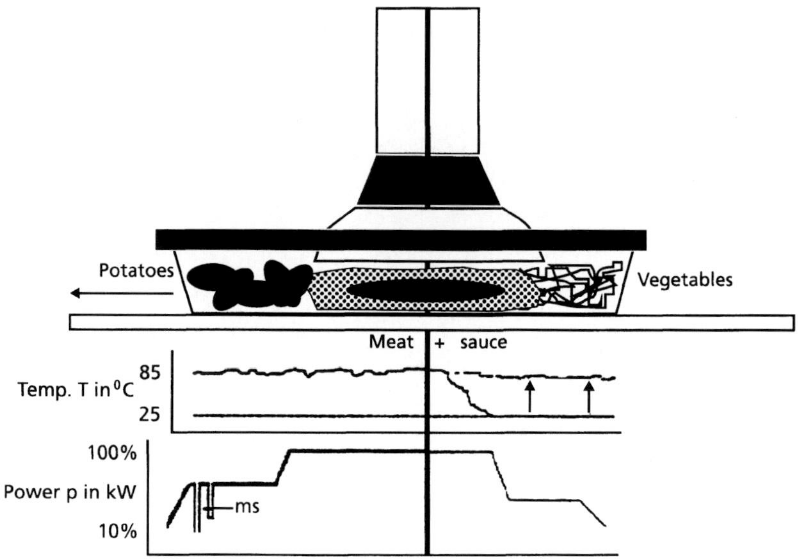

Figure 6.6 Customised heating profiles.

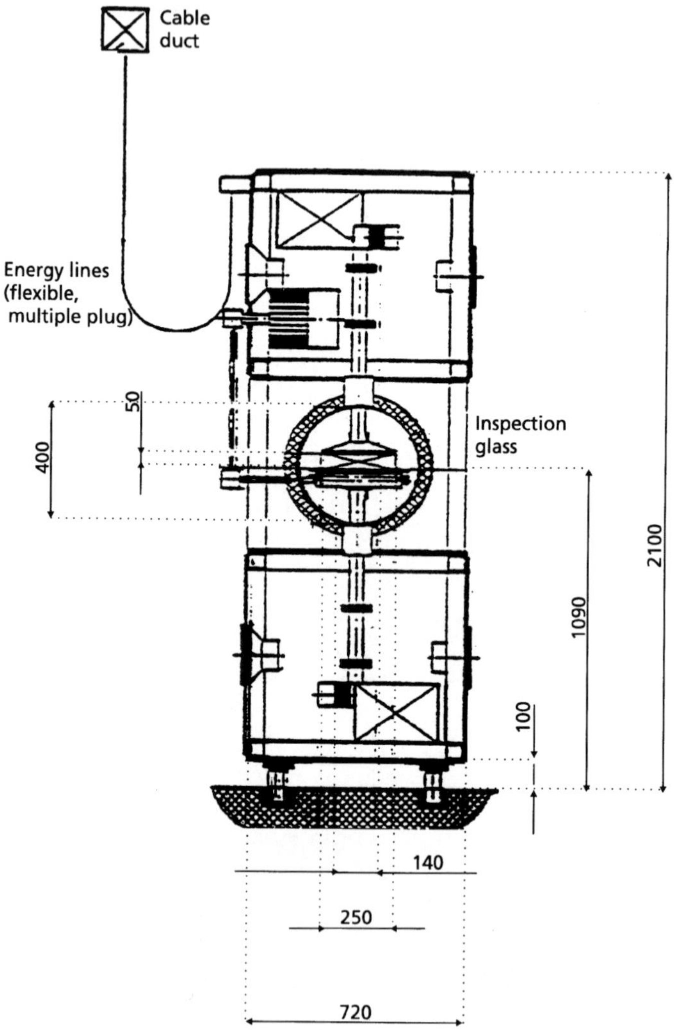

Figure 6.7 Pressurised microwave laboratory line with: two magnetrons of 1.2 kW each in hybrid or multimode configuration; production speed simulation; completely computerised (master and field PC equipment) for dosing and positioning the microwave energy in relation to the product within the tray.

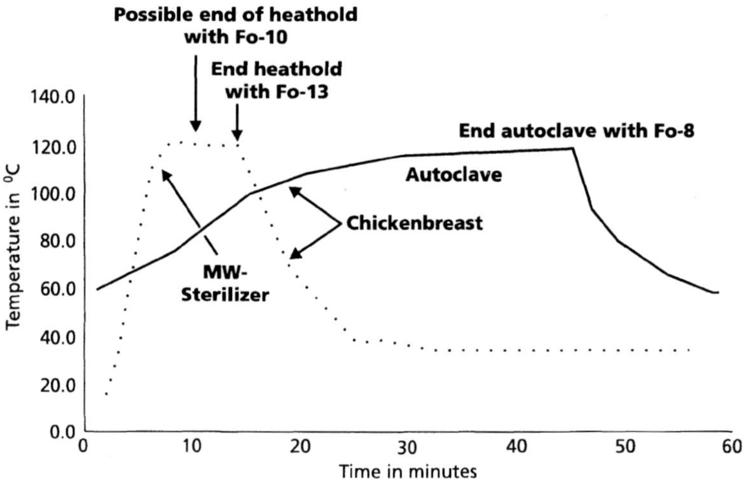

Figure 6.8 Temperature profiles for microwave treatment and conventional treatment.

6.6 The future

Microwave processing in the preservation area has made great strides over the last 10 years but has not yet made the major breakthrough. This is for a mixture of market and technical barriers. The basic handicap is the relatively high price of microwave lines and that is expected to remain the case in the medium term.

The technical barriers come from problems offered not only in equipment design but also in the interaction of microwaves with food. The dielectric behaviour of food can be complex and is not understood by the equipment manufacturers. Therefore, further advances may come through close collaboration between supplier and user. If this were to happen there is a bright future for microwave technology in the preservation area.

References

Brown, G.H. and Morrison, W.C. (1954) An exploration of the effects of strong radio frequency fields on micro-organisms in aqueous solutions. *Food Technol.*, **8(8)**, 361–366.

Brown, G.H., Hoyler, C.N. and Bierwirth, R.A. (1947) *Theory and Application of Radio Frequency Heating*. Van Nostrand-Reinhold, Princeton N.J.

Cleary, S.F. (1977) Biological effects of microwave and radiofrequency radiation. *Crit. Rev. Environ. Control*, **121**, t66.

Cole, M.B., *et al.* (1989) Destruction of *Listeria monocytogenes* during microwave cooking. *Lancet*, **i**, 218.

Correlli, J.C., *et al.* (1977) Effect of 2.6–4.0 GHz microwave radiation on *E. coli*. *J. Microwave Power*, **12**, 267–270.

Dessel, M.M., Bowersox, E.M. and Jeter, W.S. (1960) Bacteria in electronically cooked foods. *J. Am. Diet. Assoc.*, **37(9)**, 230–233.
Fleming, H. (1944) Effect of high frequency fields on micro-organisms. *Electron. Eng.*, **63(1)**, 18–21.
Fruin, J.T. and Guthertz, L.S. (1982) *J. Food Protect.* **45(8)**, 695–698.
Fung, D.Y.C. and Cunningham, F.E. (1980) Effect of microwaves on microorganisms in foods. *J. Food Protect.* **43(8)**, 641–650.
Gerling, J.E. (1986) Microwaves in the food industry: Promise and reality.
Goldblith, S.A. and Wang, D.I.C. (1967) Effect of microwaves on *Escherichia coli* and *Bacillus subtilis*. *Appl. Microbiol.*, **15(6)**, 1371–1375.
Goldblith, S.A., Tannenbaum, S.R. and Wang, D.I.C. (1968) Thermal and 2450 MHz microwave energy effect on the destruction of thiamine. *Food Technol.*, **22(10)**, 1266–1268.
Hamid, M.A.K., Boulanger, R.J., Tong, S.C., Gallop, R.A. and Periera, R.R. (1969) Microwave pasteurisation of raw milk. *Microwave Power*, **4(4)**, 272–275.
Hamid, M.A.K., Boerner, W.M. and Tony, S.C. (1970) Microwave irradiation of potato-waste water. *J. Microwave Power*, **5(1)**, 44–46.
Hamrick, P.E. and Butler, B.T. (1973) Exposure of bacteria to 2450 Mhz microwave radiation. *J. Microwave Power*, **8(3)**, 227–233.
Ingram, M. and Page, L.J. (1953) The survival of microbes in modulated high frequency fields. *Proc. Soc. Appl. Bacteriol.*, **16**, 69–87.
Lacey, B.A., Winner, H.T., McLellan, M.E. and Bagshawe, K.D. (1965) Effects of microwave cookery on the bacterial contents of food. *J. Appl. Bacteriol.*, **28**, 331–335.
Lacey, R.W., *et al.* (1988) Listeria survives microwave heating (letter to Editor). *J. Hospital Infection*, **14**, 84–86.
Lechowich, R.V., Benchat, L.R., Fox, K.I. and Webster, F.H. (1969) A procedure for evaluating the effects of 2450 MHz microwaves on *Streptococcus faecalis* and *Saccharomyces cerevisiae*. *Appl. Microbiol.*, **17(1)**, 106–110.
Lorenz, K. (1976) Microwave heating of foods – changes in nutrient and chemical composition. *CRC Critical Rev. Food Sci. Nutr.*, **339**, 370.
Lubec, *et al.* (1989) Letter to the Editor. *Lancet*, **ii**, 1392.
Nyrop, J.E. (1946) A specific effect of high frequency electric currents on biological objects. *Nature (London)*, **157**, 51.
Proctor, B.E. and Goldblith, S.A. (1951) Electron magnitude radiation fundamentals and their applications in food technology. *Adv. Food Res.*, **3**, 120–196.
Robe, K. (1996) Improve flavour of pasteurised products. *Food Process. Mark. (London)*, **27(3)**, 84–86.
Zarotschenzeff, M.T. (1944) Will dielectric heating power to way for South America meat? *Meat*, **22(11)**, 21–23.

7 Hydrostatic pressure treatment of food: equipment and processing

B. MERTENS

7.1 Introduction

Hydrostatic pressure technology is a novel non-thermal food processing technology whereby foods are subjected to high hydrostatic pressure, generally in the range of 100–600 MPa, at or around room temperature.

Just as high temperature inactivates microorganisms, so high hydrostatic pressure (HHP), whether or not in combination with moderately elevated or reduced temperature, inactivates vegetative microorganisms, spores and enzymes and, as such, allows one to achieve increases in the shelf-life of foods. Since HHP denatures proteins and polysaccharides, the technology also presents interesting applications in the area of food texturisation. In both areas, preservation and texturisation, the non-thermal character of the technology opens up unique opportunities to the food industry for the development of novel foods of superior nutritional and sensory quality, novel texture, more convenience, higher safety and increased chilled or ambient shelf-life.

Today, the European, Japanese and US food industries are investing a great deal of research and development effort in this new technology. In Europe, since early 1993, National Governments (e.g. in France, UK, Germany and Sweden) as well as the European Union have launched basic and applied collaborative research programmes; the objective of these programmes is to explore the possibilities and limitations of the HHP technology for food applications. At the same time, several of Europe's major food processors are conducting their own research and development programs in this area.

At present no commercial HHP processed foods have emerged on the European food market. In view of the present cost of the technology and its limitations in terms of capacity, it is undoubtedly true to state that HHP will not replace conventional preservation and processing technologies for high volume, commodity type foods in the foreseeable future. However, HHP is expected to offer commercially feasible alternatives to conventional technologies and find its niche applications in the growing market of high quality, high priced foods.

This chapter reviews the current status of HHP technology with a view to its possible application in the food industry. A distinction is made between those requirements that can and those that cannot be readily met by the present status of the technology. Key aspects have recently been reviewed in references [1–6].

7.2 General description of an industrial high pressure system

An industrial high pressure system consists of a high pressure vessel and its closure, a pressure generation system, a temperature control device and a material handling system which can be more or less automated.

7.2.1 *The high pressure vessel and its closure*

The heart of the system is obviously the *pressure vessel*, which is in many cases a forged monolithic, cylindrical vessel constructed in low alloy steel

Figure 7.1 Example of monobloc high pressure vessel (maximum working pressure, 200 MPa (2000 Bar); internal diameter, 940 mm; internal height, 3100 mm; internal volume, 6700 litres) (National Forge).

of high tensile strength. Figure 7.1 shows an impressive example of such a so-called monobloc vessel. The wall thickness of these monobloc vessels is determined by the maximum working pressure, the vessel diameter and the number of cycles for which the vessel is designed. Depending on the vessel internal diameter, the use of monobloc vessels is typically limited to maximum working pressures of 4000–6000 Bar. In case higher pressures are required, pre-stressed vessel designs like multilayer vessels or wire-wound vessels are used.

In these designs, vessels are deliberately fabricated with residual compressive stresses, so that the maximum stress level in the vessel wall during pressurisation is lowered, hence reducing the required vessel wall thickness. Due to the number and complexity of the processes involved in the manufacture of pre-stressed vessels, the cost of such vessels is significantly higher than for 'simpler' monolithic vessels which are forged and processed in one single piece. Therefore pre-stressed vessel technology is only applied in those areas where the boundaries of monobloc technology are exceeded. Figure 7.2 shows a schematic representation of a wire-wound vessel with wire-wound frame closure.

Depending on the application, different *closure designs* can be used. When the dwell time at working pressure is short, as is the case for CIP

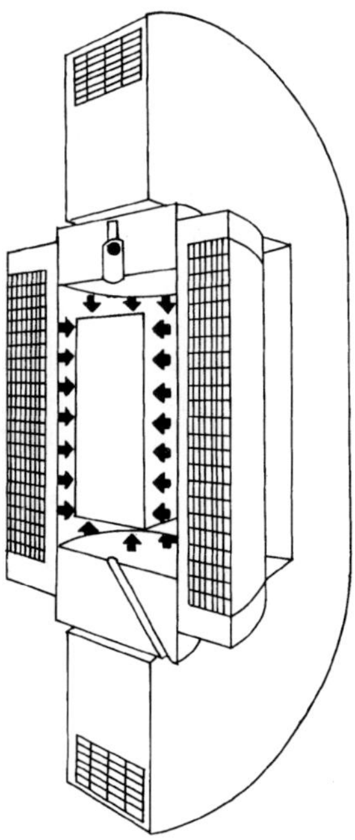

Figure 7.2 Schematic representation of wire-wound vessel with wire-wound closure (ABB).

Figure 7.3 Interrupted thread closure of large industrial cold isostatic press (National Forge).

(cold isostatic press) applications (see also Section 7.3) and food applications, fast opening and closing interrupted thread closures are used (Figure 7.3). In this way, the vessel down-time is minimised and the vessel productivity, expressed in number of cycles per unit of time or in throughput, is maximised, which is of great importance in view of the economic feasibility of the HHP process.

When the vessel down time for opening and closing is negligible as compared to the overall cycle time, like for most HIP (hot isostatic press) applications, where the dwell time at working pressure amounts to several hours, less expensive continuous thread closures are used (Figure 7.4).

Alternatively, the pressure vessels can be closed by means of a frame closure (Figure 7.5). In this case, the axial forces on the threadless closure are controlled by a wire wound frame or a frame with plates, which is moved over the vessel during pressurisation. This type of closure, however, requires more space (for the movement of the frame) than the threaded closures, while access to the vessel is hindered. In addition, the

Figure 7.4 Continuous thread closure of large industrial hot isostatic press (National Forge).

opening/closing time is longer than for the interrupted thread closure and the elastic elongation of the frame over its entire length results in important movement of the seal and, as a result, higher seal wear. Therefore, and because of the higher cost of a frame closure as compared with a threaded closure, frame closures are typically only used when the pressure-diameter boundaries of the threaded closure are exceeded.

7.2.2 Pressure generation

Once loaded and closed, the vessel is filled with a pressure transmitting medium. In most current cold and warm isostatic pressing applications, the pressure medium is simply water mixed with a small percentage of soluble oil for pump lubrication and anti-corrosion purposes. Alternatively, a low viscosity mineral or vegetable oil can be used. Also pure, distilled water can be applied, but requires an adapted use of materials (stainless steel) in the vessel, high pressure piping and valving, and the pressure intensifier. After all air is removed from the vessel by means of a low pressure, fast fill and drain pump in combination with an automatic de-aeration valve,

Figure 7.5 Hot isostatic press with frame closure (National Forge).

high hydrostatic pressure is generated by direct compression, indirect compression or through heating of the pressure medium.

In the case of *direct*, piston-type *compression* (Figure 7.6), the pressure medium in the high pressure vessel is directly pressurised by a piston, driven at its larger diameter end by a low pressure pump. Low pressure at the large surface end of the piston is multiplied by the ratio of the piston surfaces, according to the hydraulic principle, to yield high pressures at the small diameter end of the piston, i.e. in the vessel. This method of pressure multiplication allows very fast compression, but is in practice restricted to small diameter laboratory or pilot high pressure systems, due to the limitations (i.e. sensitivity to wear) of the dynamic high pressure seal between piston and internal vessel surface.

The *indirect compression* method (Figure 7.7) simply uses a high pressure intensifier which pumps the pressure medium from the reservoir into the closed and de-aerated high pressure vessel, until the desired pressure is reached. The high pressure intensifier uses the same pressure multiplica-

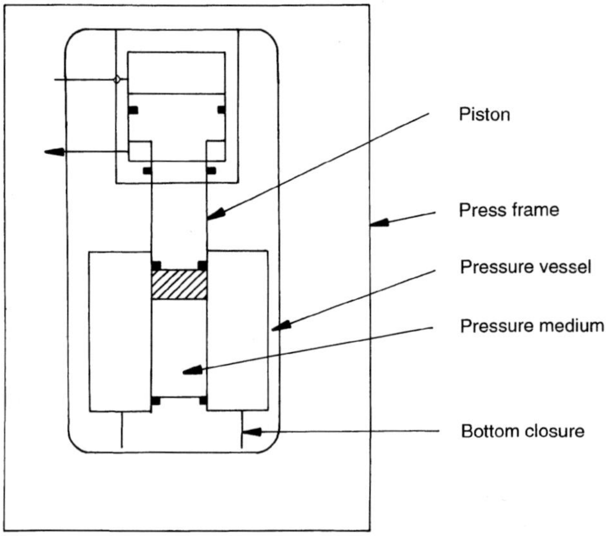

Figure 7.6 Direct pressure generating method.

tion or hydraulic principle as described above. For this method, only static high pressure seals are applied within the high pressure vessel. Most industrial cold, warm and hot isostatic pressing systems use this indirect pressurisation method.

The last pressurisation method is based on the thermal *expansion of the pressure medium* to generate high pressure. This method can obviously

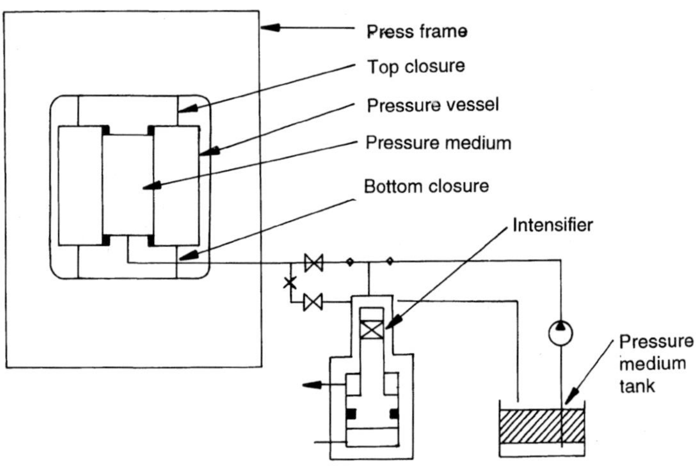

Figure 7.7 Indirect pressure generating method.

only be used for those applications where high pressure in combination with elevated temperature is required, like hydrothermal quartz growth (see also section 11.3) and some chemical reactions. It is clear that this method of pressurisation requires very accurate temperature control within the entire internal volume of the vessel. In fact, in all cases where pressure is used in combination with temperature, the effects of temperature variations on pressure, through expansion or shrinking of the pressure medium or the load, have to be taken into account and well controlled.

7.2.3 Temperature control

The temperature of the load and pressure transmitting medium inside the high pressure vessel can be controlled in two ways. By heating or cooling the entire vessel at the outside, so-called external temperature control, or by internal temperature control whereby the heat or cold source is placed inside the vessel.

The simplest execution of external temperature control uses electric heater bands wound around the vessel. This method only allows heating of vessel and its contents above room temperature; when both heating and cooling are required, the vessel can be fitted with a jacket in which a heating/cooling medium is circulated (Figure 7.8). The external temperature control method is relatively simple of design and inexpensive, but yields poor heating/cooling rates due to the enormous thermal inertia of

Figure 7.8 Temperature control by means of heating/cooling jacket (National Forge).

the pressure vessel and the small heat exchange surface between vessel and pressure medium. Consequently this method of temperature control is in practice only used to keep pressure vessels and their loads at constant temperature. For most commercial WIP (warm isostatic pressing) applications, including possible food applications, temperature cycling during pressure treatment or fast and frequent temperature changes between cycles are not required and external temperature control offers an adequate and cost effective solution.

Many laboratory HHP systems that are currently supplied to food research institutes and the food industry incorporate an internal temperature control system. This system not only allows more experimental freedom but also drastically reduces the wait between experimental cycles at different temperatures. As mentioned above, internal temperature control implies that the heat/cold source is placed inside the vessel. This can be achieved in several ways. One way is to circulate the cooling/heating medium through the top and/or bottom closure of the vessel, so that these are transformed into internal heat exchangers. The heat transfer efficiency of the closures can be further improved by fitting them with fins. Another, though technically more complex, solution is to use a cylindrical heat-exchanger which is shrink-fitted inside the high pressure vessel. A third method of internal heating is used in the area of hot isostatic pressing; in a hot isostatic press, the gaseous pressure medium is heated to temperatures of up to 2200°C by means of a cylindrical electrical furnace that is fitted inside the vessel and encloses the load.

Alternatively, the pressurising medium can also be heated/cooled in a heat exchanger outside the pressure vessel and fast circulated in the pressure vessel before and during pressurising.

In all cases of internal heating thermal losses to the vessel can be limited and heating/cooling rates can be further improved by thermal insulation of the internal vessel surface. It is clear that internal heating/cooling is a more sophisticated and expensive way of temperature control than external heating.

In addition, the heating/cooling source and thermal insulation reduce the available volume in the high pressure vessel. In this respect note again that the high pressure vessel is the most expensive part of any industrial high pressure system.

7.2.4 *Material handling*

Once the high pressure treatment is completed, the vessel is opened, unloaded, loaded with new material and closed again. Just as for batch retort systems used in the food industry, in many commercial high pressure systems currently used in the industry, all of these operations are automated in order to minimise operational cost and maximise the system

Figure 7.9 Example of integrated high pressure system for cold isostatic pressing of ceramics (National Forge).

efficiency. In some cases the material handling system also incorporates automatic mould filling, closing and washing stations, and automatic transport to and from the pressure vessel. Figure 7.9 shows an example of such an integrated high pressure system for the cold isostatic pressing of ceramics.

7.3 Current commercial applications of high pressure technology

7.3.1 *Isostatic pressing*

(a) Cold isostatic pressing (CIP). Cold isostatic pressing is essentially a forming technology. Powdered materials, like metal, ceramics, carbon/graphite and plastic (PTFE), are filled into an elastomer mould which is subsequently treated with high isostatic pressure at room temperature and in that way compacted. Depending on the type of material, pressures ranging from 50 (for PTFE) up to 600 MPa (for powder metals) are industrially applied. The dwell time at the requested pressure is typically zero, possibly a few seconds. The CIP-process can be either 'wet-bag' or 'dry-bag'. '*Wet-bag*' means that the mould is filled outside of the pressure vessel and then placed into the pressure vessel. In the case of '*dry bag*',

processing, the mould is fixed inside the pressure vessel and filled in-situ. Of today's applications, CIP, together with WIP (below), is most similar to potential future food applications. Also for food processing applications, both the dry bag (in bulk) and wet bag (in-container) process option are of interest.

(b) Warm isostatic pressure (WIP). WIP is, similar to CIP, a forming technology, whereby isostatic pressure is however, applied in combination with temperatures ranging from ambient to 200°C. WIP is normally used for powders that undergo a chemical reaction during the compression phase.

(c) Hot isostatic pressing (HIP). The basic function of a hot isostatic press is 'to uniformly heat a given workload, while gas pressure is applied to all surfaces, with accurate control of temperature and pressure'. Temperature in this case means up to 2200°C, while pressure is typically of the order of 100–400 MPa. Due to the very high temperatures that are used in HIP, the pressure medium needs to be a gas, such as argon, nitrogen, helium or even air. The high compressibility of gas, together with the time required to uniformly heat the load in the vessel, result in long cycle times: typically 6–12 hours. As CIP and WIP, HIP is mainly used to shape metallic and ceramic powders.

In addition, HIP is also frequently used for the densification of castings, i.e. to anneal their microporosity and as such improve their mechanical properties. For example, many parts used in the aeronautic and space industries are hot isostatically pressed to improve their fatigue resistance and reliability. HIP is also used to diffusion bond and densify simultaneously two different materials (e.g. ceramic and metal).

7.3.2 Quartz growing

The hydrothermal growth of quartz crystals takes place in a highly corrosive medium under conditions of high temperature (up to 420°C) and high pressure (up to 200 MPa). The entire process takes place in a forged high pressure vessel which if filled with a strong alkaline (NaOH) solution. The lower section of the vessel, the so-called dissolving section is filled with natural quartz nutrient. The upper 'growing' section of the vessel is filled with crystallisation seeds. Both sections are separated by a perforated plate and are kept at slightly different, but constantly controlled temperature; the lower section at higher temperature than the upper section. The nutrient slowly dissolves in the lower (hotter) section, and this solution rises by convection to the upper (cooler) section, where crystallisation occurs on the seeds to form large monocrystals. The quartz growing cycle takes from 24 to 36 days, depending upon the quality and size of the crystals required.

7.3.3 Chemical reactors

The rate and equilibrium of chemical reactions is influenced by pressure. Therefore some industrial chemical reactions (e.g. the synthesis of low density polyethylene) are performed under high isostatic pressure so as to increase the reaction rate and/or control the equilibrium towards maximum yield of the desired reaction product.

7.4 Current status of high hydrostatic pressure technology with a view to food processing

7.4.1 Introduction

High pressure technology is a perfectly mastered, well established technology. Table 7.1 summarises the main commercial applications of high pressure technology. Most of these applications have been applied on an industrial scale for already more than 30 years.

Just as for thermal processing, food can be HHP processed in two fundamentally different ways: *in-container*, that means that the HHP process is executed after filling and sealing of the food into its final or intermediate package, or *in bulk* followed by aseptic or ultra-clean filling and sealing. Both methods represent their own specific challenges which will be discussed in section 11.5.

This section takes a closer look at the current status of industrial HHP technology and identifies these technical requirements related to commercial HHP food processing that can be readily met.

Table 7.1 Today's commercial high pressure applications and typical process conditions

Field of application	Commercial process conditions		
	Pressure (MPa)	Temperature (°C)	Cycle time
Cold Isostatic Pressing (CIP)	50–600	Room temperature	0.3–5 min.
Warm Isostatic Pressing (WIP)	200–400	200–250	min.
Hot Isostatic Pressing (HIP)	100–400	1000–2000	5–10 hours
Quartz growing	200	430	4–6 weeks
Simulators	Various	Various	Various
Chemical reaction vessels	200–250	300	Months (continuous)

7.4.2 *HHP food process conditions: time, temperature and pressure*

The cycle times, temperatures and pressures applied in today's main commercial application of HHP technology are listed in Table 7.1. As shown, current commercial applications involve maximum working pressures ranging from 50 to 600 MPa, temperatures from room temperature up to 2200°C, and cycle times varying from as little as 20 seconds up to several weeks.

Research has shown that most commercially interesting applications of HHP technology in the food industry can be achieved by combining pressures in the range of 400–600 MPa, with temperatures from 5°C up to 90°C and cycle times of the order of 10–30 minutes. Given the above – and in the capacity range required for the niche foods markets in which HHP will be applied in the foreseeable future – the current status of HHP technology readily allows these process conditions to be met.

7.4.3 *Capacity requirements*

The HHP process is a batch process. In contrast to thermal processes, where both for in-container as well as aseptic processing continuous (hydrostatic steriliser, rotary cooker/cooler, etc.) and batch-wise (batch autoclave, orbitort, etc.) processes have been developed, HHP is, and in the foreseeable future will remain, essentially a batch-wise process: the current status of technology does not allow to transfer a packaged or even bulk food continuously from atmospheric pressure to a process pressure of several 100's MPa.

Therefore, just as for a batch autoclave system, the throughput of a HHP system is the product of three parameters: the number of cycles per vessel and per unit of time, the batch volume per vessel and the number of vessels in the system.

(a) Number of cycles per vessel and per hour. The number of cycles that one vessel can perform per hour is determined by the batch cycle time. This batch cycle time in turn is the sum of the material handling time – the time to load and unload the vessel, eventually including the time to open and close the vessel (only in the case of in-container HHP processing) – the dwell time at a given pressure and the pressurisation–decompression time.

The instantaneous character of hydrostatic pressure offers the possibility to achieve short dwell times. In contrast to thermal processing, the duration of a HHP treatment is not influenced by pressure transfer phenomena, but only by the kinetics of the inactivation of denaturation reactions contemplated at the given pressure. Recent European and Japanese scientific publications typically mention dwell times in the range of 1–30 minutes; for the current commercial applications in Japan, dwell times of the same order of magnitude are cited. Our own research has shown that dwell times as short as 5 minutes are sufficient to render some high acid foods commercially sterile.

The material handling time will vary strongly with the process applied: clearly bulk processes will require less time to replace a HHP processed batch by an unprocessed one, than in-container processes, where loading and unloading itself not only take more time but also necessitate the opening and closing of the vessel. In the case of in-container HHP processing, material handling times can be minimised by efficient automation of the loading/unloading operation and by using a fast opening/closing system. An interrupted threaded closure e.g. as used for CIP (clean in place) applications, allows opening and closing of the vessel in a few seconds.

Finally, the overall cycle time is determined by the compression/decompression time. For a given vessel volume and load compressibility, the compression time is solely a function of pump capacity. From a technical point of view the pressurisation of a vessel of several hundreds of litres to 400 MPa can be achieved within a few seconds by installing the appropriate pumping capacity.

However, economical and practical considerations will determine the optimal pumping capacity for a given high pressure system. The decompression rate is purely a technical issue; due to the low compressibility of food, decompression of a food processing HHP vessel can be achieved within a few seconds. The vessel cycle life is not affected by compression or decompression rates. The effects of fast compression/decompression on food quality, biological systems and food packages however need to be further investigated.

(b) Batch volume per vessel. The batch volume is the product of the internal vessel volume and the efficiency with which that volume is used, i.e. filled with food. At pressures as high as 600 MPa, which are required for food processing applications, there is a limit to size of the vessels that can be forged and consequently to the effective internal vessel volume. However, as Table 7.2 shows, the current state of technology allows the forging of extremely large vessels, with very significant internal volumes. Note for example, example No. 4 of a vessel with maximum working pressure of 550 MPa and internal volume of 700 litre.

As for the volume efficiency, expressed as the percentage of the internal vessel volume that is effectively filled with food, the bulk or aseptic process is more favourable than the in-container process. For the bulk process, volume efficiencies between 85 and 95% are possible; in the case of the in-container process, our studies have shown that volume efficiencies are situated between 45% and 75%.

Table 7.2 Examples of existing high pressure vessels (National Forge)

Example number	Maximum working pressure (Bar)	Diameter (mm)	Length (mm)	Internal volume (1)
1	1000	1700	4000	9000
2	2000	1000	4000	3150
3	4100	600	4500	1250
4	5500	600	2500	700
5	6900	250	750	37
6	10300	100	1000	8.5
7	13800	90	550	3.5

(c) Number of vessels in the system. Just as for batch autoclave systems, the throughput of a HHP system can be increased by multiplying the number of vessels. The multiple vessel concept is not new and already industrially used for 'dry bag' isostatic presses to reach mass production of technical ceramics. Technical elements, investment cost and operational cost considerations will determine the optimal configuration, i.e. the number and dimensions of the vessel(s) of a HHP system for a given capacity requirement.

(d) Conclusion. Given the above remarks concerning the three elements of the capacity equation, it is clear that capacities of several tonnes per hour can readily be met by today's status of HHP technology. Such capacities are adequate for the commercial HHP processing of the high quality, niche market foods to which HHP technology will be applied in the foreseeable future.

7.4.4 Fast cycling in combination with three shifts per day, 300 days per year operation

A short cycle time is essential for the economic feasibility of HHP food processing. Economic considerations also dictate an intensive use of this high capital cost equipment. A short cycle time in combination with three shifts per day, 300 days per year operation, means up to a hundred or more cycles per day and thousands of cycles per year per vessel.

Today, more than 30 years of industrial experience, especially in the materials industry where CIP systems with very short cycle times are in operation in very harsh conditions, allow the manufacture of high pressure vessels with long cycle lives. One system supplied by National Forge, for example, is composed of three vessels, each with an internal volume of more than 20 litres, each vessel cycling at 80 cycles per hour (45 seconds per cycle). Some of these fast cycling systems currently used for the production of technical ceramics and refractories operate 3 shifts per day, 7 days per week, as could be the case for the food industry.

7.4.5 *Process control*

Safe food processing with high pressure requires accurate and reliable pressure and temperature control inside the vessel. Pascal's law states that '. . . pressure exerted at any point upon a confined liquid is transmitted undiminished in all directions'. Consequently, by measuring and controlling the pressure in one point of the vessel, the pressure in all other points within the vessel is, by definition, controlled. As for temperature, today's applications of warm and hot isostatic pressing, quartz growing and chemical reactors show that temperature within a vessel can be measured and controlled with great accuracy. Synthetic quartz growing, for example, requires a temperature accuracy to within 1°C at temperatures in the range of 350–420°C, during a period of 4–5 weeks.

7.4.6 *Safety*

Two important points need to be mentioned with respect to safety, First, that the energy enclosed in a high pressure vessel at several 100's of MPa, filled with water or food, is very limited. The energy is stored in two ways: as elastic strain energy of the vessel and as the compression energy of the pressure medium. For all practical purposes the elastic strain energy is only a fraction of the total stored energy and can therefore be neglected.

The compression energy contained in a high pressure vessel can be approximated using the following formula:

$$\text{Energy} = \tfrac{2}{5} \times c \times P \times V_0$$

where c = compressibility of the pressure medium (i.e. water in the case of food processing), P = pressure, V_0 = initial volume.

Table 7.3 lists the compression energy contained in a high pressure vessel as a function of internal vessel volume for water at 400 MPa pressure. The table shows that the energy stored in a high pressure vessel during food processing is minimal. In fact, 1 kg of saturated steam at 121°C contains more energy than a high pressure vessel with an internal volume of 100 litres, filled with water at 400 MPa and room temperature.

Secondly, it is important to point out again that HHP technology is not a new technology. As high pressure systems are now commonly used in many industries – thousands of pieces of high pressure equipment are in operation worldwide – safety rules have been defined to assure that the

Table 7.3 Compression energy contained in a high pressure vessel filled with water at a 400 MPa, as a function of the vessel volume

Internal vessel volume (1)	Energy* (kJ)
10	192
50	960
100	1920
250	4800
1000	19200

1 kg of saturated steam at 120°C contains 2706 kJ.
1 six-cart FMC/Surdry retort (1400 mm diameter) filled with saturated steam at 120°C contains 29766 kJ.

pressure vessels as well as the main components of a high pressure system are perfectly safe for continuous operation.

7.5 The challenges of the commercial application of high pressure technology in the food industry

7.5.1 Technical challenges

(a) Introduction. Similarly to thermal processing, two basically different methods for HHP processing of foods have been proposed: bulk processing followed by aseptic filling and sealing and in-container processing. Figures 7.10 and 7.11 summarise respectively concepts for bulk and in-container processing that have been proposed in scientific literature today. Table 7.4 lists the specific advantages and drawbacks of each method. Obviously, both methods represent very different technical challenges. In the case of in-container processing, the main technical challenge lies in the material handling part of the high pressure system; in the case of the bulk-aseptic method the main technical challenge to be addressed in the aseptic design of the high pressure vessel(s). These and other technical challenges are briefly discussed below.

(b) Sanitation, cleaning, disinfection. Distinction needs to be made here between two methods. The bulk method, where there is direct contact between the pumpable food and the internal vessel surface, high pressure valves, sensors and tubing. And the in-container method where no direct contact between food and high pressure systems occur.

For the in-container process the same sanitation requirements and GMP-guidelines as for thermal in-container sterilisation systems apply. Namely that all water coming into contact with the food container during

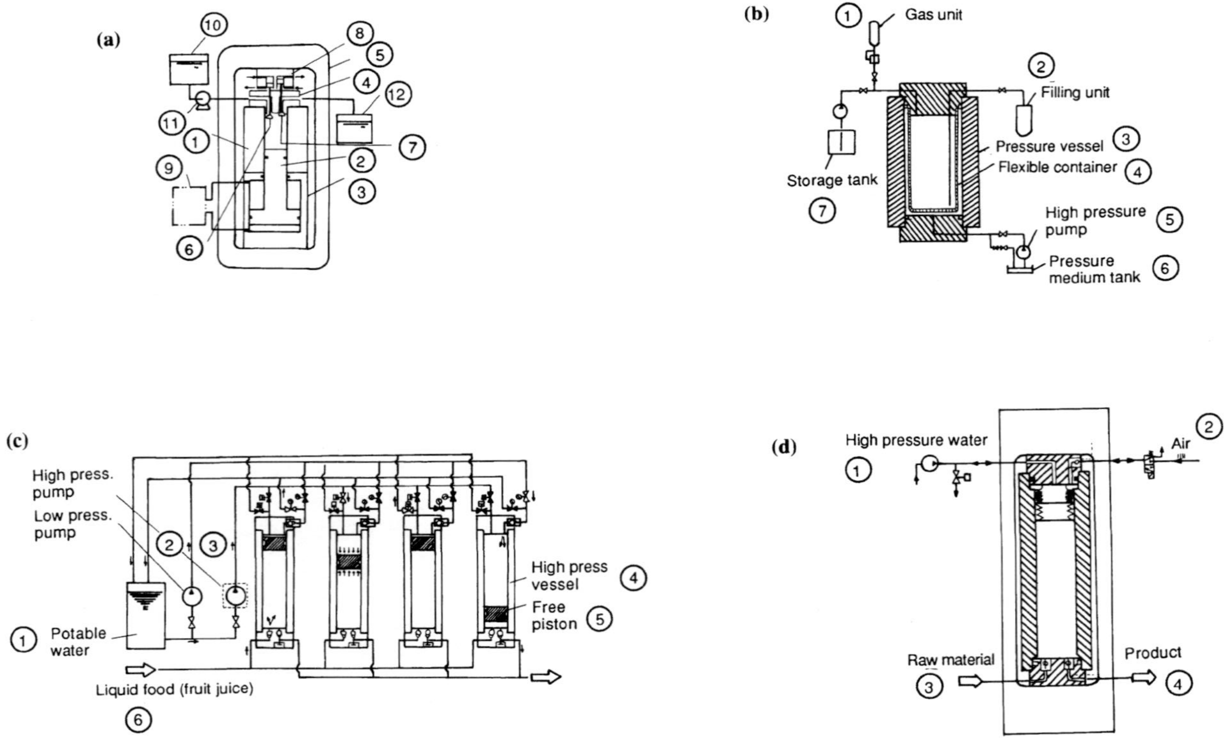

Figure 7.10 Concepts for the bulk HHP processing of pumpable foods. (a) Pressure multiplying piston (direct compression). 1. Pressure cylinder, 2. Pressure piston, 3. Hydraulic cylinder, 4. Top closure, 5. Yoke frame, 6. Inlet valve, 7. Outlet valve, 8. Actuator, 9. Hydraulic power unit, 10. Supply tank, 11. Feed pump, 12. Aseptic charge tank. (b) Membrane concept. 1. Gas unit, 2. Filling unit, 3. Pressure vessel, 4. Flexible container, 5. High pressure pump, 6. Pressure medium tank. (c) Free floating piston. 1. Potable water, 2. Low pressure pump, 3. High pressure pump, 4. High pressure vessel, 5. Free piston, 6. Liquid food (fruit juice), 7. Storage tank. (d) Bellow concept. 1. High pressure water, 2. Air, 3. Raw material, 4. Product.

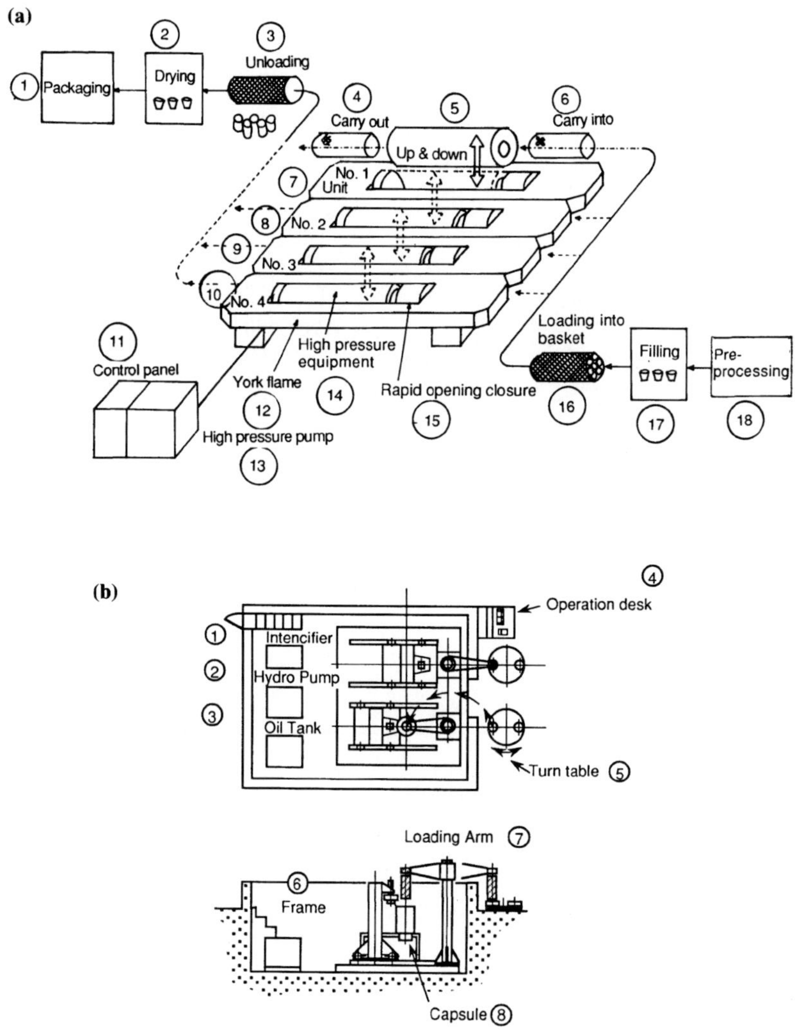

Figure 7.11 Concepts for in-container HHP processing of foods. (a) Horizontally placed vessels. 1. Packaging, 2. Drying, 3. Unloading, 4. Loading, 5. Up and Down, 6. Unloading, 7. Unit 1, 8. Unit 2, 9. Unit 3, 10. Unit 4, 11. Control panel, 12. Yoke frame, 13. High Pressure pump, 14. High Pressure equipment, 15. Rapid opening closure, 16. Loading into baskets, 17. Filling, 18. Pre-processing. (b) Vertically placed vessels with loading arm. 1. Intensifier, 2. Hydro pump, 3. Oil tank, 4. Operation desk, 5. Turn table, 6. Frame, 7. Loading arm, 8. Capsule.

Table 7.4 The advantages and drawbacks of bulk and in-container high pressure food processing

Advantages	
In-container	Bulk
Applicable to all foods, both liquid and solid	Simple material handling (pumps, pipes, valves, etc.)
Minimal risk of post-treatment contamination	High container flexibility (e.g. glass, metal containers are possible)
The high pressure part of the system is readily available. No major developments required	Maximum HP vessel volume efficiency (> 90%)
	Minimum vessel dead time (fast loading/unloading, no opening/closing of vessel required)
Disadvantages	
In-container	Bulk
Complex material handling	Only applicable to pumpable foods
Low volume efficiency (50–70%)	Still need aseptic filling step, increasing post treatment contamination risk
Low container flexibility	Aseptic design of all high pressure components in contact with food required
High high pressure vessel dead-time (loading/unloading, fill, vent, open, close)	

or after the HHP treatment must be of drinking water quality and that chemicals used for disinfection, lubrication and corrosion inhibition must be food compatible. Since there is no direct food contact, no specific material requirements apply.

The above requirements can easily be met by today's high pressure systems. If necessary, high pressure systems can be designed such that periodic cleaning and steam sterilisation of the internal vessel surface and other parts of the system are possible. When using adapted materials (stainless steel) for the construction of piping, internal vessel surface, valves and intensifier, pure distilled water can be used as pressure transmitting medium.

In case of bulk treatment, the high pressure system becomes part of an aseptic line and all of its food contact elements (vessel, tubing, valves . . .) must be CIP-able (Clean-In-Place) and SIP-able (Sterilise-In-Place). In order to allow aseptic or ultra-clean transfer of the processed food from the vessel to a buffer bank, at least one aseptic or ultra-clean high pressure valve is required.

(c) Package design. This technical challenge only applies to the in-container process. Literature shows that existing multilayer plastic and aluminium foil packages can be used in combination with the HHP process. Our own research has demonstrated that any container-lid combination, with sufficient overall flexibility to compensate the complete compression of the headspace and the limited volume reduction of the food inside the container, can be HHP treated without permanent physical deformation. Several authors have reported that the mechanical and barrier properties of multilayer plastic and aluminium foil materials, as well as the heat seal strength or integrity, are not affected by the pressure treatment [7–9].

For rigid packaging materials, like glass and metal, it has been suggested to improve the overall package flexibility by incorporating compression zones or combining glass or metal containers with flexible plastic closures. The feasibility of these solutions, however, still needs to be proven.

Package design is important for volume efficiency and hence for the economical feasibility of the technology. Container and lid should indeed be designed such that headspace is minimal. This is not only essential for achieving maximum volume efficiency, but also for minimising compression/decompression time and container stresses during pressurisation. In addition, volume efficiency can be optimised by adapting package shape and dimensions to the cylindrical shape of the pressure vessel in general and the diameter of the given vessel(s) in particular. Figure 7.12 shows how the use of hexagonally shaped one-litre packages, of which the dimensions have been adapted to the vessel dimensions, results in a volume efficiency of 75%.

(d) Material handling. The material handling system of a high pressure system for in-container food processing and of a conventional automated batch retort system contains the same elements.

The basic functions of a material handling system are as follows: a loader receives the flow of individual untreated containers and groups them into cylindrical stacks. A transport system picks up the cylindrical stacks of untreated containers and places them in the high pressure vessel; after pressure treatment the same transport system removes the treated stack from the vessel and transfers it to the unloader. The unloader individualises the treated containers again and places them on a take-away conveyor to secondary packaging. Empty trays or pallets are automatically transferred from the unloader to the loader.

All the technology mentioned above has already been developed for automated batch retort systems and can be applied to in-container high

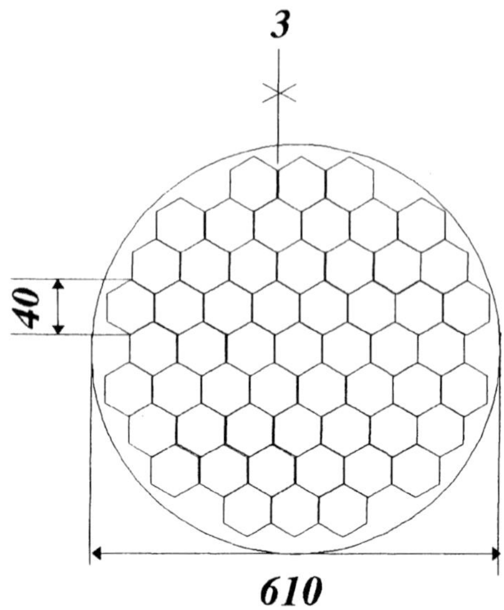

Figure 7.12 Example of container design, dimensions and configuration yielding high volume efficiency. Vessel internal diameter: 610 mm. Vessel internal length: 3000 mm. Vessel working pressure: 400 MPa. Horizontal clearance: 3 mm. Vertical clearance: 5 mm. Package shape: Hexagonal. Package dimensions: 40 × 40 × 40 × 240 mm. Package volume: 1 litre. 55 bottles per layer: 78% efficiency per layer. 12 layers per vessel: 660 bottles, 75% total volume efficiency.

pressure processing systems with the following adaptations:

(i) Stacks of containers are cylindrical instead of square or rectangular. This requires an adaptation of the container loading and unloading system.
(ii) Stacks for high pressure vessels typically have a length:diameter ratio of 1:3 to 1:4 whereas typical retort stacks are more or less cubic. This requires an adapted transport system.
(iii) high pressure vessels are mostly placed in a vertical position, most modern retorts in a horizontal position. This requires an adapted vessel loading/unloading system.

7.5.2 *Economical and commercial challenges*

(a) Investment cost. Whether or not high pressure technology will be commercially applied widely in the food industry will largely depend on the economical feasibility of the process; in particular the capital cost associated with the installation of a commercial high pressure system is an important barrier to its commercial implementation in the food industry.

The cost of the high pressure vessel represents the main fraction of the total cost of any industrial high pressure system and, at constant cycle life, is determined by the maximum working pressure and the required internal vessel volume. It is therefore important to minimise these two parameters.

Here lies a formidable challenge for food technologists and engineers, namely to optimise process conditions (pressure, temperature, time) such that both working pressure and the required effective volume for a given throughput are minimised, while still obtaining the desired unique food properties.

In addition to the process conditions, volume efficiency plays an important role in the economic feasibility of the high pressure process. Packages used for high pressure processing should be designed towards minimum headspace and maximum filling efficiency of the available cylindrical volume: a higher volume efficiency means less internal volume for a given throughput and consequently lower investment cost. In this phase also the bulk treatment – aseptic filling alternative should be evaluated since this processing method allows very favourable volume efficiencies.

Finally, once optimal process conditions and the required internal volume for the given throughput are defined, the high pressure system configuration (number of vessels, vessel dimensions, vessel position horizontal or vertical, loader/unloader configuration, material handling system, hoist and closure type, number of pumps, pump power) as well as the degree of automation have to be optimised towards minimum investment cost and operational cost.

(b) Identification of commercially feasible applications. This is possibly the most difficult of all challenges. First, it requires the identification of technically and economically feasible food applications: the identification of those foods for which distinctly unique properties are obtained, at a reasonable cost, by using high pressure processing instead of conventional food processing technologies. In this respect the need for a better basic understanding of the effects of high pressure on biological systems and for a better insight in the possibilities and limitations of the technology through fundamental research cannot be stressed enough.

Second, once technically and economically feasible applications have been identified, marketeers need to determine if the unique properties of the high pressure food justify the extra cost associated with the process or, in other words, if the consumer will accept the product and be prepared to pay a premium for the high pressure processed food.

7.6 Outlook

High pressure technology offers the food industry a unique opportunity to develop new foods of high nutritional and sensory quality, novel texture, more convenience, higher safety and increased shelf-life. The status of high pressure technology today is such that capacity, operating, process control and safety requirements for high pressure food processing can readily be met. Commercial high pressure food processing however, poses specific requirements in relation to sanitation and cleaning, package design, material handling and operational cost. The most difficult challenge of the commercial application of high pressure food processing probably lies in the uncertain field of marketing, i.e. identifying those niche applications for which the higher high pressure processing cost is justified by superior or unique product properties.

Acknowledgements

The author expresses his gratitude to G. Deplace and P. Colman (National Forge) and to E. Kurtz (FMC) for their valuable input in this publication.

References

1. Sasaki, I., Hori, K. and Inove, A. (1989) A high pressure equipment for food processing. In *Use of High Pressure Equipment in Food*, ed. by R. Hayashi. Japan, p. 191.
2. Kanda, T. (1990) Recent trend of industrial high pressure equipment and its application to food processing. In *Pressure Processed Food, Research and Development*, ed. by R. Hayashi. Japan, p. 335.
3. Hori, K., Manabe, Y., Kaneko, M., Sekimoto, T., Sugimoto, Y. and Yamane, T. (1992) The development of high pressure processor for food industries. In *High Pressure and Biotechnology*, ed. by Balny, C., Hayashi, R., Hereman, K. and Masson, P. France, p. 449.
4. Hori, K. (1990) Prospects and technical tasks as high pressure processor for food production. In *Pressure Processed Food, Research and Development*, ed. by R. Hayashi. Japan, p. 351.
5. Manabe, Y., Kaneko, M., Hori, K. and Sugimoto, Y. (1991) High pressure food processor and its system. In *High Pressure Science for Food*, ed. by R. Hayashi. Japan, p. 449.
6. Yamauchi, Y. and Onizuka, K. (1991) High pressure equipments for food processing. In *High Pressure Science for Food*, ed. by R. Hayashi. Japan, p. 436.
7. Kohno, M. and Nakagawa, Y. (1990) Packaging for high pressure food processing. In *Pressure-Processed Food, Research and Development*, ed. by R. Hayashi. Japan, p. 303.
8. Ochiai, S. and Nakagawa, Y. (1991) Package for high pressure food processing. In *High Pressure Science for Food*, ed. by R. Hayashi. Japan, p. 381.
9. Mertens, B. (1993) Packaging aspects of high pressure food processing technology. *Packaging Technology and Science*, **6(1)**, 31–36.

8 Hydrostatic pressure treatment of food: microbiology

D. KNORR

8.1 History and key issues of high pressure application

The first data in history on the effects of high hydrostatic pressure on organisms were most likely compiled by Certes (1883) who found viable bacteria in water samples obtained from 5100 m depth (approx. 50 MPa). The author then continued high pressure studies using a high pressure apparatus, with which hydrostatic pressures of 600 MPa could be obtained (Certes, 1884).

Hydrostatic pressure treatment (HPT) of food was applied for the first time by Hite (1899). This publication also represents the first reported case of the reduction of spoilage organisms by high hydrostatic pressure. Subsequent work by Hite and co-workers expanded from treatment of milk and meat products to fruit and vegetables (Hite *et al.*, 1914). Peaches and pears treated at about 400 MPa, at room temperature and for 30 min were still reported "good after 5 years of storage".

Chlopin and Tammann (1903) using pressures of about 300 MPa suggested that high pressure acted by reducing the vital functions of microorganisms and indicated that some of the organisms tested changed under high pressure into "a condition of faint from which they do not recover until some time".

Higher pressure resistance of spores than of vegetative cells was shown (Timson and Short, 1965) and pressure induced germination of bacterial spores was demonstrated by Gould and Sale (1970). Sale *et al.* (1970) offered a mechanism for pressure inactivation of bacterial spores by pressure induced initiation of germination followed by inactivation of the germinated forms. However, no further actions were taken at that time on the use of high pressure treatment in food processing.

In 1982 high pressure research on food systems was reinitiated at the University of Delaware (Hoover *et al.*, 1989) in the USA and in 1986 food-related high pressure activities were begun at Kyoto University (1989) in Japan. This resulted in the formation of the Japanese Society for High Pressure and in the launching of the first commercial products in 1990. Subsequently numerous academic and industrial activities have been carried out, especially in Japan and in Europe (e.g. Hayashi, 1990, 1991, 1992; Balny *et al.*, 1992).

High pressure is attractive to the food scientist and food processor because it possesses a number of unique advantages over conventional thermal processes (Knorr, 1993; Overview, 1993). Among those are the instant transmittance of high pressure throughout the food system which makes HPT independent of the size and geometry of the sample and shortens processing times (no 'heating' and 'cooling' periods are required). Pressure can be applied at ambient temperatures and the retention of many important quality factors of the raw material can be accomplished (Cheftel, 1992). Consequently it seems feasible that HPT can expand unit operations currently available to food processors and can result in the development of new products and processes (Eshtiaghi and Knorr, 1993).

Important factors that need to be considered during high pressure treatment are volume changes (e.g. decrease in the case of water, increase in the case of proteins), changes in the phase transition of water, decrease in pH due to enhancement of ionic dissociation resulting from electrostriction, and increase in temperature (Cheftel, 1992).

8.2 Current applications

A listing of current industrial applications of high pressure treated foods is given in Table 8.1. Various additional product developments are under way but due to the confidential nature of the projects data are not available.

Table 8.1 Current industrial applications of high pressure treated food products (after Cheftel, 1993)

Company	Product	Processing conditions
Meidi-ya	Jams, fruit dressing, fruit sauce (topping) yoghurt, fruit jelly	400 Mpa, 10–30 min, 20°C
Pokka Corp	Grapefruit juice	120–400 Mpa 2–20 min 20°C
Wakayama Food Industries	Mandarine juice	300–400 Mpa, 2–3 min 20°C
Nishin Oil Mills	Non-frozen tropical fruits ('freeze' at −18°C)	50–200 MPa
Fuji Ciku Mutterham	Beef (tenderization)	100–150 Mpa 30–40 min 20°C
Kibun	'Shiokara' (control of autolysis)	?

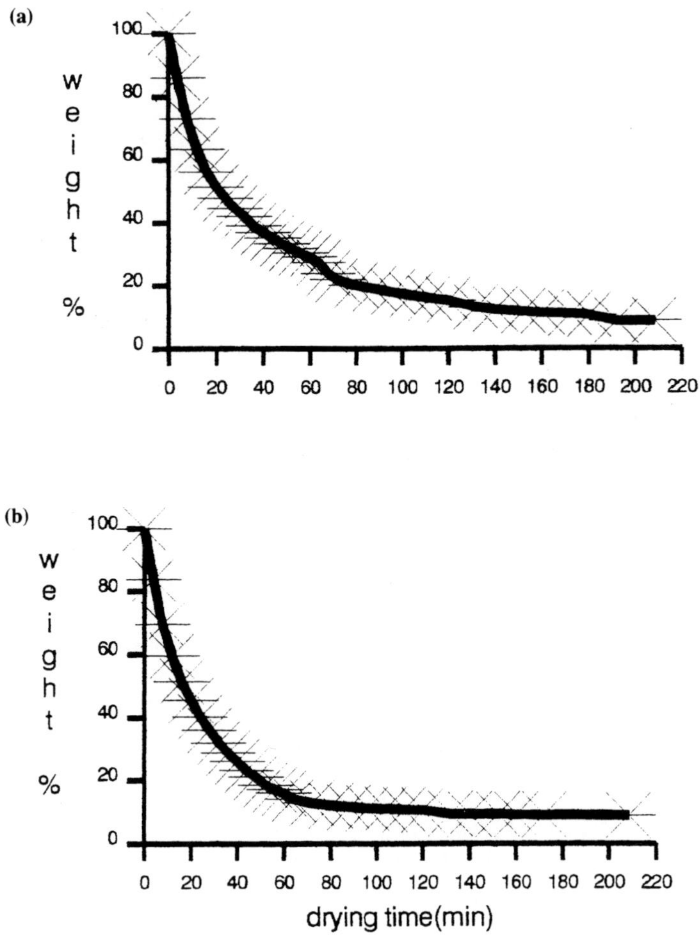

Figure 8.1 Drying curves of hot water blanched (a) and pressure pretreated apple cylinders (b). (After Estiaghi and Knorr, 1993, unpublished observations.)

Besides product developments process developments should also be mentioned. Among those are a combined freeze concentration/high pressure process (Watanabe *et al.*, 1991a), pressure shift freezing (Kanda and Aoki, 1993) and the development of a high pressure blanching process with less environmental pollution effects than conventional hot water blanching processes (Eshtiaghi and Knorr, 1993). Recently drying experiments have been carried out with pressure blanched fruits and vegetables. Data in Figure 8.1 represent a typical drying curve obtained for pressure blanched products.

8.3 Pressure effects on microorganisms

8.3.1 Possible mechanisms of action

According to Hoover *et al.* (1989) high hydrostatic pressure induces a number of changes to the morphology, biochemical reactions, genetic mechanisms, and cell membrane and wall of microorganisms.

Cellular morphology can be affected by high pressure. Walsby (1972) demonstrated that at pressures of 0.6 MPa intracellular gas vacuoles can collapse. Osumi *et al.* (1992) reported that in the case of yeasts above 200 MPa the cell wall was damaged and the subcellular structure altered with mitochondrial cristae being damaged. According to these authors areas with high electron density were noted in the matrix, membrane pores in the nucleus opened and were disrupted. Separation of the cell wall and disruption in the homogeneity of the intermediate layer between the cell wall and the cytoplasmic membrane was shown to occur (Kriss *et al.*, 1969). ZoBell (1970) found that bacteria from the early log phase of growth are normally more barosensitive than cells from the stationary, dormant or death phase. This is in agreement with our data where barotolerance of microorganisms at water activities below $a_w = 0.90$ was proven high, most likely due to a transition of the vegetative organisms into a dormant condition (Oxen and Knorr, 1993).

Proteins are important targets in pressure inactivation of cells and the fluidity of the membrane seems to play an important role in pressure inactivation. For example, *Rhodotorula rubra* cells suspended in 40–55% sucrose solutions were barotolerant up to 30°C of treatment temperature but could be inactivated at higher temperatures (Figure 8.2), most likely due to changes in the fluidity of the membrane.

Since the activity of Na/K ATPase is reduced by high pressure and is dependent on the associated bilayer (Chong *et al.*, 1985), and since a correlation exists between the activity of ion pumps and the decrease of bilayer fluidity and high pressure, the following hypothesis has been formulated by Smelt (1993): To maintain homeostasis the internal pH should be kept at value which should be sufficient to maintain a proton gradient across the cell membrane. If pressure is applied, several sites within the bacterial cell can be damaged. Among others, the following events may occur. Membrane-bound ATPase can no longer perform its function, either because of direct denaturation or because of dislocation in the membrane. ATP is no longer hydrolysed and hence is no longer available to perform the efflux of protons. The internal pH drops and the cell finally dies. Experimental work is underway to substantiate this hypothesis.

Spores have proven pressure resistance (Gould and Sale, 1970; Sale *et al.*, 1980; Timson and Short, 1965). Timson and Short (1965) discussed the mechanisms of the lethal effects of hydrostatic pressure and suggested that the primary mechanism involved is the ionization and concomitant precipitation of protein complexes. These authors went on to suggest that bacterial spores are more resistant to pressure than vegetative cells because the spore proteins are protected against solvation and excessive ionization by dipicolinic acid. Heden (1964) found that the melting temperature of DNA from *Bacillus subtilis* was 86–92°C, above which biological activity disappeared rapidly. However the melting temperature *increased* at a pressure of 270 MPa, which protected the biological activity of DNA against thermal inactivation.

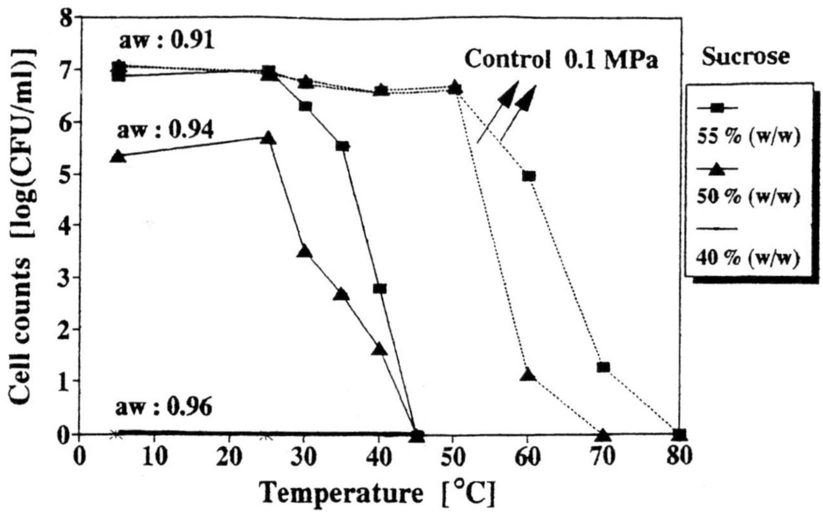

Figure 8.2 Effects of water activity, temperature of treatment and pressure (400 MPa, 15 min) on the viability of *Rhodotorula rubra* cells (after Oxen and Knorr, 1993).

Gould (1986) suggested the possibility of a highly viscous glassy state existing within the core (of a spore) and contributing to resistance and dormancy. The author further questioned whether, like raised temperature and increased hydration, raised pressure could tend to move such a system through its glass transition zone. Pressure and temperature certainly interact strongly in their destabilizing effects on spores. Sapru and Labuza (1993) recently used the polymer glass-transition theory to gain information about a possible general mechanismus to explain the high heat resistance of bacterial spores. Within this context it seems also of interest that germination of plant seeds was also affected by HPT (Davies, 1926). However, here too, far more evidence is required than is available yet.

8.3.2 Pressure inactivation of vegetative cells in food systems

Preservation of jam preparations by high pressure treatment could be performed (Hori *et al.*, 1991). However the destruction of *Saccharomyces cerevisiae* incorporated into strawberry jam (pH 3.3–3.4) was found to be less effective, and required higher pressure levels, when the sugar content was increased from 20 g/100 g to 50 g/100 g. The authors concluded that pressure-resistant *Candida* strains were present (Horie *et al.*, 1992). However, based on the data compiled by Oxen and Knorr (1993), it seems likely that the pressure resistance observed by Horie *et al.* (1992) was due to the physiological condition of the organism ('dormancy') at the reduced

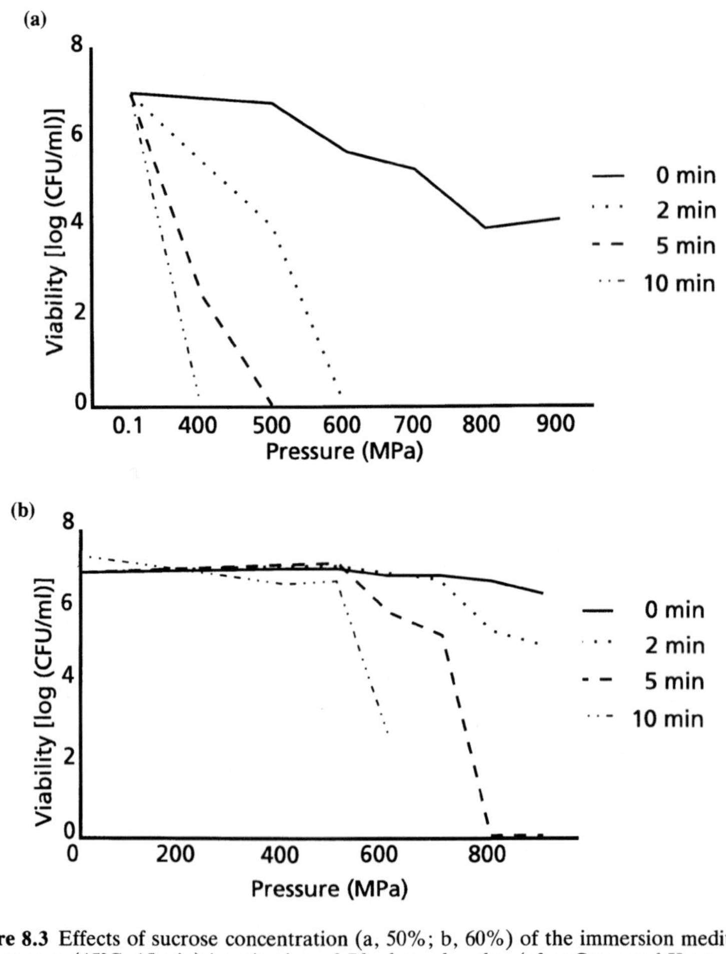

Figure 8.3 Effects of sucrose concentration (a, 50%; b, 60%) of the immersion medium on the pressure (45°C, 15 min) inactivation of *Rhodotorula rubra* (after Oxen and Knorr, 1993, unpublished observations).

water activity of $a_w = 0.94$. This effect, which could be overcome by an increase in temperature and time of HPT, is demonstrated in Figure 8.3.

It also seems noteworthy that the effectiveness of ethanol as an immersion medium on the reduction of *Rhodotorula rubra* was also hindered at high sucrose concentrations (Figure 8.4).

Pressure inactivation of yeasts and moulds has been reported in citrus juices (Ogawa *et al.*, 1990, 1992). Juices pressurized at 400 MPa for 10 min and at 40°C did not spoil during 2–3 months of storage. A population

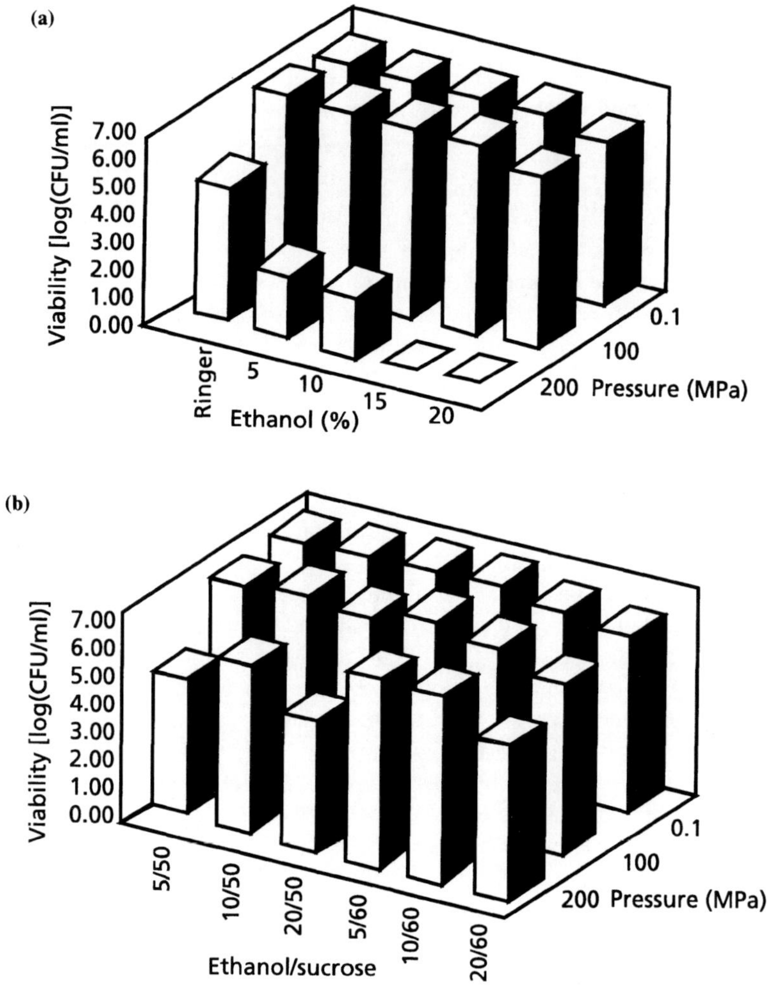

Figure 8.4 Effects of various ethanol concentrations (0–20%) in water (a) and various ethanol concentrations (5–20%) in 50% or 60% sucrose solutions (b) on the pressure inactivation (25°C, 15 min) inactivation of *Rhodotorula rubra*.

of 10^6 CFU/ml *Listeria monocytogenes* was inactivated by exposure to 340 MPa at 23°C in UHT milk and the same population of *Vibrio parahaemolyticus* was killed by exposure to 170 MPa within 10 min at 23°C in clam juice (Styles *et al.*, 1991). *Salmonella senftenberg* 775W, a very heat resistant vegetative microorganism could be reduced in chicken puree by high pressure treatment but was shown to recover in chicken-based medium (Hoover *et al.*, 1989). An interesting food application of high pressure treatment has been suggested by Watanabe *et al.* (1991b) who pressure inactivated cultures of *Erwinia ananas* which then could be safely used, in contrast to temperature-treated organisms, as ice-nucleation active cells in food systems.

8.3.3 Pressure effects on bacterial spores

Spore counts can be lowered by exposure to HPT but pressure treatment alone may not be sufficient for substantial reduction of viable spore counts (Hoover, 1993). This became evident with the very first pressure experiments with food systems (Hite, 1899; Hite *et al.*, 1914). Based on this evidence, pressure induced germination of spores has been examined (Clouston and Wills, 1969; Gould and Sale, 1970) and subsequent pressure treatment of the germinated/germinating spores was shown to be an effective means of reducing spore counts. However, 'super dormant' spores which may be present in the tail of a germination curve, may also be more resistant to germination induced by pressure (Russell, 1982). Consequently *combination processes* in conjunction with HPT may be required if sterilization of food products is sought. The effectiveness of a combined pressure induction of germination and subsequent HPT is demonstrated in Figure 8.5.

Based on these data, *Bacillus subtilis* spores may be among the most pressure resistant bacterial spores. Enfors and Molin (1978) examined the effects of carbon dioxide at 0.1–5.5 MPa on the germination of bacterial spores. On the one hand germination of *Bacillus cereus* spores was inhibited at atmospheric pressure but carbon dioxide strongly enhanced the germination rate of the *Clostridia*; on the other hand germination of *Clostridium sporogenes* and *Clostridium perfringens* was inhibited completely at 1 MPa and at 2.5 MPa respectively.

Pressurization cycles have recently been suggested by Hayakava *et al.* (1994). The authors reported that *Bacillus stearothermophilus* spores could be reduced by four to six log cycles when four to six cycles of oscillatory pressurization (600 MPa, 70°C, 5 min/cycle) were applied. However, Seyderhelm and Knorr (1992) showed similar results when applying static high pressure treatment for similar total times of treatment.

An interesting additional application of HPT was recently reported by Nakatomi (1993) who suggested high pressure as an effective means

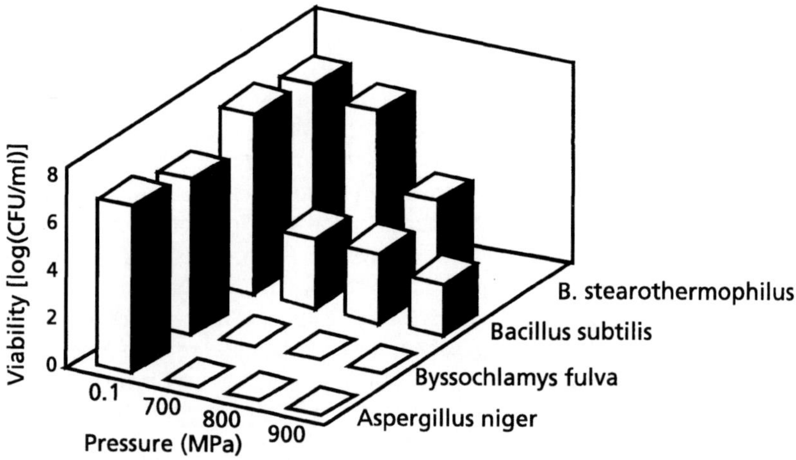

Figure 8.5 Effects of pressure induced germination of spores (100 MPa, 60°C, 20 min), followed by 45 min at 20°C and 0.1 MPa and then HPT (45°C, 2 min) on the inactivation of bacterial spores (after Oxen-Bodenhausen, 1994, unpublished observations).

for separating ascospores from sporulating cultures in *Saccharomyces cerevisiae*.

8.4 Combination treatments

Combination of mild heat treatment and HPT has been shown to be an effective means of reducing microbial counts in food systems (Takahashi et al., 1991). This can be demonstrated by data provided in Figure 8.2. At reduced levels of water activity of the immersion medium during HPT an increase in temperature from 20°C to 45°C clearly increased its effectiveness in reducing microbial counts. It is also important to note that the same level of inactivation of *Rhodoturola rubra* cells ($a_w = 0.91$) was achieved at 45°C at 400 MPa as was reached at 80°C, at atmospheric pressure.

Several recent studies indicate that it is possible to reduce bacterial spores through combinations of mild heat and HPT (Kinugasa *et al.*, 1992; Kowalski *et al.*, 1992; Seyderhelm and Knorr, 1992; Hayakawa *et al.*, 1994). In addition, it has been shown that the pressure-induced germination of spores is very temperature dependent. For example, pressurization of spores at low temperatures (e.g. near 0°C) even at pressures as high as 1000 MPa was relatively non-germinative and non-lethal, whereas, as the temperature was raised, pressure-induction of germination became very effective, even at pressures below 100 MPa (Gould and Sale, 1970). The pressure-germinated spores could then be inactivated by the pressure itself, if sufficiently high, or by some other treatment such as heat in combination with HPT (Clouston and Wills, 1969; Gould and Sale, 1970; Sale *et al.*, 1970; Gould, 1989; see also Figure 8.5).

The mode of action of high hydrostatic pressure on spores in these combination treatments has been suggested to be via electrostriction, i.e. a pressure-induced rise in the ionization of spore components. Analysis of the kinetics of pressure induced germination by Clouston and Wills (1970) indicated that it acted in this way to disturb a Donnan equilibrium in the spore so as to allow the central protoplast to begin to hydrate (Gould and Sale, 1970; Murrell and Wills, 1977). Gould (1989) indicated that the location and identity of the ionizable components remain speculative. Since dielectric studies have shown that the ions in the central, relatively dehydrated, protoplasts of spores are remarkably immobilized in some manner (Carstensen *et al.*, 1971), Gould (1989) further suggested that an increase in ionization could clearly raise their solubilities, and therefore osmotic effectiveness, with consequent germination, heat and radiation sensitization, by initiating osmotically driven hydration of the protoplast.

As with heat, ionizing radiation has been shown to reduce the resistance of spores to the subsequent application of hydrostatic pressure (Sale *et al.*, 1970). The effect, however, was an additive one, rather than a synergistic one (Wills, 1974, 1975). Gould (1989) indicated that the mechanisms of the additive effects are not known for certain but probably involve firstly the spore-germinative effect of pressure, and secondly the effect of irradiation on spore cortex peptidoglycan, allowing partial rehydration of the core with consequent reduction in radiation resistance.

Compressed gases including krypton, xenon, N_2O and CO_2 have been effective in inactivating vegetative cells (Thom and Marquis, 1984; Kamihira *et al.*, 1987; Stahl and Rau, 1985; Haas *et al.*, 1988; Kühne and Knorr, 1990; Lin *et al.*, 1992). Generally, it is considered that the major sites of action for these gases are within the hydrophobic regions of the cell membrane. The gases are highly hydrophobic and could dissolve in and distort these regions. They could also act by upsetting hydrophobic interactions in proteins (Thom and Marquis, 1984). As for the antimicrobial effect of supercritical carbon dioxide two possible mechanisms have been suggested. Lin *et al.* (1992) related the inactivation process to the diffusion of CO_2 into, and the extraction of vital biomaterials from, the cells. Haas *et al.* (1988) proposed diffusion of the highly diffusive and soluble CO_2 in its supercritical state, into the microbial cell and, if the water activity is sufficiently high, a conversion of CO_2 into carbonic acids which leads to a drop in internal pH and subsequent cell death.

The effectiveness of such a combination is demonstrated in Figure 8.6. However, no effects of supercritical CO_2 on bacterial spores could be observed (Titze *et al.*, 1994, unpublished data).

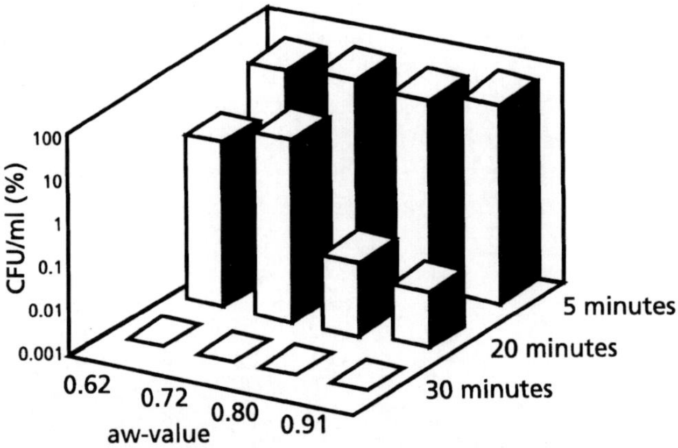

Figure 8.6 Effects of water activity of the samples and time of treatment on the inactivation of *Saccharomyces rouxii* with dense (static, 32 MPa, 45°C) carbon dioxide (after Titze *et al.*, 1994, unpublished observations).

Pretreatment with biopolymers prior to high pressure treatment has also proven successful. Papineau *et al.* (1991) reported antimicrobial effects of chitosan with high hydrostatic pressure on *Escherichia coli*, *Staphylococcus aureus* and *Saccharomyces cerevisiae*. However, no synergistic effects could be observed. Data in Figure 8.7 represent an example of the pretreatment of microbial suspensions followed by HPT.

Popper and Knorr (1990) and subsequently Baldwin and Robinson (1994) used chitosan or lysozyme and zymolase respectively as pretreatments prior to high pressure homogenization of vegetative cells. Popper and Knorr (1993) recently also demonstrated the effectiveness of enzyme

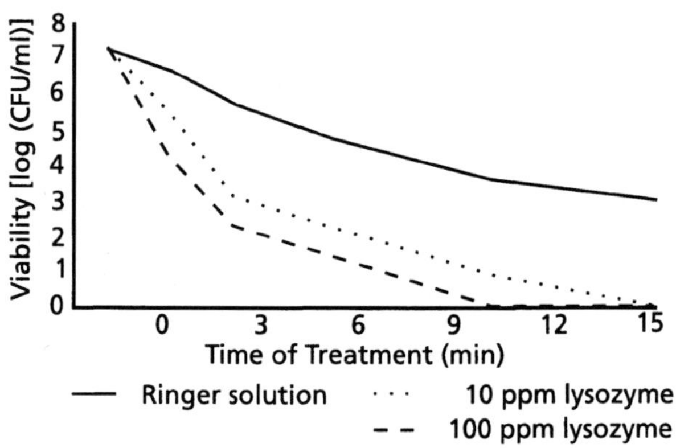

Figure 8.7 Inactivation of *E. coli* with high hydrostatic pressure (400 MPa, 25°C) with and without pretreatment with lysozyme.

cocktails (e.g. lysozyme, lactoperoxidase, glucose oxidase) on the inactivation of vegetative microbial cells in liquid foods at atmospheric pressure. It might be of interest to combine such a treatment with HPT.

Chemical agents such as ethanol or sodium sulphite as immersion media during HPT were also proven to be effective (Figure 8.8).

Ultrasonic waves have been shown to possess antimicrobial potential which is most likely due to dramatic and extremely short pressure and

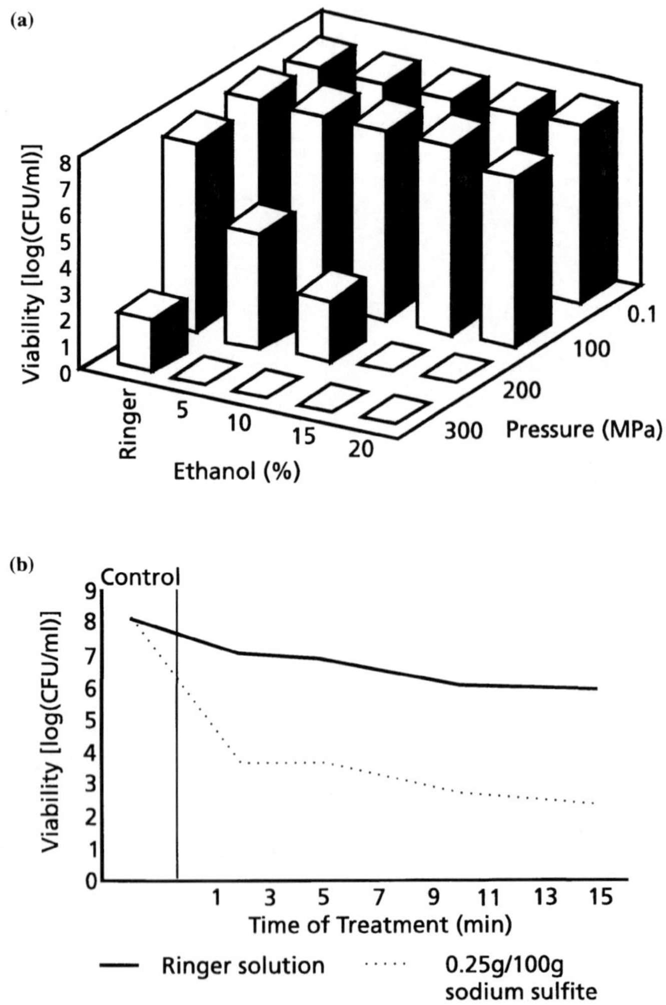

Figure 8.8 Effects of high pressure treatment (25°C, 25 min) in combination with ethanol (a) or sodium sulphite (b) as immersion media on the viability of *Listeria innocua* cells (after Oxen-Bodenhausen and Knorr, 1994, unpublished observations).

temperature changes as a consequence of cavitation (Heglein, 1987; Riesz, 1991). The potential of ultrasonic treatment in conjunction with HPT becomes evident when evaluating the data in Figure 8.9. While ultrasonic treatment alone was not highly effective in activating *Rhodoturola rubra* cells suspended in an isotonic solution or in 50% and 60% sucrose solutions, its combination with HPT proved successful. While *Rhodoturola rubra* immersed in 60% sucrose solution could not be completely inacti-

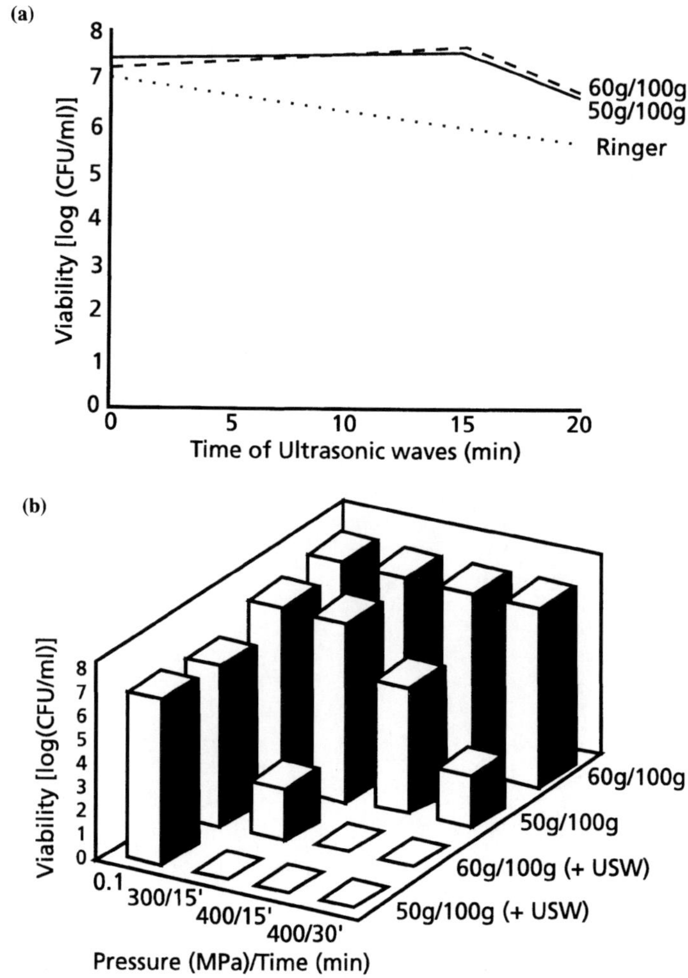

Figure 8.9 Effects of ultrasonic waves (100 W/cm^2, 25°C, 25 min) without (a) and with (b) subsequent pressure treatment on the viability of *Rhodotorula rubra* cells suspended in an isotonic solution or in 50% and 60% sucrose solutions (after Oxen-Boolenhausen and Knorr, 1994, unpublished observations).

vated even at 800 MPa for 15 min at 25°C (see Figure 8.3), pretreatment with ultrasonic waves (100 W/cm^2, 25°C, 25 min) made it possible to achieve inactivation at 400 MPa.

Decrease in the pressure tolerance of resting cells of *Escherichia coli* by pretreatment with alternating current (200 mA/cm^2) was reported by Shimada and Shimahara (1991) and shown to be dependent on the composition of the solutions surrounding the cells during pressure treatment (up to 400 MPa, room temperature, 10 min). We are actively involved in the application of high electric field pulses for food processing and preservation (Knorr *et al.*, 1994) and work is underway to identify the effectiveness of combinations of treatment with high electric field pulses and HPT on vegetative cells as well as on bacterial spores.

8.5 Conclusions

The evidence provided in this chapter suggests that high pressure has the potential to inactivate vegetative cells and, if pressure-induced germination can be achieved, to also kill bacterial spores. However, unless it becomes possible to completely understand pressure effects on microbial systems it is not likely that complete safety requirements can be assured. However, data need to be collected on whether super dormant spores are capable of germination during storage of pressure treated food systems or whether (and under which condition) they remain dormant. Data are also required on storage-dependent changes in food systems *after* HPT, and shelf-life-limiting factors need to be determined. It is also essential that increasing knowledge is gained on the impact of HP on enzyme activities in order to make high pressure processing an even more successful food preservation option.

Acknowledgements

Parts of the work presented have been supported by the German Research Association (DFG-Kn/260-3), the German Industrial Foundation (AIF-FV-8774), the European Community (AIR-296) and by CPC-Europe.

References

Baldwin, C.V. and Robinson, C.W. (1994) Enhanced disruption of *Candida utilis* using enzymatic pretreatment and high-pressure homogenization. *Biotechnol. Bioeng.*, **43**, 46–56.

Balny, C., Hayashi, R., Heremans, K. and Masson, P. (1992) *High Pressure and Biotechnology*. John Libbey Eurotext, Montrouge.

Carstensen, E.L., Marquis, R.E. and Gerhardt, P. (1971) Dielectric study of the physical state of electrolytes and water within *Bacillus cereus* spores. *J. Bacteriol.*, **107**, 106–113.
Certes, A. (1883) Sur la culture, a l'abri des gerunes atmospheriques, des eaux et des sediments rapportes par l'expedition du "Travailleur" et du "Talisman". *Compt. Rend.*, **98**, 690.
Certes, A. (1884) De l'action des hautes pressions sur les phenomenes de la putrefaction et sur la vitalité des microorganismes d'eau donce et d'eau de mer. *Compt. Rend.*, **99**, 385–388.
Cheftel, J.C. (1992) Effects of high hydrostatic pressure on food constituents: an overview. In Balny *et al.* (eds), *High Pressure and Biotechnology*. John Libbey Eurotext, Montrouge, p. 195.
Cheftel, J.C. (1993) *Hautes pressions on Japon*. Biosynthese Japon No. 5, Service pour la Science et la Technologie de l'Ambassade de France au Japon.
Chlopin, G.W. and Tammann, G. (1903) Ueber den Einfluss hoher Drucke auf Mikroorganismen. *Z. Hygiene Infektionskrank*, **45**, 171–204.
Chong, P.L., Fortes, P.A. and Jameson, D.M. (1985) Mechanisms of inhibition of Na/K ATPase by hydrostatic pressure studied with fluorescent probes. *J. Biol. Chem.*, **260**, 14484–14490.
Clouston, J.G. and Wills, P.A. (1969) Initiation of germination and inactivation of *Bacillus pumilus* spores by hydrostatic pressure. *J. Bacteriol.*, **97**, 684–690.
Clouston, J.G. and Wills, P.A. (1970) Kinetics of initiation of germination of *Bacillus pumilis* spores by hydrostatic pressure. *J. Bacteriol.*, **103**, 104–143.
Davies, P.A. (1926) Effect of high pressure on germination of seeds (*Medicago sativa* and *Melilotus alba*). *J. Gen. Physiol.*, **9**, 805–809.
Enfors, S.O. and Molin, G. (1978) The influence of high concentrations of carbon dioxide on the germination of bacterial spores. *J. Appl. Bacteriol.*, **45**, 279–285.
Eshtiaghi, M.N. and Knorr, D. (1993) Potato cubes response to water blanching and high hydrostatic pressure. *J. Food Sci.*, **58**, 1371–1374.
Gould, G.W. (1986) Water and the survival of bacterial spores. In Leopold, A.C. (ed.), *Membranes, Metabolism and Dry Organisms*. Comstock Publishing, Cornell University Press, Ithaca.
Gould, G.W. and Jones, M.V. (1989) Combination and synergistic effects. In Gould, G.W. (ed.), *Mechanisms of Action of Food Preservation Procedures*. Elsevier, London, p. 401.
Gould, G.W. and Sale, A.J.H. (1970) Initiation of germination of bacterial spores by hydrostatic pressure. *J. Gen. Microbiol.*, **60**, 335–346.
Haas, G.J., Prescott, H.E., Duddley, E., Dik, R., Hintlian, C. and Keane, L. (1988) Inactivation of microorganisms by carbon dioxide under pressure. *J. Food Safety*, **9**, 253–261.
Hayakawa, I., Kanno, T., Yoshiyama, K. and Fujio, Y. (1994) Oscillatory compared with continuous high pressure sterilization of *Bacillus stearothermophilus* spores. *J. Food Sci.*, **59**, 164–167.
Hayakawa, I., Kanno, T., Tomita, M. and Fijio, Y. (1994) Application of high pressure for spore inactivation and protein denaturation. *J. Food Sci.*, **59**, 159–163.
Hayashi, R. (1989) *Use of High Pressure in Food*. San-Ei Shuppan Co., Kyoto.
Hayashi, R. (1990) *Pressure-Processed Foods: Research and Development*. San-Ei Shuppan Co., Kyoto.
Hayashi, R. (1991) *High Pressure Science for Food*. San-Ei Shuppan Co., Kyoto.
Hayashi, R. (1992) *High Pressure Bioscience and Food Science*. San-Ei Shuppan Co., Kyoto.
Heden, C.G. (1964) Effects of hydrostatic pressure on microbial systems. *Bacteriol. Rev.*, **28**, 14–29.
Hegelein, A. (1987) Sonochemistry: historical developments and modern aspects. *Ultrasonics*, **25**, 6–16.
Hite, B.H. (1899) The effects of pressure in the preservation of milk. *West Virginia Agric. Exp. Sta. Bull.*, **58**, 15–35.
Hite, B.H., Giddings, N.J. and Weakley, C.E. (1914) The effect of pressure on certain microorganisms encountered in the preservation of fruits and vegetables. *West Virginia Agric. Exp. Sta. Bull.*, **146**, 2–67.
Hoover, D.G. (1993) Pressure effects on biological systems. *Food Technol.*, **47(6)**, 150–155.

Hoover, D.G., Metrick, C., Papineau, A.M., Farkas, D.F. and Knorr, D. (1989) Biological effects of high hydrostatic pressure on food microorganisms. *Food Technol.*, **43(3)**, 99–107.
Horie, Y., Kimura, K., Ida, M., Yosida, Y. and Ohki, K. (1991) Identification and pressure sterilization of a *Candida* strain resistant to high pressure. *Nippon Nogeikagaki Kaishi*, **68**, 713–718.
Horie, Y., Kimura, K., Ida, M., Yosida, Y. and Ohki, K. (1991) Jam preparation by pressurization. *Nippon Nogeikagaki Kaishi*, **65**, 975–980.
Kanda, Y. and Aoki, M. (1993) Development of pressure-shift freezing: Part I. Observations of ice crystals of frozen tofu. In Hayashi, R. (ed.), *High Pressure Bioscience and Food Science*. Scan-Ei, Kryoto, p. 27.
Kanihira, M., Taniguchi, M. and Kobayashi, T. (1987) Sterilization of microorganisms with supercritical carbon dioxide. *Agric. Biol. Chem.*, **51**, 407–412.
Kinugasa, H., Takeo, T., Fukumoto, K. and Ishihara, M. (1992) Changes in tea components during processing and preservation of tea extracts by hydrostatic pressure sterilization. *Nippon Nogeikagaku Kaishi*, **66**, 707–712.
Knorr, D., Geulen, M., Grahl, T. and Sitzmann, W. (1994) Food application of high electric field pulses. *Trends Food Sci. Technol.*, **5**, 71–75.
Knorr, D. (1993) Effects of high-hydrostatic-pressure processes on food safety and quality. *Food Technol.*, **47(6)**, 156–161.
Kowalski, E., Ludwig, H. and Tauscher, B. (1992) Hydrostatic pressure to sterilize food. 1. Application to pepper (*Piper nigrum* L.). *Deutsche Lebensm. Rundsch.*, **88**, 74–75.
Kriss, A.E., Mitskevich, I.N. and Cherni, N.E. (1969) Changes in the ultrastructure and chemical composition of bacterial cells under the influence of high hydrostatic pressure. *Mikrobiologiya*, **38**, 108–113.
Kühne, K. and Knorr, D. (1990) Effects of high pressure carbon dioxide on the reduction of microorganisms in fresh celery. *ZFL* (*J. Food Industry*), **41(10)**, EFS55–EFS57.
Lin, H., Yang, Z. and Chen, L. (1992) Inactivation of *Saccharomyces cerevisiae* by supercritical and subcritical carbon dioxide. *Biotechnol. Prog.*, **8**, 458–461.
Murrell, W.G. and Wills, P.A. (1977) Initiation of *Bacillus* spore germination by hydrostatic pressure: effect of temperature. *J. Bacteriol.*, **129**, 1272–1280.
Nakatomi, Y., Hamada, K. and Shimada, S. (1993) An efficient method for separating ascospores from sporulating cultures in *Saccharomyces cerevisiae* by hydrostatic pressure. *Biosci. Biotech. Biochem.*, **57**, 170–171.
Ogawa, H., Fukuhisa, K., Kubo, Y. and Fukumoto, H. (1990) Pressure inactivation of yeasts, molds, and pectinesterase in Satsuma mandarin juice: effects of juice concentration, pH, organic acids, and comparison with heat sanitation. *Agric. Biol. Chem.*, **54**, 1219–1225.
Ogawa, H., Fukuhisha, K. and Fukumoto, H. (1992) Effect of hydrostatic pressure on sterilization and preservation of citrus juice. In Balny *et al.* (eds), *High Pressure and Biotechnology*. John Libbey Eurotext, Montrouge.
Osumi, M., Yamada, N., Sato, M., Kobori, H., Shimada, S. and Hayashi, R. (1992) Pressure effects on yeast cell ultrastructure: changes in the ultrastructure and cytoskeleton of the dimorphic yeast, *Candida tropicalis*. In Balny *et al.* (eds), *High Pressure and Biotechnology*. John Libbey Eurotext, Montrouge, p. 9.
Overview (1993) Use of hydrostatic pressure in food processing. *Food Technology*, **47(6)**, 149–172.
Oxen, P. and Knorr, D. (1993) Baroprotective effects of high solute concentrations against inactivation of *Rhodotorula rubra*. *Lebensm. Wiss. Technol.*, **26**, 220–223.
Papineau, A.M., Hoover, H.G., Knorr, D. and Farkas, D.F. (1991) Antimicrobial effect of water-soluble chitosans with high hydrostatic pressure. *Food Biotechnol.*, **5**, 45–57.
Popper, L. and Knorr, D. (1990) Application of high pressure homogenization for food preservation. *Food Technol.*, **44(7)**, 84–89.
Popper, L. and Knorr, D. (1993) Nicht-thermische Inaktivierung von Mikroorganismen durch antimikrobielle Enzymsysteme. *Bioengineering*, **9**, 27–34.
Riesz, P. (1991) Free radical generation by ultrasound in aqueous solutions of volatile and non-volatile solutes. *Adv. Sonochem.*, **2**, 23–64.
Russell, A.D. (1982) *The Destruction of Bacterial Spores*. Academic Press, London, p. 259.
Sale, A.J.H., Gould, G.W. and Hamilton, W.A. (1970) Inactivation of bacterial spores by hydrostatic pressure. *J. Gen. Microbiol.*, **60**, 323–334.

Sapru and Labuza, T.A. (1993) Glassy state in bacterial spores predicted by polymer glass transition theory. *J. Food Sci.*, **58**, 445–448.

Seyderhelm, I. and Knorr, D. (1992) Reduction of *Bacillus stearothermophilus* spores by combined high pressure and temperature treatments. *ZFL* (*J. Food Industry*), **43(4)**, 17–20.

Shimada, K. and Shimahara, K. (1991) Decrease in high pressure tolerance of resting cells of *Escherichia coli* K-12 by pretreatment with alternating current. *Agric. Biol. Chem.*, **55**, 1247–1251.

Smelt, J.P.P.M. (1993) *High Pressure Inactivation of Microorganisms: Possible Mechanisms of Inactivation*. Progress Report, EC-AIR project No. 296, Bruxelles.

Stahl, E. and Rau, G. (1985) Hochdruck-Behandlung von Mikroorganismen. *Naturwissenschaften*, **72**, 144–145.

Styles, M.F., Hoover, D.G. and Farkas, D.F. (1991) Response of *Listeria monocytogenes* and *Vibrio parahaemolyticus* to high hydrostatic pressure. *J. Food Sci.*, **56**, 1404–1407.

Takahashi, K., Ishi, H. and Ishikawa, H. (1991) Sterilization of microorganisms by hydrostatic pressure at low temperature. In Hayashi, R. (ed.), *High Pressure Science for Food*. San-Ei, Kyoto, p. 225.

Thom, S.R. and Marquis, R.E. (1984) Microbial growth modification by compressed gases and hydrostatic pressure. *Appl. Environ. Microbiol.*, **47**, 780–787.

Timson, W.J. and Short, A.J. (1965) Resistance of microorganisms to hydrostatic pressure. *Biotechnol. Bioeng.*, **7**, 139–159.

Walsby, A.E. (1972) Gas filled structures providing buoyancy in photosynthetic organisms. In Sleigh, M.A. and Macdonald, A.G. (eds), *The Effects of Pressure on Living Organisms*. Academic Press, New York, p. 233.

Watanabe, M., Arai, E., Kumeno, K. and Honma, K. (1991a) A new method for producing non-heated jam sample: The use of freeze concentration and high pressure sterilization. *Agric. Biol. Chem.*, **55**, 2175–2176.

Watanabe, M., Makino, T., Kumeno, K. and Arai, S. (1991b) High-pressure sterilization of ice nucleation-active bacterial cells. *Agric. Biol. Chem.*, **55**, 291–292

Wills, P.A. (1974) Effects of hydrostatic pressure and ionising radiation on bacterial spores. *Atomic Energy* (*Austral.*), **17**, 2–10.

Wills, P.A. (1975) Inactivation of *Bacillus pumilus* spores by combination hydrostatic pressure–radiation treatment of medical products. In *Radiosterilization of Medical Products*, International Atomic Energy Agency, Vienna, p. 45.

ZoBell, C.E. (1970) Pressure effects on morphology and life processes of bacteria. In Zimmermann, A.M. (ed.), *High Pressure Effects on Cellular Processes*. Academic Press, New York, p. 85.

9 Effect of heat and ultrasound on microorganisms and enzymes

F.J. SALA, J. BURGOS, S. CONDÓN, P. LOPEZ and J. RASO

9.1 Historical perspective

Heat treatment and low temperature storage are at present amongst the main methods of food preservation.

Cold storage has the advantage of leaving nutritive value and organoleptic properties of foods almost unchanged. However refrigeration does not necessarily ensure food safety.

Heat, with its destructive effect on enzymes and microorganisms ensures safety and long preservation, but the effects of heat on foods can also be very detrimental. Some heat preservation processes cause significant reductions in nutritive value and drastic changes in organoleptic properties.

As no other method of food preservation to replace heat has yet been developed, a new approach in food preservation has arisen. This is the combination of heat with other methods that together enhance the lethal effect of heat on microorganisms and enzymes, and result in a lowering of the intensity of conventional heat treatments. This approach has lately revived the interest of scientists in what has become known as the 'preservation of foods by combined processes'.

Although the preservation of foods by combined processes is not new, the advance of scientific knowledge has opened new possibilities: the influence of different microenvironmental parameters on the heat resistance of microorganisms and enzymes is now much better known and new lethal effects of some physical phenomena have been reported. For example, the combination of heat and reduced pH to lower microbial heat resistance allowed the use of milder heat treatments and was one of the first combined processes to be employed. Other combinations widely used are those with sodium chloride, nitrite, etc. More recently, technological advances have led to the investigation of other possible combinations: These include the combination of heat with ultrahigh pressures (Gould, 1973), which is currently attracting the interest of many research groups worldwide and, most recently, the combination of heat with ultrasound.

The idea of using ultrasound to speed up chemical reactions (known as sonochemistry) and to inactivate microorganisms is not new. It was first suggested at the beginning of this century (Harvey and Loomis, 1929). But the idea of enhancing the effect of heat by ultrasound, in order to reduce the severity of current heat processes, is new, and is now the main subject of our research group.

9.1.1 *Heat inactivation of microorganisms and enzymes*

(a) Microorganisms

Although the preservation of foods by heat was first performed in France by Nicholas Appert around 1810, this remained for a long time an empirical practice until the scientific knowledge on the mechanism of the preservation effect began to accumulate. The works of Bigelow (1921) finally established the sound basis on which, still today, current methods of heat preservation rely.

The observation by Bigelow that the death of microorganisms followed a first order reaction kinetic pattern was essential for the future development of the technology of food preservation by heat. Bigelow showed that every unit of heating time of a microbial population at a given temperature reduced the number of viable cells by a constant proportion. By plotting the log of the number of survivors as a function of heating times, a straight line is therefore obtained. In this plot (survival curve), the minutes needed to reduce the number of viable cells to 1/10 (one log cycle) of its original value is now known as the 'decimal reduction time', or D_t value. When log D_t values are plotted vs their corresponding heating temperatures (Decimal Reduction Time Curve; DRTC), again a straight line is obtained. The number of degrees Celsius of temperature increase for the log D_t value to decrease by one log cycle is known as z value. The heat resistance of microorganisms is defined by these two parameters.

Once a D_t value is known, the kinetics of death (Survival curve) allows prediction of the numbers of survivors after a given heating time. Furthermore, as z value allow the calculation of the lethal effect of each temperature, the total lethal effect of any given process, including heating and cooling phases, can be estimated, thus avoiding undue overprocessing.

Work carried out during the last decades by food microbiologists on the influence of different factors on heat resistance of microorganisms and on the kinetics of death, has led to some authors to question the validity of some published heat resistance data and that of the Bigelow's kinetics.

It appears that heat resistance data (D_t and z values), once considered as well defined and constant parameters are in fact very variable, being influenced by many factors. For example, the pH of heating menstruum is one of the most important and one of the first to be known. But many others, such as the water activity (a_w) (Alderton *et al.*, 1980); sporulation temperature (Beaman and Gerhardt, 1986; Condón *et al.*, 1992b) and growth medium (Donnelly and Busta, 1980), composition of heating medium (Blocher and Busta, 1983; Condón and Sala, 1991) and incubation temperature and medium after heat treatment (Cook and Gilbert, 1968; Feeherry *et al.*, 1987) have also been investigated.

Some heat resistance data reported in literature should therefore be accepted with caution as factors influencing these data were unknown to the authors or not taken into account. The effect of some influencing factors can be so big as to make the differences in heat resistance between two populations of the same strain, bigger than those between two unrelated species (Put and Aalbersberg, 1967).

The capacity of different parameters to strongly influence the heat resistance of microorganisms is today an important issue in thermobacteriology, as are the deviations from theoretical death rate kinetics, such as those reported by some authors on survival curves and DRTC. Among different deviations reported on the patterns of survival curves (Moats *et al.*, 1971; Brown and Ayres, 1975), 'tails' and 'shoulders' are the most frequent and best characterized. A combination of both can explain most, if not all deviations of linearity of survival curves.

'Tails' are end portions of survival curves that appear with a decreasing slope at the final stages of heating. A comprehensive review of the 'tail' phenomenon is that of Cerf (1977). The different hypotheses to explain this phenomenon have been classified by Cerf into two groups of theories: 'vitalistic' and 'mechanistic'. 'Vitalistic theories' try to explain deviations from linearity of survival curves by postulating a different heat resistance for each individual cell in a population. 'Mechanistic theories' assume a logarithmic death rate and blame the appearance of deviations to methodological artefacts (Stumbo, 1973) or to the development of a higher heat resistance *during* heat treatments (Mackey and Derrick, 1986a). The development of a higher heat resistance of vegetative cells during heat treatments, reported by different authors, would be due to metabolic changes (Mackey and Derrick, 1986b) or to interactions with the heating menstruum. Although the fraction of the population having a higher heat resistance is normally very small, its heat resistance can be so high that it can become the factor determining the intensity of heat treatments (Moats *et al.*, 1971; Condón *et al.*, 1992a). Current concern to reduce the intensity of heat treatments in order to improve the quality of food products has led to a more detailed study of death rate kinetics and to model heat treatments in such a way as to take into account deviations of logarithmic death rate (Cole *et al.*, 1993).

'Shoulders', appear in the first portion of the survival curves, with different shapes. In 'shoulders' the slope of survival curves is always smaller and sometimes the number of survivors not only does not decrease, but can even increase. This phenomenon is less frequent in vegetative cells and some authors have related it to cell clumps disgregation (Hansen and Riemann, 1963). 'Shoulders' are much more frequent in spore suspensions.

In some authors' opinions (Shull *et al.*, 1963; Lewis *et al.*, 1965), in spores, 'shoulders' are often caused by a lack of 'activation'. A high proportion of a spore population is often unable to germinate. The spores are in a latent ('dormant') state. Some chemical/physical treatments can 'activate' them, restoring their germination capacity. Heat is a well-known activation agent. During the first moments of a heat treatment two opposite phenomena therefore take place, each at a different rate and both catalysed by heat: the activation of 'dormant spores' and the concurrent and subsequent inactivation of all spores. It is now believed that activation follows, as does heat inactivation, first-order reaction kinetics (Abraham *et al.*, 1990; Sapru *et al.*, 1993). The profile of shoulders would be determined by the balance between both rate constants.

In the last 30 years attempts have been made to develop mathematical models of death rates that would include 'shoulders'. Shull *et al.* (1963), in his attempt to develop the first model, postulated that activation and heat inactivation were two separate and successive phenomena. This opinion was also shared by other authors (Abraham *et al.*, 1990) who also postulated that activation is in fact the limiting factor of the inactivation phenomenon. On the contrary, other authors (Rodriguez *et al.*, 1991; Sapru *et al.*, 1992) believed that both phenomena are simultaneous and a prior 'activation' is not necessary for spores to be inactivated by heat. There is no agreement about whether the heat resistance of activated and dormant spores is the same. While some have developed mathematical models assuming equal heat resistances (Rodriguez *et al.*, 1991), the models of others assume that they are different (Sapru *et al.*, 1993).

Other deviations of linearity of death kinetics in DRTC have also been reported. However, these are less known and there is no agreement among authors. While some investigators have reported that z values increase at higher temperatures of treatment (DRTC curves bend upwards) (Wang *et al.*, 1964; David and Merson *et al.*, 1990) others have reported that they decrease (Cerf and Hermier, 1973; Hermier *et al.*, 1975). In some authors' opinion (Cerf and Hermier, 1973) the decrease of D_t values at high temperatures of treatment could be due to a thermal shock that would be greater the higher the temperature of treatment, causing the DRTC to bend downwards (decreasing z values). According to these authors the higher D_t and z values at higher temperatures reported in literature would be explained by methodological errors. Difficulties in the measurement of the very short heating times involve in high temperature treatments would result in poor estimations of heating times. Much therefore remains uncertain about heat resistance and death kinetics.

Despite substantial efforts carried out during the last half of this century, the mechanism(s) of heat inactivation are not yet clear. According to the first interpretation, the strict logarithmic order of death, as postulated by Bigelow, could most easily be explained by a mechanism involving the destruction of one single or a small number of vital molecules per cell. As the kinetics of cell death have become better known many authors have attempted to explain deviations from linearity by implying in these mechanism different vital molecules and/or structures. Heat has been reported to damage different cell structures, including damage to cell membranes, ribosomes, DNA, RNA and enzymes. DNA is still considered the most likely lethal target molecule, but damage occurring at the same time in different molecules and/or structures may also result in heat inactivation. Some of these injuries can be repaired and ultimately it is the balance of intensity of injury/capacity to repair that determines cell viability. A more detailed review of the mechanisms of microbial heat inactivation and injury is that of Gould (1989).

(b) Enzymes. Although enzyme inactivation is often required for food stabilization, the food industry is generally not concerned with it because heat treatments for microbial destruction exceed those required for the inactivation of enzymes. However, there are some food industries, such as those involved in the production of citric juices, tomato pastes and juices, and those engaged in vegetable drying or freezing, which are very concerned with the inactivation of endogenous deleterious enzymes. For example, a small fraction of orange pectinesterase that is highly heat resistant is responsible for orange juice cloud loss and serious negative effects on juice reconstitution from concentrates (Versteeg, 1979; Versteeg *et al.*, 1980). Extracellular proteases and lipases from psychrotrophic bacteria can be a problem in the dairy industry as they can become the limiting factor for long-term preservation of UHT milk (Cogan, 1977; Law, 1979; Burton, 1988). For these food industries any combined treatment which would add to the enzyme inactivation effect of heat with little or no damage to the nutritive value or sensory quality of the food would be valuable.

Active centres of enzymes consist of amino acid residues, usually far apart from each other in the primary structure, but brought together in the native three dimensional structure of the macromolecule. A number of physical and chemical agents, such as heat, acidic or alkaline pH, proteases, oxygen, chaotropic agents, etc. are capable of unfolding the molecule (Putman, 1954; Joly, 1965) disassembling the active centre and inactivating the enzyme. From the practical standpoint heat is the most important inactivating agent (Klibanov, 1983).

The native catalytically active structure is maintained by a delicate balance of different non-covalent forces, mainly ionic and hydrophobic interactions and hydrogen bonds. An increase in temperature diminishes all these forces (except hydrophobic interactions, whose strength increases up to approximately 110°C; Privalov *et al.*, 1986; Baldwin, 1986) inducing protein unfolding. Protein unfolding is thought to be the first step in heat enzyme inactivation (Tanford, 1968; Lapanje, 1978). While heat inactivation is an irreversible phenomenon, protein unfolding is a reversible one. The irreversibility of enzyme inactivation by heat is brought about by secondary events which are highly specific for individual enzymes. These secondary events can be divided into two groups (Klibanov, 1983): covalent and non-covalent changes. Covalent changes result in chemically altered enzymes. Non-covalent changes consist of two types of transformations: polymolecular (aggregation) and monomolecular (incorrect folding). Non-covalent changes seem to be prevalent at low temperatures and close to neutral pH, whilst at high temperatures, long heating periods and extreme pH values, chemical changes appear to be involved. Aggregation takes place mainly in concentrated solutions.

Due to differences in the type and number of bonds and interactions involved in maintaining the native conformation of the active centre and other parts of the enzyme molecule and to the fact that they are not equally affected by heat and other protein unfolding agents, in a number of enzymes, inactivation occurs before noticeable conformational changes can be detected (Liang *et al.*, 1990; Kelly and Price, 1991; Zhou *et al.*, 1993). On the contrary, in some instances, heat inactivation of enzymes may be preceded by extensive molecular unfolding (Dominguez *et al.*, 1992).

Enzyme resistance to heat inactivation is very dependent on environmental factors such as pH, ionic strength, soluble solids and presence of other denaturing agents. All these factors may, by themselves, induce important conformational changes which may not affect catalytic activity (Dominguez *et al.*, 1992). However, they can modify the bonds and interactions playing the prevalent role of maintaining the macromolecular structure. These conformational changes influence heat resistance and the nature of the changes that heat can induce.

The heat inactivation of enzymes is generally considered as being a single-step, two-state (active–inactive) process. An alternative kinetic model has been proposed (Ray and Koshland, 1961) in which partially inactivated states, which retain fractional activity, have to be taken into account.

Single-step two-state processes follow first-order kinetics and there are very few instances in which heat enzyme inactivation has been shown to be a true non first-order process. Therefore two approaches can be followed to analyse data obtained in enzyme heat inactivation experiments. Residual enzyme activity can be plotted, either as it is done with survival curves in thermobacteriology (see above) to obtain D_t and z values or, alternatively, data can be processed in accordance with the classical equations used in the study of first order chemical reactions. This way the timecourse of heat inactivation at constant temperature is expressed in terms of rate constants (K) and the temperature dependence of the inactivation reaction(s) in terms of activation energy (*Ea*). The rate constant of heat inactivation (k) is obtained from plots of log of residual activity as a function of time (slope = $-k/2.303$) and the *Ea* from Arrhenius plots (Slope = $-Ea/R$).

There are a number of reports of enzyme heat inactivation in which the plots of log of residual activity vs time appear to be nonlinear. Most of the time the deviation of linearity is due to the presence of more than one enzyme (or isoenzyme) catalysing the same reaction. In some instances, the plot of log of activity *vs.* time gives a good fit to second-order kinetics (Dominguez *et al.*, 1992) due to the formation of enzyme aggregates.

9.1.2 Destructive effect of ultrasound waves on microorganisms and enzymes

Although ultrasound has been known for more than a century and the first scientific mathematical approach was reported in 1917 by Lord Raleigh, the development of the use of ultrasound has been relatively slow. The lethal effect of ultrasound in microorganisms was first reported by Harvey and Loomis in 1929. This was the beginning of intense activity in this field during the 1930s.

The development of the investigations on the effect of ultrasound on microorganisms was parallel to that of the technology of instrumentation. The first ultrasonic apparatuses were piezoelectric generators of quartz submerged in oil, that generated ultrasonic waves of a very high frequency but low intensity (10 W/cm^2 approx.). Modern ultrasonic instruments consist essentially of a piezoelectric generator having a crystal of zirconate titanate that changes its shape under the effect of an electric field of 1000 V and a frequency of 20 kHz, supplied by a standard 50–60 Hz, 120–220V converter. This electric energy, transformed into mechanical energy of the same frequency, is transmitted to a titanium alloy disruptor horn. The horn transmits and amplifies this energy onto its tip, which is submerged into the menstruum being ultrasonicated. Amplification is dependent on the volume and shape of the disruptor. The oscillation of the horn's tip generates the sonic wave which is transmitted into the liquid. The availability of these new instruments in the last decade has revived interest in the effect of ultrasound on the kinetics of chemical reactions, and as catalysts of new ones, promoting the remarkable advance in sonochemistry. The renewed interest has also led to new attempts to optimize existing instruments and to design new ones. A comprehensive review of generators available in 1991 is that of Berlan and Mason (1992).

Ultrasound is currently used for cleaning, in liquid dispersion, liquid degasification, homogenization, welding, etc. In Food Technology it is used for the mixture and emulsification of unmiscible liquids, to regulate crystallization and to speed up ageing of meat, liquors and wines, etc.

The physics of ultrasound is very complex and despite the information available (Neppiras, 1980; Suslick, 1988), many aspects of its mechanisms remain obscure. Only basic aspects will be described here.

When a sonic wave hits a liquid medium it originates, in the hitting spot, longitudinal waves that propagate into the liquid creating alternating compression and expansion zones that, under certain circumstances, provoke stable or transient cavitation. Stable cavitation originates in small bubbles, the sizes of which oscillate slightly during compression and expansion cycles. After thousands of cycles, the interchange of gas that takes place can make them grow (Atchley and Crum, 1988). During oscillation, stable cavitation can, under given circumstances, change to a transient cavitation and vice versa. Stable cavitation has normally been related to small gas bubbles dissolved in the liquid in a stable manner. Ultrasound makes them vibrate, creating shock waves that cause strong currents in the adjacent liquid (Scherba *et al.*, 1991). Other bubbles are attracted into the sonic field by these small bubbles and these coalescence at high speeds, originating new very small bubbles that create microcurrents around themselves. This phenomenon is known as 'microstreaming'. Normally, all these phenomena take place at pressures in the range of 1–100 kPa (Hughes and Nyborg, 1962). Currents generated are of such intensity that they can catalyse chemical reactions and disrupt erythrocytes or even some bacteria, if a minimum speed of particles flowing off the sonication focus is reached (Kinsloe *et al.*, 1954). This cavitation occurs when the sonic waves are of high frequency (Scherba *et al.*, 1991; Mason, 1993) or of very low amplitude (Hughes and Nyborg, 1962).

In transient cavitation the sizes of bubbles change very quickly, in only a few oscillation cycles, and they finally collapse with different intensities. Under normal conditions liquid molecules remain very close to each other due to the liquid tensile strength. Under ultrasonication, during the compression cycle, they approximate and during the expansion cycle, they separate. When negative pressure in the liquid created by the expansion cycle is low enough to overcome the tensile strength, small bubbles are formed. These bubbles become nuclei of cavitation. Tensile strength is, in pure liquids, very high (i.e. in pure water it is about 10^5 kPa; Suslick, 1989) and not even with modern generators is it possible to overcome it. However, pure liquids are very difficult to obtain and even the purest water always has small bubbles adhered to contaminating particles or to the wall of the container. So, in fact, with modern instruments, cavitation can be obtained when the negative pressure during the expansion cycle is very low.

During the expansion/compression cycles the gas bubbles expand and contract and an equilibrium is reached between the vapour in and out of the bubble. The surface area of the bubble is greater during the expansion than during the compression cycle. This makes gas diffusion during the expansion cycle also bigger, and the size of the bubble increases in each cycle. The ratio between the rate of diffusion during expansion and the rate of diffusion during compression grows in each cycle until a resonance size is reached. When this occurs the bubble cavity grows very fast in just one cycle and as the amount of energy supplied by the ultrasound is not enough to retain the vapour phase, a sudden condensation occurs. Condensed molecules hit each other violently and, as a result, shock waves originate, creating spots of very high temperature and pressure. According to some authors (Suslick, 1988), during implosion, temperatures and pressures increase in these spots at very fast rates (10^6 °C/s) attaining, in only 1 microsecond, values as high as 5500°C and peaks of pressure of 5×10^4 kPa. It is currently believed that most of phenomena taking place in liquids during sonication are due to cavitation.

The cavitation threshold (minimum oscillation of pressure for cavitation to occur: amplitude of pressure) depends on many factors (Atchley and Crump, 1988). Dissolved gas is the main one. In liquids saturated with gases, the threshold is very low and increases linealy as vapour pressure in the liquid decreases. Also, hydrostatic pressure determines the threshold of cavitation. This should never be too high as to prevent gas diffusing out of the bubble. Other factors such as specific heat of the gas in the bubble, that of the liquid and the tensile strength also influence the cavitation threshold. Temperature is a very important factor. A lineal relationship has been observed between the cavitation threshold and the reciprocal of temperature. It decreases as temperature increases and becomes zero at the boiling point of the liquid.

Cavitation threshold increases as solids concentration decreases or when the ion concentration increases. But it depends, above all, on the frequency of the sonic waves (Berlan and Mason, 1992).

The effects of ultrasound depend on the intensity of implosion. Energy liberated by cavitation depend on the ratio of radius of bubble at maximum size to initial radius. This ratio is determined by the acoustic intensity and is influenced by many different factors, including: sonication menstruum, sonication parameters and sonication chamber. Sonication menstruum and ultrasonication parameters are interrelated. Any factor hampering cavitation draws more energy from the generator to keep the amplitude constant. However, if cavitation is nevertheless attained under these conditions, its effects are bigger (Berliner, 1984).

Cavitation nuclei and bubbles form and grow more quickly the higher the temperature of the liquid because vapour pressure increases and tensile strength decreases. However, although more bubbles are formed, the violence of collapses are smaller, because high vapour tension acts as a cushion (Alliger, 1975). The thermal conductivity of dissolved gas is also important as it determines maximum temperatures attained in the liquid hot spot during implosion and thus the effects of ultrasonication. Furthermore, the increase in temperature reduces viscosity of the liquid and favours cavitation. However, at this lower viscosity the violence of the implosion is smaller.

The viscosity of the menstruum also influences the diffusion of the acoustic waves. The range of transmission of ultrasound is inversely proportional to the kinematic coefficient of viscosity. High hydrostatic pressures tend to prevent cavitation as more output is needed. But if this is available, the violence of cavitation is even more intense.

The characteristics of the ultrasound generator, and specially its output, and the frequency and amplitude of ultrasonic waves, are also determinants of the intensity of cavitation. As any parameter hampering cavitation (pressure, tensile strength, low temperature, etc.) will require a bigger output under the same conditions, more powerful generators will ensure bigger ultrasonication effects.

The frequency of ultrasonication in the usable range is an important parameter as it determines the maximum bubble size before implosion (Suslick, 1989). The lower the frequency, the bigger the critical size of the bubbles and the bigger the intensity of the collapse. At very high frequencies (1 MHz approx.) cavitation is more difficult and above 2.5 MHz cavitation does not occur (Alliger, 1975). The intensity of cavitation is, however, mainly determined by the amplitude of the sonication waves and thus it is the amplitude parameter that has to be especially controlled and maintained during ultrasonication (Berliner, 1984).

The geometry of the chamber and that of the sonicating horn is also important. The effect of any given ultrasonication treatment is inversely related to the volume of the ultrasonication chamber (Davies, 1959). The maximum energy transfer of ultrasound through the wall of the sonication vessel is obtained when the wall thickness is a multiple of half the wave length. As cavitation is restricted to a zone close to the transducer, the effect of ultrasonication will also depend on the volume and the shape of the container. Also the way the sonic energy is coupled to the cavitation region has to be decided after careful consideration. Cavitation should not take place too close to the horn's tip as the bubbles generated could attenuate the effect by dispersing sonic waves. Adequate focusing devices help to solve this problem (Berlan and Mason, 1992). Size and geometry of the horn also determine the amount of energy transmitted (Berliner, 1984).

As can be seen, many different parameters influence the effect of ultrasonication. Almost all influencing parameters seem to have an optimum value in which maximum cavitation will occur. Ultrasonication conditions therefore have to be carefully chosen and controlled (Berlan and Mason, 1992) for maximum effect.

The effects of ultrasound on cells and parasites (Harvey and Loomis, 1929) and the improvements in ultrasonic generators have stimulated investigations on the eventual use of ultrasound for other microbiological purposes. Ultrasound has been used to disaggregate clumps of cells in liquid suspensions (Williams *et al.*, 1970) and to free bacteria adhered to surfaces to facilitate accurate total counting of contaminating flora (Dewhurst *et al.*, 1986). It has also been used to break cell walls in investigations of cellular components (Mett *et al.*, 1988) and its usefulness has been suggested for the pasteurization or sterilization of foods (Gaboriaud, 1986; Scherba *et al.*, 1991) and surgical instruments (Boucher, 1978).

The bactericidal effect of ultrasound was already reported in the 1920s. At that time the effect was attributed to the compression that ultrasound would generate in liquid media. It was thought later that it was due to mechanical disruption of cells by the very intense currents generated in the media by ultrasound (Harvey and Loomis, 1932). This hypothesis was shared by other investigators who also related it to the 'microstreaming' produced during stable cavitation, as it occurs during high frequency ultrasonication (Hughes and Nyborg, 1962; Williams *et al.*, 1970). Now most authors agree that, amongst other factors, it is mainly cavitation that produces the lethal effect (Kinsloe *et al.*, 1954; Davies, 1959). Although it is known that highly reactive chemical radicals and reaction products (e.g. H_2O_2) (Suslick, 1988), of well-known lethal capacity, are liberated in the aqueous media during cavitation, it is generally agreed that the lethal effect of ultrasound is due to extreme pressure variations caused by implosion. Implosion generates very high temperatures in hot spots. However, as the volume of menstruum heated at these temperatures is very small, only a very small number of cells would be affected. It has been mathematically demonstrated that practically the whole of the lethal effect is due to the pressure changes responsible for the disruption of cellular structures (Scherba, 1991).

It has been observed that microorganisms can withstand high pressures but they are incapable of withstanding the quick alternating pressures produced during cavitation. Mechanical disruption seems to be the most probable mechanism of action of ultrasound. This has been observed by different authors (Harvey and Loomis, 1929; Davies, 1959). However a lethal effect on microorganisms has been reported, with no apparent mechanical disruption (Lepeschkin and Golman, 1952; Kinsloe *et al.*, 1954).

Resistance of different species of microorganisms to ultrasonication varies widely. Data on resistance to ultrasound of different microbial species has been reported by different authors (Davies, 1959; Alliger, 1975). With the exception of *Mycobacteriaceae* (Alliger, 1975), it is generally admitted that cells of a bigger size are more sensitive to ultrasound (Kinsloe, 1954; Ahmed and Russell, 1975). Coccal forms are more resistant than rod-shaped bacteria (Jacobs and Thorney, 1954; Alliger, 1975), Gram-positive more than Gram-negative and aerobic more than anaerobic species (Ahmed and Russell, 1975). Also, the age of cells is an important factor. It has been reported (Kinsloe *et al.*, 1954) that young (4 hours) *Saccharomyces cerevisiae* cells were more sensitive than old ones (24 hours). Sporulated microorganisms are *much* more resistant than vegetative ones to ultrasound, and some authors (Paci, 1953; Sanz *et al.*, 1985) even question whether sonication can disrupt them. The resistance of microorganisms to ultrasonication depends also on the physical characteristics of the menstruum. However it is not clear whether what changes is the resistance to ultrasound or the intensity of cavitation. The influence of the temperature of medium on the lethal effect of ultrasonication seems to depend on the microorganism (Kinsloe *et al.*, 1954). The effect of pH is not clear. While some authors (Kinsloe *et al.*, 1954) reported no influence of pH on resistance, others reported a decrease in resistance at acidic pH values (Utsunomiya and Kosaka, 1979).

Comparison of data from different authors is difficult as sometimes some parameters determining the intensity of the ultrasonication are unknown. Furthermore, disagreements between the results of different authors could be explained by intrinsic differences between different species, or differences developed by different culture conditions, etc. The influence of the composition of medium on the resistance of some microorganisms to ultrasonication has also been demonstrated. It has been reported with different species of yeast and bacteria, that resistance to ultrasound was always greater in food than in nutritive broth (Jacobs and Thorney, 1954; Lee *et al.*, 1989). Resistance of *Escherichia coli* was greater in milk with higher fat content and homogenization decreased it (Jacobs and Thorney, 1954).

It is generally admitted that the numbers of survivors of an ultrasonic treatment is an exponential function of time (Kinsloe *et al.*, 1954; Davies, 1959; Mett *et al.*, 1988). However, as in heat inactivation, many deviations from linearity have also been reported. Some authors have even questioned that the kinetics of death is logarithmic (Jacobs and Thornley, 1954; Berger and Marr, 1960).

Sometimes ultrasound survival curves also show, at the start, a 'shoulder' that has also been attributed to a cellular disaggregation (Kinsloe *et al.*, 1954; Ahmed and Russell, 1975). Other authors have interpreted 'shoulders' as the result of two simultaneous first-order reactions. The first one being that of the disruption of the exosporium (of spores) and the second the true inactivating reaction of the ultrasound upon the unprotected cell (Berger and Marr, 1960).

The occurrence of 'tails' is an even more frequent and relevant abnormality. In some authors' opinions tails are not a cell recovery problem, nor due to a more resistant fraction of the population, but to a progressive loss of efficacy of the ultrasonication (Jacobs and Thornley, 1954; Lee *et al.*, 1989). According to these authors bubbles of dissolved gas lose size and stop being cavitation nuclei. Also, bacteria would stop acting as cavitation nuclei due to the destructive effect of ultrasound. The effect of ultrasound on polymers could lower the viscosity of the menstruum and thus also decrease the intensity of implosion.

Although ultrasound has been proposed for food pasteurization and sterilization, and even some processes have been patented (Gaboreaud, 1986) the probability of its use in the future for these purposes is very slim. Bacteria, especially spores, are very resistant and would require hours of ultrasonication (Sanz *et al.*, 1985; Mason, 1993).

The effects of ultrasound on *enzymes* and other food components began to be investigated about 60 years ago and was reviewed by El'Piner (1964). It was soon clear that acoustic waves could produce depolymerization of macromolecules. Depolymerization induced by ultrasonic waves has been extensively studied and deserved a number of recent reviews (Price, 1990; Brenner, 1990; Pethric, 1991). Among the first observations made were:

(i) The reversible reduction of viscosity of aqueous solutions of starch, gum arabic, gelatin and other macromolecules.
(ii) The depolymerization of starch and dextrans to high molecular weight products.
(iii) The breakdown of DNA to fragments retaining the native configuration. The rate of many of these degradations now seems to be proportional to a power of the molecular weight.

It was soon demonstrated as well that ultrasounds induced different kinds of changes, the nature of these changes being dependent on the nature of the gas dissolved in the aqueous solution. The replacement of oxygen by hydrogen can lead to an increase in the molecular weight of various kinds of macromolecules. In fact, ultrasonic irradiation induced polymerization, even from monomers (Kruus, 1991).

The effects of ultrasonic waves on proteins are very complex. Polymeric globular proteins are broken down into their subunits and, if the medium contains oxygen, the quaternary structure may not be recoverable. Lipoproteins may undergo partial delipidation (Luca *et al.*, 1970). In haemoproteins, the splitting of the haeme group from the globin is promoted (Weisler, 1960). If the ultrasonic irradiation is long enough, protein hydrolysis may occur and polypeptide chains may be broken, releasing large molecular fragments. Detachment of low molecular weight peptides and amino acids, breakdown of cyclic amino acids, oxidation of aromatic residues and aggregation of large polypeptide fragments have also been reported (Prudhomme and Grabar, 1947; El'Piner and Surova, 1954). From the pioneer studies it is known that these changes are strongly dependent on protein concentration and on the nature of the gas dissolved.

The decrease in the depolymerization effects of ultrasounds at high protein concentrations is attributable to a lowering of the cavitation effect, due to the high viscosity of the medium. This affects the two main mechanisms which are thought to be involved in depolymerization: (i) the frictional force between polymeric and solvent molecules; and (ii) the high intensity shock waves (Price, 1990). The ultrasonic field effects on enzymes depend not only on parameters related to the ultrasonic field, but also on their molecular structure and on factors related to the sonicating medium, especially on the nature of the dissolved gas. In some conditions, ultrasonically treated enzymes undergo fairly extensive changes which do not affect their catalytical functions, whilst in other conditions the enzymatic function is lost (El'Piner, 1964).

Some enzymes, such as catalase, are not inactivated during ultrasonic irradiation at low temperature. Yeast invertase is fairly resistant and only slightly inactivated after long irradiation at very low concentration (Matsudaira and Sato, 1933; Oparin *et al.*, 1954). Pepsin is also very resistant unless the irradiation takes place in very dilute solutions (Naimark and Mosher, 1953). Ribonuclease is not inactivated by irradiation in the presence of either oxygen or hydrogen, something similar occurs to serum aminopeptidase (Dubs, 1966). Lysozyme, alcohol dehydrogenase, hyalurodinase, lactate dehydrogenase, malate dehydrogenase, polyphenoloxidase and other oxidases are much more sensitive (Grabar *et al.*, 1949; Santamaria *et al.*, 1952; McCleod and Dun, 1967. Coarkley *et al.*, 1973; Kashkooli *et al.*, 1980). Even different isozymes may be differently affected (Dubs, 1966).

Inactivating effects generally require long irradiation periods and the presence of oxygen. The loss of activity of pepsin and trypsin in these circumstances has been attributed to the destruction of cyclic amino acid and sulfhydryl groups (El'Piner *et al.*, 1958; El'Piner, 1964). Enzyme inactivation is strongly reduced if hydrogen substitutes for oxygen in the sonicating medium or when antioxidants are present (Coarkley *et al.*, 1973; Santamarìa *et al.*, 1952; Grabar *et al.*, 1949). The presence of argon in the ultrasonicating medium induces ribonuclease inactivation (El'Piner, 1964). The influence of the dissolved gas on the intensity of enzyme inactivation has been related to the formation of free radicals by cavitation (Dunn and Maccleod, 1968). However, some investigators (DeGrois and Baldo, 1969) have emphasized the importance in activation of the mechanical damage. Others (Kashkooli *et al.*, 1980) have proved that cavitation is not always necessary for enzyme inactivation to occur, as it can be caused by 'microstreaming'. In fact it has been thoroughly proved that shear stress can cause degradation of high molecular weight polymers, even in the absence of bubble collapse (Price, 1990).

To increase the effectiveness of ultrasound for microbial and enzyme inactivation, different combined processes have been attempted by different authors.

The combination of ultrasound with chemical agents such as benzolbarium chloride, glutaraldehyde and H_2O_2, amongst others, has been reported to be very effective for the sterilization of surgical instruments (Ahmed and Russell, 1975). Some of these methods have been patented (Sierra, 1972). Ultrasound probably acts by increasing cell permeability to these substances.

Other uses of ultrasound have also been investigated with success. For instance, as the sterilization of effluents by ozonization is a rather long procedure (Burleson *et al.*, 1975), the ozonization has been combined with a simultaneous ultrasonic treatment. The disaggregating effect of ultrasound upon the solid matter and on the gas bubbles improves efficacy by increasing surface area. Furthermore, ultrasound accelerates the sedimentation of oxidizable organic matter, thus decreasing ozone demand. Ultrasonic treatments prior to irradiation have been reported to increase the efficacy of ionizing radiation (Dharkar, 1964). Some combined processes including the use of ultrasound are rather complex. A water potabilization method (Davies *et al.*, 1992) has been patented that includes the use of ozone, H_2O_2, radiation, UV and ultrasound.

The combination of ultrasound with heat treatment for the inactivation of microorganisms and enzymes has been investigated in the last few years with very promising results.

9.2 Destructive effect of combined treatments of heat and ultrasound under pressure: Mano-Thermo-Sonication (MTS)

Data on the combined effects of heat and ultrasound have been, until very recently, very scarce.

In the first experiments on the lethal effect of ultrasound on microorganisms, carried out during the 1930s and 1940s, researchers generally performed ultrasonication under refrigeration to avoid interference by the potentially lethal effects of heat. This is perhaps why the increase of lethality of heat treatment when combined with ultrasound remained unveiled for so long.

Although a decrease in bacterial heat resistance by a previous ultrasonication treatment was already mentioned by Paci in 1953, the first investigations to confirm these observations were not carried out until some years later. In 1972 it was reported that heat resistance at 105°C of *Bacillus cereus* and *Bacillus licheniformis* decreased after a previous ultrasonication treatment at 20 kHz (Burgos *et al.*, 1972).

In further reports (Sanz *et al.*, 1985) it was shown that *Bacillus stearothermophilus* was also made more sensitive to heat by a previous ultrasonication. Its heat resistance was reduced to ½–⅓ its original value. This effect was constant regardless of the heating temperature and the storage time between both treatments. Thermoduric streptococci were also sensitive to this combined treatment. It was later observed that this effect was even bigger when the combined heat/ultrasonic treatment was carried out simultaneously (Ordoñez *et al.*, 1984). This combined and simultaneous heat/ultrasonication treatment to inactivate bacteria was then designated 'thermoultrasonication' (TU) (Ordoñez *et al.*, 1987).

It was first seen that TU at 50°C was very effective in inactivating *Staphylococcus aureus*. However, this increase in lethality became smaller the higher the temperature of treatment (z = 64°C for heat treatments increased to z = 75.7°C for TU treatments). A similar increase in lethality of mild temperatures when combined with ultrasound had also been reported and a patent was granted for a procedure to sterilize surgical instruments dipped in surfactant and detergent solutions (Boucher, 1978). Other authors (Alliger, 1975) also reported that a mild increase in the temperature of the medium during ultrasonication, increased its lethality for vegetative cells, but was ineffective for the inactivation of spores. However, other investigators (Garcia *et al.*, 1989) did find the combination of heat with ultrasound to be effective in inactivating the spores of two strains of *Bacillus subtilis*. However, this effect became smaller the higher the temperature of treatment (Figure 9.1), disappearing at the boiling point of the menstruum (approx. 100°C). When milk or glycerol were used, this effect, although decreased, could still be observed even at this high temperature: resistance to TU was, at 100°C, 40–60% that of heat treatment alone.

As is known, high temperatures increase vapour pressure, decrease viscosity and as a consequence, reduce the intensity of cavitation. In 1992 a thermoresistometer was designed and built in our laboratory (Sala *et al.*, unpublished observations), to investigate the effect of TU under pressure in an attempt to retain the effect of ultrasound even at high temperatures. This instrument allowed heat or ultrasonication, and combined heat/ultrasonication treatments under pressure, at a wide range of temperatures (up to 140°C), pressures (up to 1000 kPa) and intensities of ultrasonication (up to 340 W, 145 μm of amplitude at a constant frequency of 20 kHz). During the treatments, the temperature and pH of the medium could be constantly monitored and temperature controlled in a narrow range (±0.05°C). An automatic system of sampling (32 samples/s) allowed the determination of very low D_t values (lower than 0.0009 min). This instrument allowed the investigation of the effects of different parameters on heat resistance and the efficacy of combined heat/ultrasound treatments under pressure over a wide range of temperatures, including temperatures higher than the boiling point of medium, and even at UHT temperatures.

The first experiments that were carried out were very promising. This new combined method of food preservation was called Mano-Thermo-Sonication (MTS-process) (Spanish Patent Number 9200686) and those reported here are some of the first results obtained with this procedure.

9.2.1 Effects of MTS on microorganisms

The heat resistance of *Bacillus subtilis* spores at pH 7, to an MTS treatment, was observed to decrease to about 1⁄10 that of the simply heated control in the range 100–112°C (Figure 9.2). Figure 9.2 shows survival curves at these two temperatures, for heat and MTS treatments respectively. As seen in this figure, when the simultaneous heat/ultrasonic

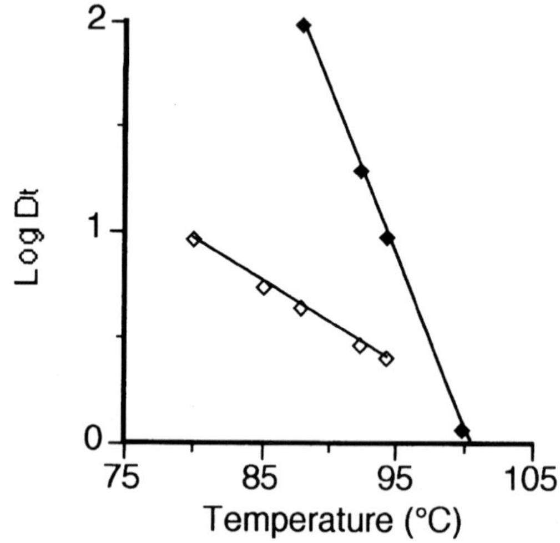

Figure 9.1 Heat resistance of *B. subtilis* spores at pH 7 to thermoultrasonication treatment (□–□) compared with heat control (■–■).

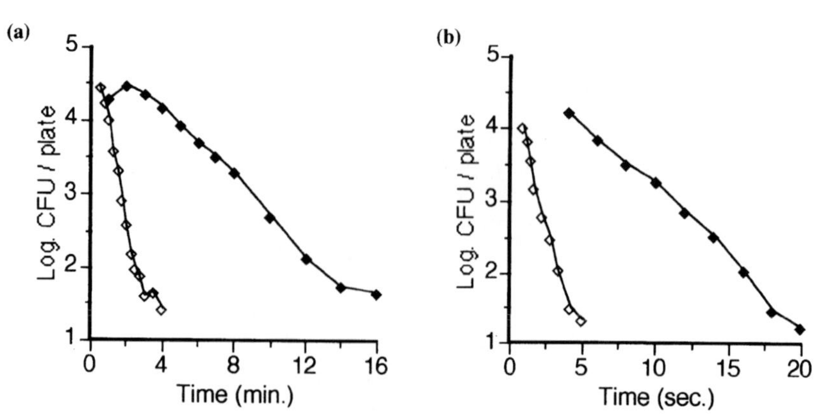

Figure 9.2 Survival curves at (a) 100°C and (b) 112°C of *B. subtilis* spores to MTS treatment (□–□) compared with heated controls (■–■).

treatment was carried out under pressure the sensitizing effect could be retained even at temperatures above the boiling point of medium.

It was shown that MTS was also effective with other microorganisms such as different sporeformers, vegetative cells, and yeasts. Figure 9.3 includes survival curves of *Aeromonas hydrophila*, *S. cerevisiae*, *B. coagulans* and *B. stearothermophilus* and shows the lethality of MTS treatment as compared to heat treatments at the same temperature.

The lethality of MTS treatments was 6–30 times greater than that of the corresponding heat treatment at the same temperature, dependent on the microorganism. The increase in lethality of MTS treatments as compared with heat treatments at the same temperature was greater for yeasts than for spores.

It was observed that the effect of MTS depended on the intensity of the ultrasonication (sonication time, amplitude and instrument output) and the pressure being applied (Sala *et al.*, unpublished observations).

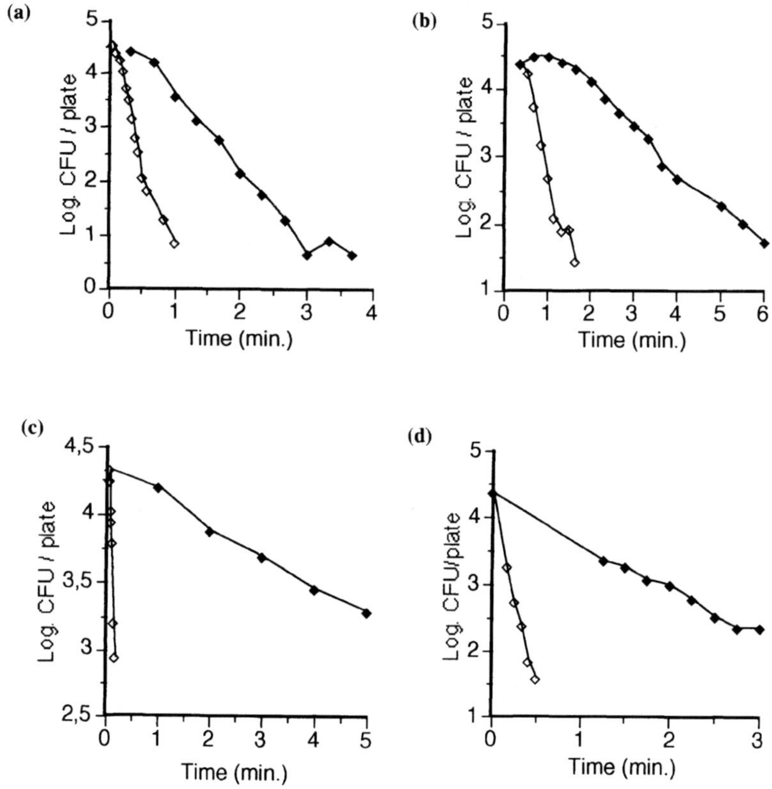

Figure 9.3 Effects of MTS treatment (□–□) and heat treatment (■–■) on survival of (a) *B. stearothermophilus*, (b) *B. coagulans*, (c) *S. cerevisiae* and (d) *Aeromonas hydrophila*.

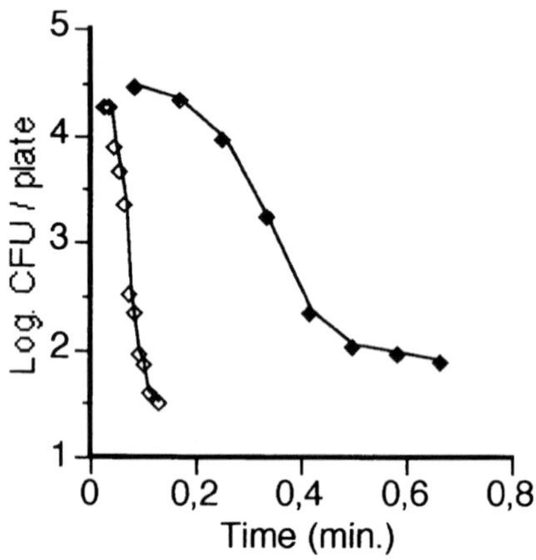

Figure 9.4 Effects of MTS treatment (□–□) and simple heating (■–■) on *B. subtilis* survival in milk (104°C).

Lethality of MTS treatments was observed in some liquid foods (Sala *et al.*, unpublished observations). Figure 9.4 shows, as an example, the survival curve corresponding to a heat and an MTS treatment of *B. subtilis* in milk.

The effect of MTS is not an additive but a synergistic one. An ultrasonication treatment at 45°C under pressure that failed to show any lethal effect on the spores of *B. subtilis var. niger*, reduced heat resistance at 112°C from 0.1 to 0.015 min when the ultrasonication was performed during heat treatment under pressure (20 kHz, 117 μm and 300 kPa).

The kinetics of death of microorganisms under MTS are, as seen in the figures, the same as for heat inactivation, and 'shoulders' and 'tails' may also appear. If the fraction of the microbial populations responsible for 'tails' were less sensitive to MTS treatments the advantages of this method could be questioned. But this proved not to be the case. In Figure 9.5, 'tails' of survival curves corresponding to a heat and an MTS treatment of *B. cereus* at 100°C and 104°C are shown. As seen in this figure the effectiveness of the MTS treatments was the same for the fraction of more sensitive microorganisms as for the 'tail' fraction.

As the kinetics of microbial death caused by MTS is the same as that caused by heat, calculation of adequate MTS processes can be carried out by the same procedure.

(a) Mechanism of microbial inactivation by MTS. As the death rate resulting from MTS is logarithmic, as for heat treatments, it is possible that the intrinsic mechanisms of death are the same. Physical effects of ultrasound would sensitize target molecules and/or structures, to heat.

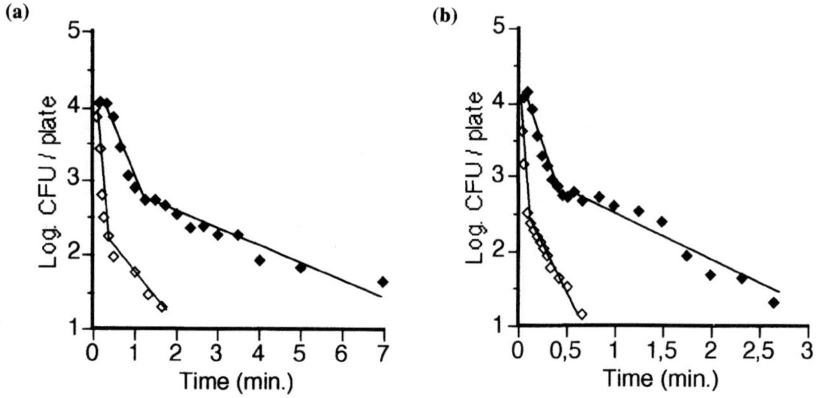

Figure 9.5 'Tails' of survival curves for *B. cereus* at (a) 100°C and (b) 104°C on MTS treatment (□–□) and heat treatment (■–■).

It is generally believed that the effects of ultrasound are due to cavitation. Hot spots at extremely high temperatures developed during cavitation are minute, and would seem to be incapable of causing heat inactivation (Scherba *et al.*, 1991). It is thought that the destructive effect of ultrasound is due to a physical damage caused by extreme pressure changes and shock waves. It has been reported that ultrasonication disrupted the spore exosporium (Berger and Marr, 1960) and released dipicolinic acid and low molecular weight polypeptides from the cortex of some bacterial spores (Palacios *et al.*, 1991). This cortex degradation would lead to the rehydration of the protoplast which, as postulated by some authors (Gould and Dring, 1975), would result in a loss of heat resistance.

The effect of MTS on vegetative cells could possibly be explained by a comparable but different mechanism that would also affect the integrity of the cell wall.

9.2.2 Effect of MTS on enzymes

The effect of MTS on some enzymes (lipoxygenase, peroxidase, polyphenoloxidase and *Pseudomonas fluorescens* extracellular protease and lipase) is being investigated by our research group and a summary of the results obtained is here included.

MTS inactivation of the enzymes has been observed to follow first-order kinetics. During heat inactivation of horseradish peroxydase (SIGMA Type VI) a biphasic course was obtained, the second phase starting after 70% inactivation was attained (López *et al.*, 1994). Deviation from first order kinetics in the heat inactivation of peroxidase has often been described (Chang *et al.*, 1988; Ganthavorn *et al.*, 1991). This biphasic timecourse of peroxidase inactivation is usually attributed to the presence of isozymes with different heat stability. However, some authors think that it is due to the formation of enzyme aggregates (Winter, 1971).

Under MTS no signs of biphasic curves were observed. If the deviation of peroxidase heat inactivation from first-order kinetics was due to enzyme aggregation, this change in the time course could be due to the well-known dissociation effects of ultrasound on aggregates (López *et al.*, 1994).

As observed with microorganisms (Sala *et al.*, unpublished observations), enzyme inactivation by a combined treatment of heat and ultrasound under pressure is not an additive but a synergistic phenomenon (Table 9.1). The enzyme inactivation efficacy of heat is increased by a factor which depends on the nature of the enzyme and the working conditions.

The role of the different bonds and interactions involved in protein structure stabilization are not equally important in maintaining the native structure of the catalytic centre of each of the enzymes and they are not equally affected by heat and ultrasonic waves. Therefore, the magnitude of the synergistic effect must reflect differences in the bonds and interactions which stabilize the active centres. On the other hand, since at low

Table 9.1 Rate constant of enzyme inactivation by heat, ultrasonic waves (uw) and MTS in 20 mM potassium phosphate buffer, pH 6.5

Enzyme	Treatment		$k(min^{-1})$
Lipoxygenase	uw*	(37°C)	0.0007
	Heat	(67°C)	0.023
		(70°C)	0.105
		(74°C)	0.89
		(76°C)	3.4
	MTS*	(67°C)	1.15
		(69°C)	1.84
		(71.3°C)	5.23
		(74.2°C)	8.86
		(77°C)	32.9
Peroxidase	uw†	(37°C)	0.005
	Heat	(110.4°C)	0.124
		(124.5°C)	0.475
		(134.8°C)	1.16
		(140.8°C)	2.19
	MTS†	(121.5°C)	0.69
		(126.5°C)	0.84
		(136.6°C)	2.02
		(142.6°C)	4.26
Polyphenol oxidase	uw‡	(37°C)	0.0435
	Heat	(70.7°C)	2.35
	MTS‡	(70.7°C)	5.52

*Pressure 4 kg/cm^2; amplitude 76 μm
†Pressure 4.8 kg/cm^2; amplitude 145 μm
‡Pressure: 4 kg/cm^2; amplitude 35 μm.

temperature there seems to be more than one mechanism involved in the enzyme inactivation by ultrasound, it seems reasonable to assume that differences in the synergistic efficiency may be also related to the nature of the groups participating in the catalysis and to the molecular weight of the protein.

The activation energies of heat and MTS destruction of peroxidase (Table 9.2) are of a comparable magnitude. This means that the simultaneous application of heat and ultrasonic waves remain equally efficient at any temperature. This probably indicates that one single inactivating mechanism is operating both in heat and MTS treatments.

For peroxidase heat inactivation, a mechanism involving the release of the haem moiety of the enzyme, previous to the denaturation of the liberated apoprotein, has been suggested (Lu and Whitaker, 1974). Our data (López and Burgos, unpublished observations) are in agreement with this mechanism operating as well during MTS inactivation. It has to be clarified whether the potentiating effects of ultrasonic waves on peroxidase destruction are related to an increase in the dissociation rate or to the hindering of the haem binding to apoprotein. This could be due to an increase in denaturation rate or in the rate of haem destruction.

On the contrary, the inactivation of lipoxygenase and the extracellular protease of *P. fluorescens* by MTS have lower energy requirements than heat inactivation. This means that the lower the temperature, the higher the potentiation effect of ultrasound. This, most likely, means that two different inactivating mechanisms are in operation. Our data on lipoxygenase inactivation by MTS (López and Burgos, unpublished observations) are partly in agreement with a mechanism implying, as the main pathway of inactivation, the production of ·OH radicals.

Differences in the intensity of the potentiation effect induced by changed working conditions must reflect the influence of these conditions on some physicochemical parameters (e.g. number of cavitational collapses produced, energy released, net absorbed power, hot spots temperature at collapse, free radical production, shock wave intensity etc.).

The synergistic effect of MTS at constant pressure decreases for lipoxygenase and *P. fluorescens* protease when the temperature is increased. This will probably occur with most enzymes, because it is what can be expected from any mechanism involving direct or indirect participation of a sonochemical reaction. In sonochemical reactions, rates are slowed down as the temperature is increased (Suslick, 1988; Mason and Lorimer, 1988), because:

(i) A higher bulk temperature results in an increase of the water vapour pressure in the cavitation bubble (which make the collapse less effective).
(ii) At higher temperature the number of cavitation collapses available to release energy will be a lower.

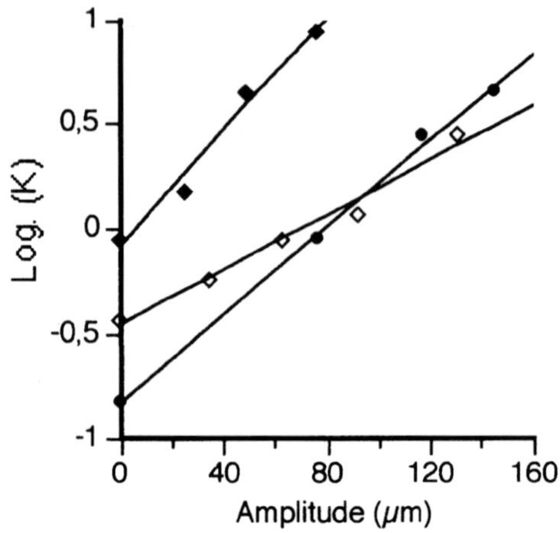

Figure 9.6 Plot of enzyme inactivation rate (k) in MTS vs. ultrasound amplitude. Effect of the amplitude of ultrasonic waves at 3 kg/cm^2 gauge pressure. Treatment medium 20mM phosphate buffer, pH 6.5. Effect on lipoxygenase at 74°C (◆); peroxidase at 123°C (◇); polyphenoloxidase at 60°C (●).

It was to be expected that an increase in the amplitude would lead to higher MTS efficacy. This has proved to be the rule obeyed by the three enzymes for which it has been studied. Figure 9.6 shows that the log of the inactivation rate is linearly related to amplitude. In lipoxygenase, changes in the amplitude of the ultrasonic waves also modify the E_a of the MTS inactivation reaction(s) (Table 9.2), which diminishes progressively as the amplitude is increased (López and Burgos, unpublished observations) revealing that the relative contribution to total enzyme inactivation by the mechanism added to heat, by the simultaneous application of ultrasonic waves, grows as the amplitude is increased. The effect of pressure on MTS efficacy on lipoxygenase was studied (Lõpez *et al.*, 1994)

Table 9.2 Activation Energy (*Ea*) for enzyme destruction by heat and MTS of peroxidase, lipoxygenase and *Pseudomonas fluorescens* extracellular protease in 20 mM potassium phosphate buffer, pH 6.5

Enzyme	Treatment	*Ea* (J. mol^{-10} k^{-1}) × 10^{-5}
Protease	Heat	0.96
	MTS	0.57
Lipoxygenase	Heat	5.43
	MTS	3.2
Peroxidase	Heat	1.18
	MTS	1.23

at a fixed amplitude (63 μm) and temperature (70°C). No significant change in the inactivation rate was detected between 150 and 500 kPa, but the inactivation rate multiplied by a factor of 3 from 500 kPa to 7500 kPa.

There are many other factors related to the treatment medium which may substantially affect MTS efficiency for enzyme inactivation. An investigation of some of them performed in the authors' laboratory using lipoxygenase as a model (López and Burgos, unpublished observations) revealed that the synergistic effect of heat and ultrasonic waves:

(i) Is almost independent of the ionic strength in the range 0–1.
(ii) Increases as pH decreases in the pH range 5–8, by a factor which is temperature dependent.
(iii) Diminishes with increasing enzyme concentration.
(iv) Increases with soluble solids concentration.

The influence of some of these factors, like the concentrations of soluble solids and enzyme, will probably be common to all enzymes because they influence the intensity of cavitation (El'Piner and Surova, 1954; Dognon and Simonot, 1948). Furthermore the influence of enzyme concentration on the efficacy of enzyme inactivation by ultrasound has been observed at low temperatures even when cavitation does not occur (Kashkooli *et al.*, 1980). The influence of pH and ionic strength will probably be found to be quantitatively very different or even opposite with different enzymes.

It is therefore obvious that an appropriate combination of temperature, hydrostatic pressure, amplitude and other factors such as saturating gas composition, may lead to a substantial increase in MTS inactivation efficacy for particular purposes.

9.3 Conclusions

Data reported in literature on the effects of ultrasound on microorganisms, enzymes and food components, do not allow one to foresee any future utilization of ultrasonication by itself for food stabilization. The resistance of most microorganisms and enzymes to ultrasonication is so high that the intensity of treatments required would probably produce extensive changes in composition and physical properties that would impair the important organoleptic properties of foods.

The increase in lethality of heat treatments when these are combined with ultrasonication under pressure (Mano-Thermo-Sonication) introduces the long sought possibility of a substantial reduction in the intensity of conventional heat treatments for the obtention of safe foods, whilst improving quality of foods preserved by traditional heat processes.

As ultrasound may also have side-effects, it is the balance of desirable and undesirable effects that will ultimately determine the possibility of the utilization of MTS for the treatment of any given product.

References

Abraham, G., Debray, E., Candau, Y. and Piar, G. (1990) Mathematical model of thermal destruction of *Bacillus stearothermophilus* spores. *Applied and Environmental Microbiology*, **56**, 3073–80.

Ahmed, F.I.K. and Russell, C. (1975) Synergism between ultrasonic waves and hydrogen peroxide in the killing of microorganisms. *Journal of Applied Bacteriology*, **39**, 31–40.

Alderton, G., Chen, J.K. and Ito, K.A. (1980) Heat resistance of the chemical resistance forms of *Clostridium botulinum 62A* spores over the water activity range 0 to 0.9. *Applied and Environmental Microbiology*, **40**, 511–15.

Alliger, H. (1975) Ultrasonic disruption. *American Laboratory*, **10**, 75–85.

Atchley, A.A. and Crump, L.A. (1988) Acoustic cavitation and bubble dynamics. In *Ultrasounds. Its chemical, physical, and biological effects* (ed. K.S. Suslick), VCH Publishers, New York, pp. 1–64.

Baldwin, R.L. (1986) Temperature dependence of the hydrophobic interaction in protein folding. *Proceedings of National Academic of Science*, **83**, 8069–72.

Beaman, T.C. and Gerhard, P. (1986) Heat resistance of bacterial spores correlated with protoplast dehydration, mineralization and thermal adaptation. *Applied and Environmental Microbiology*, **52**, 1242–46.

Berger, J.A. and Marr, A.G. (1960) Sonic disruption of spores of *Bacillus cereus*. *Journal of General Microbiology*, **22**, 147–157.

Berlan, J. and Mason, T.J. (1992) Sonochemistry: from research laboratories to industrial plants. *Ultrasonics*, **30**, 203–212.

Berliner, S. (1984) Application of ultrasonic processors. *International Biotechnology Laboratory*, **2**, 42–49.

Bigelow, W.D. (1921) The logarithmic nature of thermal death-time curves. *Journal of Infectious Diseases*, **28**, 528–532.

Blocher, J.C. and Busta, F.F. (1983) Bacterial spore resistance to acid. *Food Technology*, **11**, 87–99.

Boucher, R.M.G. (1978) Process for ultrasonic pasteurization. *United States Patent*, **4**, 211, 744.

Brenner, D. (1990) Historical introduction to Sonochemistry. In *Advances in Sonochemistry*, Vol. 1 (ed. T. Mason), Jai Press, London, pp. 1–37.

Brown K.L. and Ayres, C.A. (1985) Thermobacteriology of UHT processed foods. In *Developments in Food Microbiology* (ed. R. Davies), Applied Sciences Publishers, London, pp. 119–52.

Burgos, J., Ordoñez, J.A. and Sala, F.J. (1972) Effect of ultrasonics waves on the heat resistance of *Bacillus cereus* and *Bacillus licheniformis* spores. *Applied Microbiology*, **24**, 497–98.

Burleson, G.R., Murray, T.M. and Pollard, M. (1975) Inactivation of viruses and bacteria by ozone, with and without sonication. *Applied Microbiology*, **29**, 340–4.

Burton, H. (1988) *Ultrahigh-temperature Processing of Milk and Milk Products*, Elsevier Applied Science, London.

Cerf, O. and Hermier, J. (1973) Thermoresistance anormale de spores bacteriennes chauffees par injection directe dans la vapeur. *Le lait*, **43**, 23–29.

Cerf, O. (1977) Tailing of survival curves of bacterial spores. *Journal of Applied Bacteriology*, **42**, 1–19.

Chang, B.S., Park, K.H. and Lund, D.B. (1988) Thermal inactivation kinetics of horseradish peroxidase. *Journal of Food Science*, **153**, 920–23.

Coarkley, W.T., Brown, R.C., James, C.J. and Gould, R.K. (1973) The inactivation of enzymes by ultrasonic cavitation. *Archives of Biochemistry and Physics*, **159**, 722–29.

Cogan, T.N. (1977) A review of heat resistant lipases and proteinases and the quality of dairy products. *Journal of Food Science and Technology*, **1**, 95–105.

Cole, M.B., Davies, K.W., Munro, G., Holyoak, C.D. and Kilsby, D.C. (1993) A vitalistic model to describe the thermal inactivation of *Listeria monocytogenes*. *Journal of Industrial Microbiology*, **12**, 232–39.

Condon, S. and Sala, F.J. (1991) Heat resistance of *Bacillus subtilis* in buffer and foods of different pH. *Journal of Food Protection*, **55**, 605–8.

Condon, S., Garcia, M.L., Otero, A. and Sala, F.J. (1992a) Effect of culture age, pre-incubation at low temperature and pH on the thermal resistance of *Aeromonas hydrophila*. *Journal of Applied Bacteriology*, **72**, 322–6.
Condon, S., Bayarte, M. and Sala, F.J. (1992b) Influence of the sporulation temperature upon the heat resistance of *Bacillus subtilis*. *Journal of Applied Bacteriology*, **73**, 251–6.
Cook, A.M. and Gilbert, R.J. (1968) Factors affecting the heat resistance of *B. stearothermophilus* spores. *Journal of Food Technology*, **3**, 385–93.
David, J.R. and Merson, R.L. (1990) Kinetic parameters for inactivation of *Bacillus stearothermophilus* at high temperatures. *Journal of Food Science*, **55**, 488–93.
Davies, R. (1959) Observations on the use of ultrasound waves for the disruption of microorganisms. *Biochimica et Biophysica Acta*, **33**, 491–93.
Davies, P.W., Greenhalgh, S.H., Donnelly, J.K. and Stentiford, E.I. (1992) Treatment of water. *European Patent Application* 0567225 A1.
DeGrois, X. and Baldo, X. (1969) Explanatory hypothesis of the absence of erosion, of chemical effects and of sonoluminescence in true ultrasonic cavitation. *Acustica Internat.* **21(4)**, 222–8.
Dewhurst, E., Rawson, D.M. and Steele, G.C. (1986) The use of a model system to compare the efficiency of ultrasound and agitation in the recovery of *Bacillus subtilis* spores from polymer surfaces. *Journal of Applied Bacteriology*, **61**, 357–363.
Dharkar, S.D. (1964) Sensitization of microorganisms to radiation by previous ultrasonic treatment. *Journal of Food Science*, **29**, 241–3.
Dognon, A. and Simonot, Y. (1948) Actions des ultrasounds sur les suspensions. Influence de la concentration des particles. *Comptes Rendue Academic Sciences*, **227**, 1234–42.
Dominguez, J.M., Acebal, C., Jimenez, J., Mata, A., Macarron, R. and Castillon, M.P. (1992) Mechanisms of thermoinactivation of endoglucanase I from *Trichoderma roseiii QM 9414*. *Biochemistry Journal*, **287**, 583–88.
Donnelly, L.S. and Busta, F.F. (1980) Heat resistance of *Desulfotomaculatum nigrificans* in soy protein infant formula preparations. *Applied and Environmental Microbiology*, **40**, 727–5.
Dubs, C.A. (1966) Ultrasonic effects on isoenzymes. *Clinical Chemistry*, **12**, 181–86.
Dunn, F. and Macleod, R.M. (1968) *Journal of Acoustic Society of America*, **40**, 932–40.
El'Piner, I.E. (1964) *Ultrasounds: Physical Chemical and Biological Effects*, Consultants Bureau, New York, pp. 149–229.
El-Piner, I.E. and Surova, M.D. (1954) Acceleration of protein degradation process in an ultrasonic field. *Doklady Akad Nauk SSSR*, **94**, 243–50.
Feeherry, F.E., Munsey, D.T. and Lowley, D.R. (1987) Thermal inactivation and injury of *Bacillus sterarothermophilus* spores. *Applied and Environmental Microbiology*, **53**, 365–70.
Gaboriaud, P.L.F. (1986) Stérilisation de liquides par ultrasons. *French Patent* 2 575 641 A1.
Ganthavorn, C., Nagel, C.W. and Powers, J.R. (1991) Thermal inactivation of asparagus lipoxygenase and peroxidase. *Journal of Food Science*, **56**, 47–9.
Garcia, M.L. (1985) *Acción de los tratamientos ultrasonicos y térmicos en los esporos de B. subtilis*. Doctoral Thesis. Facultad de Veterinaria, Universidad Complutense, Madrid.
Garcia, M.L., Burgos, J., Sanz, B. and Ordoñez, J.A. (1989) Effect of heat and ultrasonic waves on the survival of two strains of *Bacillus subtilis*. *Journal of Applied Bacteriology*, **67**, 619–28.
Gould, G.W. (1973) Inactivation of spores in food by combined heat and hydrostatic pressure. *Acta Alimentaria*, **2**, 377–83.
Gould, G.W. and Dring, G.J. (1975) Heat resistance of bacterial endospores and concept of an expanded osmoregulatory cortex. *Nature*, **258**, 402–5.
Gould, G.W. (1989) Heat-induced injury and inactivation. In *Mechanisms of Action of Food Preservation Procedures* (ed. G.W. Gould), Elsevier Applied Science, London, pp. 11–42.
Grabar, P., Voinovitch and Prudhome, R.O. (1949) Action des ultrasonides sur une oxidase. *Biochemica et Biophysica Acta*, **3**, 412–17.
Hansen, N.J. and Riemann, H. (1963) Factors affecting the heat resistance of nonsporting organisms. *Journal of Applied Bacteriology*, **20**, 314–18.
Harvey, E. and Loomis, A. (1929) The destruction of luminous bacteria by high frequency sound waves. *Journal of Bacteriology*, **17**, 373–9.
Harvey, E. and Loomis, A. (1932) High speed photomicrografy on living cell subjected to supersonic vibrations. *Journal of General Physiology*, **15**, 147.

Hermier, J., Begue, P. and Cerf, O. (1975) Relationship between temperature and sterilising efficiency of heat treatments of equal duration. Experimental testing with suspensions of espores in milk heated in an ultra-high-temperature sterilizer. *Journal of Dairy Research*, **42**, 437–44.

Hughes, D.E. and Nyborg, W.L. (1962) Cell disruption by ultrasound. *Science*, **138**, 108–14.

Jacobs, S.E. and Thornley, M.J. (1954) The lethal action of ultrasonic waves on bacteria suspended in milk and other liquids. *Journal Applied Bacteriology*, **17**, 38–55.

Joly, M. (1965) *A Physicochemical Approach to Denaturation of Proteins*, Academic Press, New York.

Kashkooli, H., Roony, J. and Rooxby, R. (1980) Effects of ultrasound on catalase and malatedehydrogenase. *Journal of Acoustic Society of America*, **67**, 1798–1801.

Kelly, S.M. and Price, N.C. (1991) The unfolding and refolding of pig heart fumarase. *Biochemical Journal*, **275**, 745–49.

Kinsloe, H., Ackerman, E. and Reid, J.J. (1954) Exposure of microorganisms to measured sound fields. *Journal of Bacteriology*, **68**, 373–80.

Klibanov, A. (1983) Stabilization of enzymes against thermal inactivation. *Advances in Applied Microbiology*, **29**, 1–28.

Kruus, P. (1991) Sonochemical initiation of polymerization. In *Advances in Sonochemistry*, vol. 2 (ed. T.J. Mason), JAI Press, London, pp. 2–21.

Lapanje, S. (1978) *Physicochemical Aspects of Protein Denaturation*, Wiley, New York.

Law, B. (1979) Reviews of the progress of dairy science. Enzymes of psychotrophic bacteria and their effects on milk and milk products. *Journal of Dairy Research*, **46**, 573–88.

Lee, B.H., Kermasha, S. and Baker, B.E. (1989) Thermal ultrasonic and inactivation of *Salmonella* in thin films of aqueous media and chocolate. *Food Microbiology*, **6**, 143–42.

Lepeschkin, W.W. and Golman, D.E. (1952) Effects of ultrasound on cell structure. *Journal of Cellular Composition and Physiology*, **40**, 393–97.

Lewis, J.C., Snell, N.S. and Alderton, G. (1965) Dormancy and activation of bacterial spores. In *Spores III* (eds L.L. Campbell and H.O. Halvorson), American Society for Microbiology, Washington, D.C., pp. 47–55.

Liang, S.J., Lin, Y.Z., Zhou, J.M., Tsou, C.L., Wu, P.Q. and Zhou, Z.M. (1990) Comparison of inactivation and conformational changes of D-glyceraldehyde-3-phosphate dehydrogenase during thermal inactivation. *Biochimica et Biophysica Acta*, **1038**, 247–52.

Lopez, P., Sala, F.J., Fuente, J.L., Condon, S., Raso, J. and Burgos, J. (1994) Inactivation of peroxidase, lipoxygenase and polyphenoloxidase by manotermosonication. *Journal of Agriculture and Food Chemistry*, **42**, 552–56.

Lu, A.T. and Whitaker, J.R. (1974) Some factors affecting rates of heat inactivation and reactivation of horseradish peroxidase. *Journal of Food Science*, **39**, 1173–78.

Luca, R. and Zamfirescu-Georgiu, M. (1970) Ultrasonic effect on some physico-chemical properties of serum proteins and enzymes. *Revue Rumaine Medicine Interne*, **7**, 421–25.

Mackey, B.M. and Derrick, C.M. (1986a) Elevation of the heat resistance of *Salmonella typhimurium* during heating at rising temperatures. *Letters in Applied Microbiology*, **4**, 13–16.

Mackey, B.M. and Derrick, C.M. (1986b) Elevation of the heat resistance of *Salmonella typhimurium* by sublethal heat shock. *Journal of Applied Bacteriology*, **61**, 389–93.

Mason, T.J. and Lorimer, J.P. (1988) *Sonochemistry: Theory, Application and Uses of Ultrasound Chemistry*, Ellis Horwood, Chichester, pp. 42–47.

Mason, T.J. (1993) Sonochemistry: A technology for tomorrow. *Chemistry and Industry*, 47–50.

Matsudaira, M. and Sato, A. (1933) Effect of supersonic ray on enzymes. *Tohoku Journal of Experimental Medicine*, **22**, 412–16.

McCleod, R.M. and Dun, F. (1967) Effects of ultrasonic cavitation on trypsin, chymotrypsin and lactate dehydrogenase solutions. *Journal of Acoustic Society of America*, **42**, 527–29.

Mett, H., Schacher, B. and Wegman, L. (1988) Ultrasonic disintegration of bacteria may lead to irreversible inactivation of lactamase. *Journal of Antimicrobial Chemotherapy*, **22**, 293–98.

Moats, W.A., Dabbah, R. and Edwards, V.M. (1971) Interpretation of nonlogarithmic survivor curves of heated bacteria. *Journal of Food Science*, **36**, 523–6.

Naimark, G.M. and Mosher, W. A. (1953) Effects of sonic vibration on the proteolytic activity of pepsin. *Journal of Acoustic Society of America*, **25**, 289.

Neppiras, E.A. (1980) Acoustic cavitation. *Physics Reports*, **61**, 159–251.
Oparin, A.I., Bardinskaya, M.S. and El'Piner, I.E. (1954) Action of ultrasonic waves on yeast invertase. *Doklady Akad Nauk SSSR*, **99**, 423–432.
Ordoñez, J.A., Sanz, B., Hernandez, P.E. and Lopez-Lorenzo, P. (1984) A note on the effect of combined ultrasonic and heat treatments on the survival of thermoduric *streptococci*. *Journal of Applied Bacteriology*, **56**, 175–77.
Ordoñez, J.A., Aguilera, M.A., Garcia, M.L. and Sanz, B. (1987) Effect of combined ultrasonic and heat treatment (thermoultrasonication) on the survival of a strain of *Staphylococcus aureus*. *Journal of Dairy Research*, **54**, 61–7.
Ordoñez, J.A., Burgos, J., Raso, J., Lopez, P., Condon, S. and Sala, F.J. (1992) Procedimiento para la destrucción de microorganismos y enzimas: Proceso-MTS. *Spanish Patent* No. 93/00021.
Paci, C. (1953) L'emploi des ultra-sons pour l'assainissement du lait. *Le Lait*, **33**, 610–15.
Palacios, P., Burgos, J., Hoz, L., Sanz, B. and Ordoñez, J.A. (1991) Study of substances released to ultrasonic treatment from *Bacillus stearothermophilus* spores. *Journal of Applied Bacteriology*, **71**, 445–51.
Pethric, R.A. (1991) Ultrasonic studies of polymeric solids and solutions. In *Advances in Sonochemistry*, vol. 2 (ed. T. Mason), Jai Press, London, pp. 66–129.
Price, G.J. (1990) The use of ultrasound for the controlled degradation of polymer solutions. In *Advances in Sonochemistry*, vol. 1 (ed. T.J. Mason), Jai Press, London, pp. 231–287.
Privalov, P.L., Griko, Y.V., Venyamino, Y.S. and Kutyshenko, V.P. (1986) Cold denaturation of myoglobin. *Journal of Molecular Biology*, **19**, 487–98.
Prudhome, R.O. and Grabar, P. (1947) Etude de la denatuartio des proteides. I. Action des US sur les proteides du serum de cheval normal et sur les acides aminès cycliques. *Bulletin Société Chimie Biologie*, **29**, 122–33.
Put, H.M. and Aalbergsberg, W.I.J. (1967) Occurrence of *Bacillus subtilis* with high heat resistance. *Journal of Applied Bacteriology*, **30**, 411–19.
Putman, F.W. (1954) Protein denaturation. In *The Proteins*, Vol. 1B (eds H. Neurath and K. Bailey), Academic Press, New York, pp. 808–92.
Ray, W.J. and Koshland, D.E. Jr. (1961) A method for characterizing the type and number of groups involved in enzyme action. *Journal of Biological Chemistry*, **236**, 1973–1979.
Rodriguez, A.C., Smerage, G.H., Teixeira, A.A., Lindsay, J.A. and Busta, F.F. (1991) Population model of bacterial spores for validation of dynamic thermal processes. *Journal of Food Process Engineering*, **15**, 1–30.
Santamaria, L., Castellani, A. and Levi, F.A. (1952) Hyalurodinase inactivation by ultrasonic waves and its mechanisms. *Enzymologia*, **15**, 285–95.
Sanz, B., Palacios, P., Lopez, P. and Ordoñez, J.A. (1985) Effect of ultrasonic waves on the heat resistance of *Bacillus stearothermophilus* spores. In *Fundamental and Applied Aspects of Bacterial Spores* (eds G.J. Dring, D.J. Ellar and G.W. Gould), Academic Press, New York, pp. 251–59.
Sapru, V., Teixeira, A.A., Smerage, G.H. and Lindsay, J.A. (1992) Predicting thermophilic spore population dynamics for UHT sterilization processes. *Journal of Food Science*, **57**, 1248–57.
Sapru, V., Smerage, G.H., Teixeira, A.A. and Lindsay, J.A. (1993) Comparison of predictive models for bacterial spore population responses to sterilization temperatures. *Journal of Food Science*, **58**, 223–8.
Scherba, G., Weigel, R.M. and O'Brien, J.R. (1991) Quantitative assessment of the germicidal efficacy of ultrasonic energy. *Applied and Environmental Microbiology*, **57**, 2079–84.
Shull, J.J., Cargo, G.T. and Ernst, R.R. (1963) Kinetics of heat activation of thermal death of bacterial spores. *Applied Microbiology*, **11**, 485–7.
Sierra, G. (1972) Sterilization with gluteraldehyde. *United States Patent* 3,697,222.
Stumbo, C.R. (1973) *Thermobacteriology in Food Processing2*, 2nd edn, Academic Press, New York.
Suslick, K.S. (1988) Homogeneous sonochemistry. In *Ultrasounds. Its Chemical, Physical and Biological Effects* (ed. K.S. Suslick), VCH Publishers, New York.
Suslick, K.S. (1989) The chemical effects of ultrasound. *Scientific American*, **2**, 62–8.
Tanford, C. (1968) Protein denaturation: part C characterization of the denaturate state. In *Advance in Protein Chemistry*, vol. 23. Academic Press, pp. 121–282.

Utsunomiya, Y. and Kosaka, Y. (1979) Application of supersonic waves to foods. *Journal of Faculty of Applied Biological Science*, **18**, 225–31.

Versteeg, C., Pectinesterases from orange juice – their purification, general characteristics and juice cloud destabilizing properties. *PhD Thesis*. Agricultural University, Wageningen, The Netherlands.

Versteeg, C., Rombouts, F.M., Spaansen, C.H. and Pilnik, W. (1980) Thermostability and orange cloud destabilizing properties of multiple pectinesterases from orange. *Journal of Food Science*, **45**, 969–988.

Wang, D.I., Scharer, J. and Humphrey, A.E. (1964) Kinetic of death of bacterial spores at elevated temperatures. *Applied Microbiology*, **12**, 451–54.

Weissler, A. (1960) Effects of ultrasonic irradiation on hemoglobin. *Journal of Acoustic Society of America*, **32**, 1208–12.

Williams, A.R., Stafford, D.A., Callely, A.G. and Hughes, D.E. (1970) Ultrasonic dispersal of activated sludge flocs. *Journal of Applied Bacteriology*, **33**, 656–63.

Winter, E. (1971) Hitzebestandigkeit der peroxidase. *Z. Lebensm Unters. Forsch*, **145**, 3–6.

Zhou, H-M., Zhang, X-A., Ying, Y. and Tsou, C-L. (1993) Conformational changes at the active site of creatine kinase at low concentrations of guanidine chloride. *Biochemical Journal*, **291**, 103–7.

10 Electrical resistance heating of foods

P. FRYER

10.1 Introduction

Electrical resistance heating allows particles and liquids to heat at the same rate and permits the rapid heating of mixtures of high solids fractions. The technique has been applied to a number food processes, and has recently been developed into a commercial process for the sterilisation of food mixtures. To understand the process, it is necessary to study food electrical conductivity and predict the resulting heating patterns. Models for the process are discussed, and the ways in which electrical processing may differ from conventional thermal sterilisation suggested.

10.1.1 The thermal sterilisation of foods

The use of heating for reducing numbers of microorganisms in foods is widespread [1]. Such foods, supplied to the consumer, must be commercially sterile, and so the aim of thermal processing is to reduce the level of any microbial contamination to the point where the food will not cause a health hazard or undergo spoilage, at any point during its life. Two types of reactions occur within a food when it is heated; those which reduce the level of microbial contamination and those which result in losses in product quality, in terms of nutrition, taste and texture. The classical method of thermal preservation is the canning process, in which food is placed in the package, which is then sealed and sterilised within batch or continuous retorts. The commonest method of providing heat to the can is the use of condensing steam. Heat must then be transferred within the can, either by thermal conduction if the food is a solid or a very viscous liquid, or by convective heat transfer if the food is a less viscous liquid. When processing a can or a package, it is necessary to sterilise every part of the food. This can lead to overcooking of some or all of the material, as a result of the slowness of heat transfer.

In practice, canned food has a taste and texture significantly different to unprocessed foods. Whilst some canned products, such as tinned beans or tomato soup, have achieved wide acceptance, there is a commercial advantage in improving the quality of much canned food. Optimising the quality of packaged food is complex [2]. A number of techniques have been used to increase the quality of packaged food; these include rotating the pack to increase the amount of convective mixing, reducing the pack size to reduce the distance over which heat must be transferred, and changing the heating profile in retorts to minimise overcooking of the outer layers of the pack.

Another way of improving the product quality is to change the temperature of sterilisation. The reactions which reduce the level of bacterial spores have a higher activation energy than those which lead to reduced product quality; for example, Holdsworth [3] gives activation energies for the destruction of *C. botulinum* spores as about 300 kJ/mol whilst those for the loss of enzyme activity are about 120 kJ/mol. This implies that to maximise quality for a given level of sterility it is thus advantageous to process at as high a temperature as possible. Continuous ultra-high temperature (UHT) or high temperature short time (HTST) processes exploit this fact to produce food of a higher quality than canning. In these processes the food is rapidly heated to about 140°C, held there for a short period and then rapidly cooled prior to aseptic packaging [3]. At 140°C, food can be sterilised in a few seconds, rather than the several minutes which it takes at canning temperatures.

UHT processes require rapid heating rates to minimise time spent at the high temperatures, and ensure minimum quality loss. This is best done in a continuous process. Process flowsheets incorporate three sections:

(i) A *heating* section in which product is first heated to the required temperature.
(ii) A *holding* section, generally a horizontal or slightly inclined tube, in which it is held at temperature long enough to ensure sterility.
(iii) A *cooling* section prior to packaging.

Processing single-phase liquid foods in this way is straightforward, because fluids can be heated and cooled rapidly. It is possible to process low-viscosity liquids using forced convective heat transfer in plate or tubular heat exchangers, in which the food is contacted indirectly with hot fluid. In this equipment, high heating and cooling rates in the order of 1°C/s, are possible. This type of process offers an efficient way of sterilising liquids such as milks, fruit juices, soups and sauces. In high-viscosity products, such as creams and viscous sauces, which are shear-sensitive and may foul heavily, direct steam injection can be used to heat the material rapidly. However, it is difficult to apply these techniques to foods which contain particles, because of the slowness of thermal conduction into the solid. For particles larger than about 2 mm, the rate of heating is too great for rapid liquid heating to be feasible. If it were attempted to process a solid–liquid mixture containing large particle, by the time the particles were sterile the liquid would be enormously overcooked. Scraped surface heat exchangers can be used to process foods with a high solids fraction, but lower heating rates and long hold times may well be needed to sterilise the particulate phase, giving poor product quality.

10.1.2 Heat generation: electrical resistance heating

The need to conduct heat is the limiting factor in the sterilisation of particles. Volumetric techniques, in which heat is generated within the material, offer ways of circumventing the problem. It is possible to generate heat using microwave heating [4], in which a high-frequency electric field excites the water molecules within the material, or by electrical resistance heating ('ohmic heating'). The principle of electrical resistance heating is shown in Figure 10.1. In this process, an electric current is passed through the material, which then heats throughout its volume, as a result of its electrical resistance. The process is more energy-efficient than microwave heating, because nearly all of the electrical energy goes into the food as heat. Whereas microwave heating requires no physical contact, however, resistance heating requires electrodes in good contact with the food.

When a two-phase mixture is heated electrically, if the liquid and solid have the same electrical resistance, then the two phases will generate heat at the same rate. As a demonstration, Figure 10.2 shows the conventional and electrical heating of a food particle immersed in a liquid. In conventional heating the temperature of the particle lags behind the liquid, whilst in electrical heating the particle heats faster than the liquid, a result not possible conventionally. Using this technique, therefore, it is possible to apply HTST techniques to solid–liquid mixtures.

Electrical heating is not a new process [5]. An electrical pasteurisation process was successful in the 1930s, and a number of applications have been found in areas such as thawing and baking [6]. Before a process can

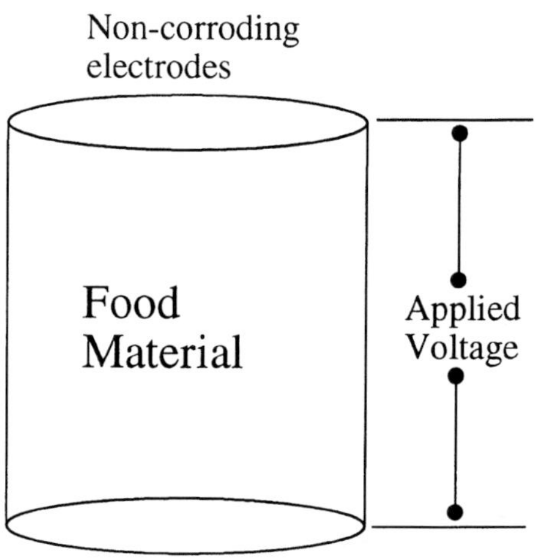

Figure 10.1 Principle of electrical resistance heating.

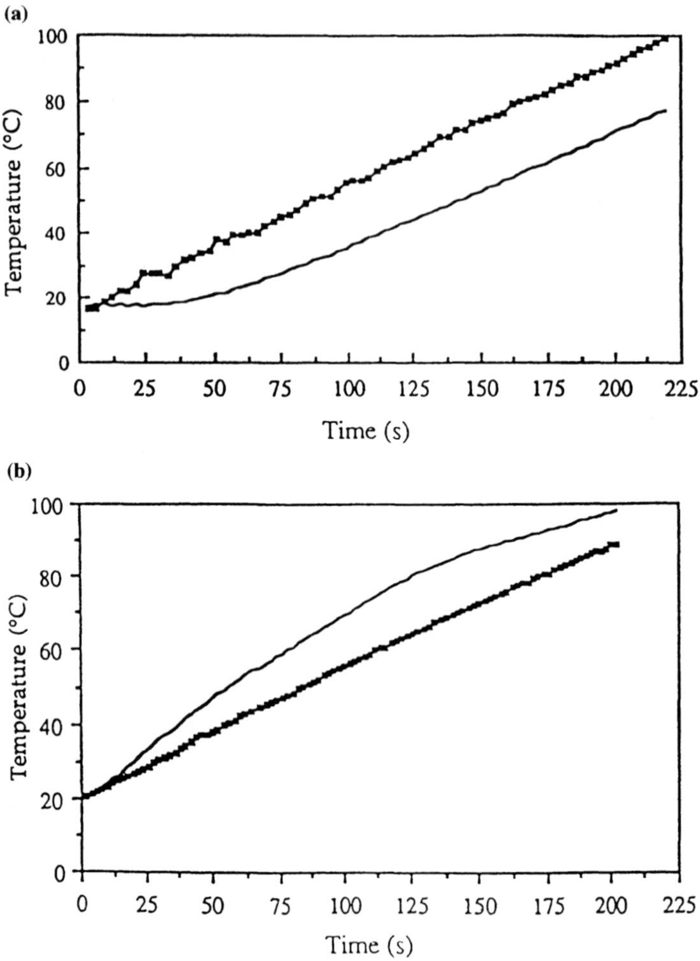

Figure 10.2 (a) Conventional and (b) electrical heating of a food particle (____) immersed in a liquid (■–■). [15]

be commercially acceptable, it must satisfy a number of process and safety criteria:

- an electrical design which avoids electrolysis and product contamination, either due to electrode breakdown or localised product scorching,
- effective control of heating and flow rates of the food,
- aseptic packaging techniques which allow a two-phase mixture to be sealed into a package under sterile conditions,
- overall cost-effectiveness.

A number of early processes did not meet these criteria, for example: the use of a DC supply led to electrolysis on the electrodes; the use of packages which contained expensive electrodes were too expensive. The current resurgence of interest in the process is due to the development of a commercial ohmic heating system by APV Baker [7–10], which incorporates solutions to the problems described above.

10.1.3 APV Baker ohmic heater

The APV Baker 'ohmic heater' was originally developed by EA Technology to Capenhurst, UK. In 1984, APV Baker secured a licence for the heating system and then developed a commercial process. Commercial scale systems are available with power outputs of 75 and 300 kW, corresponding to product capacities about 750 and 3000 kg/h. A schematic process flowsheet for the APV Baker ohmic heater is given in Figure 10.3. Food passes from a product pump into a vertical or near-vertical pipe containing a series of electrodes. Sufficient pressure is maintained in the column to ensure that the material does not boil; this can be up to 4 bar

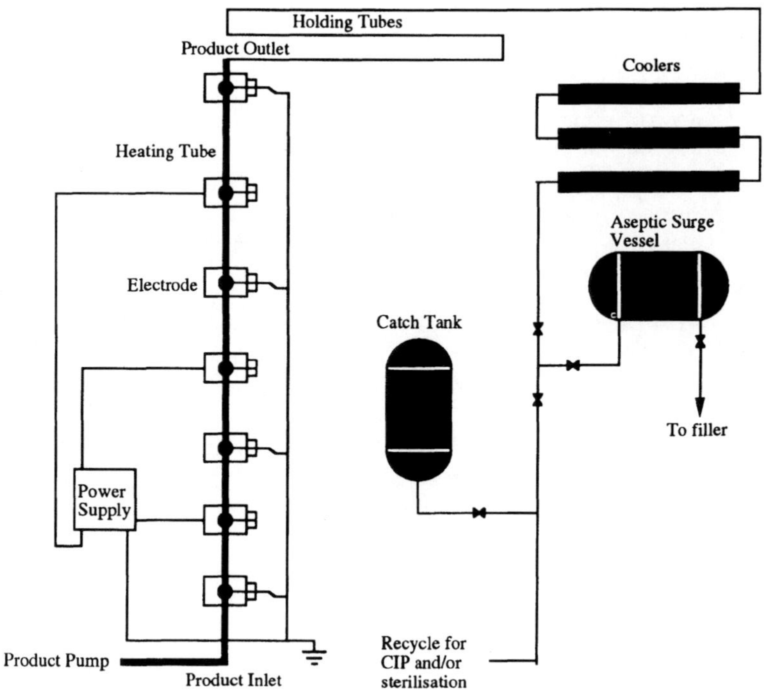

Figure 10.3 Schematic flowsheet for the APV Baker ohmic heater, after [10].

for sterilisation at 140°C. Figure 10.3 shows a 7-electrode machine, in which each of the live electrodes is surrounded by two which are earthed; 4-electrode machines have also been produced [8], in which the outer two are earthed.

Each electrode housing is machined from PTFE and contains a cylindrical cantilever electrode (supported only at one end) across the tube. Alternating current from a 3-phase supply flows between the electrodes and thus through the food. The connecting tubes are made from stainless steel lined with insulating plastic, and are made of different lengths so that the section between each pair of electrodes is always of similar impedance. As shown in the Figure, the tubes increase in length from inlet to outlet, reflecting the increase in electrical conductivity with temperature. From the heater, food passes to a holding tube, in which it is held for long enough to ensure sterility, prior to cooling and aseptic packaging. Cooling may be provided solely by water flowing on the outside of the cooling tube, or precooled liquid can be pumped into the tube after the holding section to allow an increase in the heating rate [10].

The process enables the sterilisation of foods of high solids fraction, up to about 60%, heating rates in the region of 1°C/s. Large particles, up to 25 mm in diameter, can be processed using the technique. It is now in commercial use in the UK, USA, Europe and Japan, and has been further developed for a number of specific applications for the processing of solid –liquid mixtures:

- aseptic processing of high added value ready prepared meals for storage and distribution at ambient temperature,
- pasteurisation of particulate food products for hot-filling,
- preheating of food products prior to in-can sterilisation,
- the hygienic production of high added value ready prepared meals for storage and distribution at chilled temperatures.

The overall sterilisation efficiency of the whole process has been demonstrated by passing alginate beads impregnated with spores through the heater [11]. This technique, whilst commercially vital, displays the effect of the whole process – heating, holding and cooling – rather than the behaviour of the heater alone. To understand the processes which occur during heating it is necessary to construct a model for the process. This is important both for understanding and designing the process and for confirming its acceptability by regulatory bodies.

10.1.4 Preservation by electrical heating

Electrical resistance heating is largely novel in the food industry, and requires new skills of the plant designer and operator. For example; conventional processing involves the use of steam, whose temperature will be the highest in the system. In electrical heating, as heat is generated rather than transferred, there is no theoretical limit on the temperature which can be reached. It is significantly different from microwave heating, which has a finite penetration depth of energy into any food [4] whilst electrical heating has no such limitation.

By allowing particles to be heated very quickly, the process offers significant advantages for the commercial processor. To design a food formulation to exploit the advantages of electrical heating requires an understanding of the process and the factors which affect the sterility and quality of the product. The key to the process is the rate of heat generation, which is controlled by a number of factors, the most important of which is the electrical conductivity of the food material. However, the temperature of the material is also governed by the way food flows through the heater, and thus the residence time of material within the plant. The aim of this chapter is to review published work on the electrical heating process, to describe the ways in which electrical heating is affected by the physical properties of the material, and to discuss how the quality of the product is affected by process design. Most published work has concentrated on attempting to model the process to enable output temperatures to be predicted, although some has considered the mechanism of the heating technique and the ways in which electrically cooked products may differ from conventionally processed ones.

10.2 The physics of electrical heating

10.2.1 Governing electrical equations

To produce a satisfactory product which can be successfully marketed requires that the food is sterile and of adequate quality. To be able to design a material which achieves these criteria requires the factors which affect the heating rate of the material, and the time that it spends in the heater, to be understood. A number of workers have modelled the sterilisation of flowing foods heated conventionally [12–14]. These models are not applicable to ohmic heating because the mechanism of heating is different; heat is generated within the material rather than transferred to it.

At the frequencies at which electrical heating is carried out, the heat generation mechanism is ohmic, i.e. given the familiar I^2R, where R is the resistance and I the current. Preliminary work [15] demonstrated that this equation is appropriate, i.e. that it is not required to solve Maxwell's equations within the food. Here, a more useful form of Ohm's law is one which considers voltage gradients, in which the heating rate of a food at point i is given as:

$$Q_i = \kappa_i E^2 = \kappa_i \text{ (grad } V.\text{grad } V) \quad (10.1)$$

where E is the voltage gradient and κ_i the electrical conductivity at point i. To calculate the distribution of heat generation requires that the voltage distribution is known throughout. The voltage field within a system in which the electrical conductivity varies with position can be found by solving Laplace's equation:

$$\nabla.(\kappa \nabla V) = 0 \tag{10.2}$$

Equation (10.2) must be solved throughout the material using appropriate boundary conditions, such as constant voltage on the electrodes. The equation shows that the distribution of electrical conductivities will control the local voltage distribution, and thus have a critical effect on the heating rates of solid and liquid.

This equation cannot be solved analytically except in simple situations, such as for an isolated sphere or cylinder of constant physical properties in a uniform field [15]. For example, for a sphere of radius R_p in an infinite region containing a uniform parallel field of strength E_0, equation (10.2) can be rewritten in spherical coordinates and solved by separation of variables to give the voltage distribution:

$$\text{inside the sphere:} \quad V_S = V_0 - \left(\frac{3\kappa_L}{2\kappa_L + \kappa_S}\right) E_0 r \cos\theta \tag{10.3}$$

$$\text{outside:} \quad V_L = V - (1 + \left(\frac{R_p}{r}\right)^2 \left(\frac{\kappa_L - \kappa_S}{2\kappa_L + \kappa_S}\right) E_0 \cos\theta) \tag{10.4}$$

where the subscript L denotes the liquid phase and S denotes the solids. These equations can be manipulated to give the relative heat generation rates in the two phases. This is an important quantity in determining the relative temperature changes of the two phases. Neglecting conduction, the ratio of the heat generation in the solid to that at a point in the liquid where the particle has no effect on the field is given by substitution of equations (10.3) and (10.4) into (10.1) as:

$$\frac{Q_S}{Q_L} = \frac{\rho\kappa_S\kappa_L}{(\kappa_S + 2\kappa_L)^2} \tag{10.5}$$

The above equation indicates that the relative heating rate is a function of both solid and liquid electrical conductivities. Complex heating effects are found, which can be expressed as a function of the ratio of the solid to liquid conductivity, $R_\kappa = \kappa_S/\kappa_L$ [15, 16]. A plot of equation (10.5) is given in Figure 10.4. Between $R_\kappa = 1$ and 4, the rate of heat generation in the solid is greater than that in the surrounding liquid. Outside this region, less heat is generated in the solid than in the liquid. The inset to Figure 10.4 shows the shape of the electric field, which is concentrated into the solid if it has a higher electrical conductivity. Some experimental data

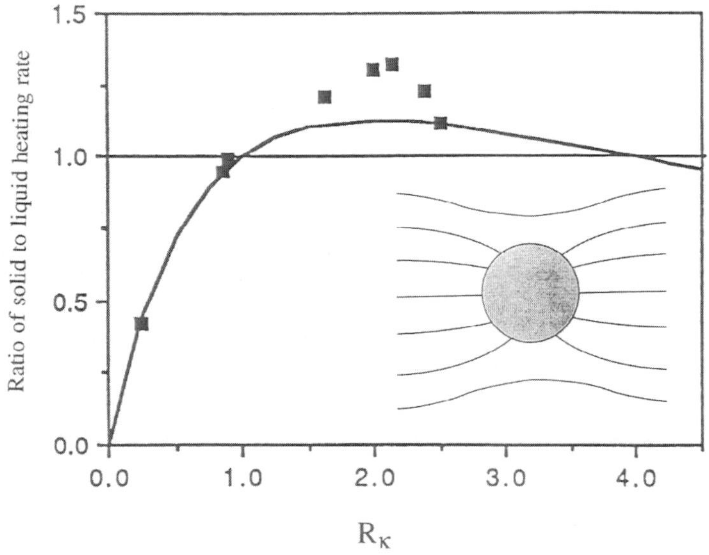

Figure 10.4 Plot of equation (10.5). Inset shows shape of electrical field [15, 19].

has been superimposed on the plot, without any attempt to distinguish between the physical properties of the two phases. The shape of the theory curve matches the experimental results, and the maxima also occur at the same point. This shows that Laplace's equation, and the Ohm's Law analysis, is indeed appropriate for this problem. It should be noted that, in the limit of infinite or zero particle electrical conductivity, the electrical heating rate of the particle will be zero; a highly conductive material will not heat, however large the current flowing through it, and at near zero conductivity no heating will occur because negligible current flows through an insulator. For conductivities in between, as shown in Figure 10.4, particles may either over- or underheat the surrounding liquid [16].

Equation (10.2) shows that the heating rate will also be a function of the shape of the particle and its orientation to the electric field – i.e. the distribution of the electrical conductivity in the system. Figure 10.5 demonstrates this clearly; the same particle will *overheat* the fluid if placed parallel to the electric field or *underheat* if placed at right angles to the field. The shape of these changes can be predicted by analogy with electrical circuits, but to represent the heating behaviour accurately it is necessary to solve equation (10.2). Commercial programs, such as ANSYS (Swanson Analysis Systems, Houston, PA) can be used. Simulations of experimental cases of Figure 10.5 show that similar over- or underheating patterns can be produced. Particles with a smaller aspect ratio, such as

(a)

(b)

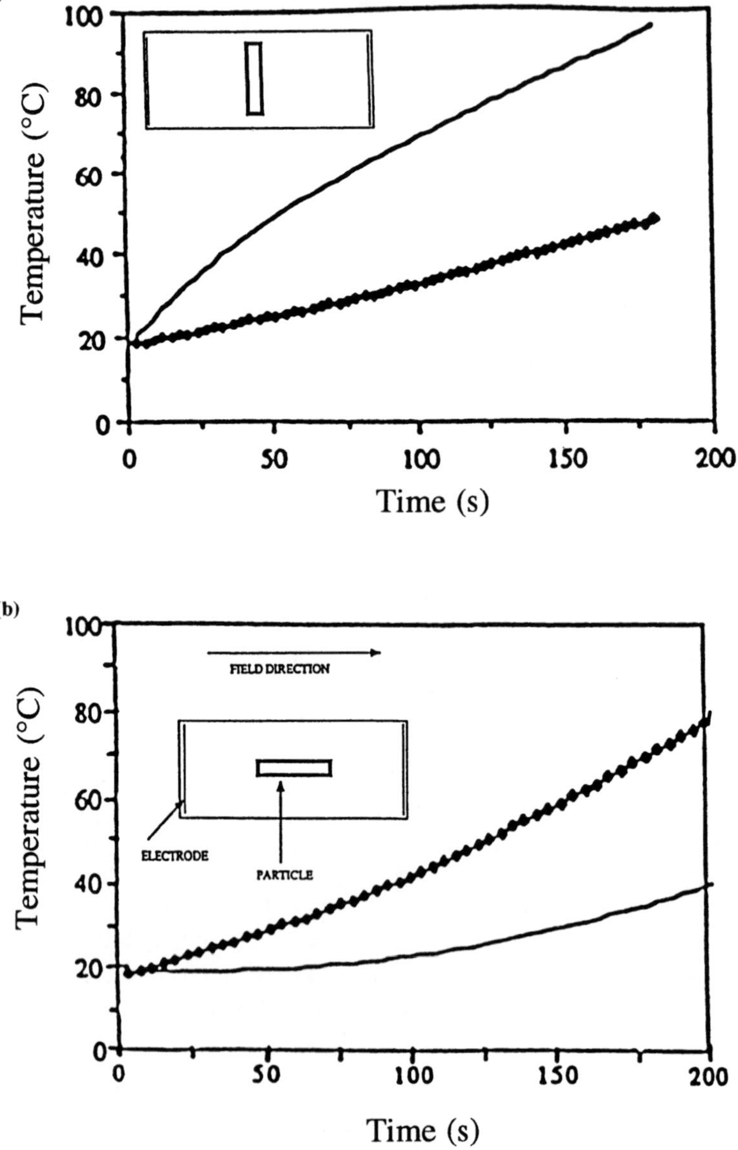

Figure 10.5 Temperature of particle (■–■) and liquid (——) according to orientation of electrical field [15].

cubes, will have less heating rate variation with orientation [17], but some effect will always be seen around sharp edges. Effects found by simulation have been verified by experiment [15, 18].

10.2.2 Thermal properties of foods

The rate of heat generation can be related to the local temperature change by the thermal properties of the solid. The factors which will affect the heating rate are:

(i) the thermal capacity of the material, the product of density ρ and specific heat c_p. It is convenient to define G, the inherent heating rate of the material in the absence of other factors, as:

$$G = \frac{dT}{dt} = \frac{Q}{\rho c_p} = \frac{\kappa E^2}{\rho c_p} \tag{10.6}$$

(ii) the rate of heat transfer from the element into the rest of the material. When thermal conduction is the only heat transfer factor, the following equation applies:

$$\frac{dT}{dt} = G + \frac{\nabla.(\lambda \nabla T)}{\rho c_p} \tag{10.7}$$

where λ is the thermal conductivity of the material. When convection acts the problem will be complicated by the presence of a velocity field; this is discussed further below.

10.2.3 Food mixtures: flow and heat generation

The commercial heating unit involves a flow of a solid–liquid mixture past a series of electrodes. Even in the case of no relative motion between the phases, i.e. plug flow of the mixture, the solution of equations (10.1), (10.2) and (10.7) to find the temperature field is not simple, since the electric and temperature field solutions are coupled by the strong thermal dependence of electrical conductivity. If there is relative motion between the liquid and solids, or a liquid velocity profile in the heater, then the Navier–Stokes equation must be solved for the velocity field along with the other equations for the voltage and temperature fields.

The flow behaviour of the material will also be a function of temperature, as the viscosity of the fluid will change with temperature. Within the liquid, the effects of forced and natural convection must also be modelled; it has been shown [19] that in a viscous liquid significant differences in temperature may arise due to the effect of heat generation. In practice, the effect of flow may even out temperature differences in the liquid.

Full solution of the three fields is not feasible given the complexity of the problems and the coupling of the equations via the viscosity and electrical conductivity variation on temperature. Approximations must be sought. The next section considers ways of predicting and modelling heating rates in real systems.

10.3 Models for electrical heating

10.3.1 Electrical conductivity of foods

As shown above, the electrical conductivity of a food and its surroundings is the key parameter which determines how rapidly it heats. It is critical to measure the conductivity of all the foods involved in the process, both when designing a formulation and, during operation, in quality control and assurance.

In comparison to other physical properties, the electrical conductivity of food materials has not been extensively studied, although some uses have been made of it, for example in process control [20]. Electrical conductivities can vary over a range of sixteen decades, from 10^8 S/m for conductors such as copper to 10^{-8} S/m for insulators such as wood. This variation is much higher than that of thermal conductivity. For a single food fluid, the electrical conductivity can increase in a linear manner with temperature [8], as

$$\kappa = \kappa_0 + \Delta\kappa\,(T - T_0) \qquad (10.8)$$

reflecting the increase ionic mobility with temperature as it increases from reference temperature T_0. This equation, although useful as a first approximation, is not accurate in all cases. Some data for conductivity have been published [21–23] showing that:

(i) Electrical conductivity can be anisotropic, i.e. different in different directions. This reflects the structure of a food; for example, it would be expected that the electrical conductivity of uncooked carrot would be different if measured parallel to or at right angles to the root structure.
(ii) Changes in electrical conductivity of a material reflect the change in its structure, for example, starch gelatinisation or cell lysis can change the conductivity by changing the ionic concentration or ionic mobility. Figure 10.6 shows data for the electrical conductivity of a potato, showing a steep increase at about 70°C [21].
(iii) The electrical conductivity of a material, originally not suitable for electrical processing, can be modified to make it suitable. For example, blanching can change the electrical conductivity of mushroom by removing air bubbles inside the structure [21].

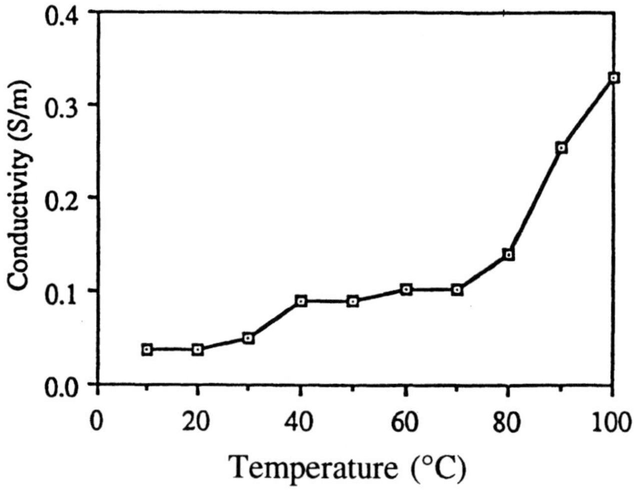

Figure 10.6 Electrical conductivity of a potato [21].

One complication is that the electrical conductivity of a food is a very strong function of the frequency of the current at which it is measured. The mobility of ions in a solution depends on the surrounding ionic atmospheres [24], which take a finite time to form; if the frequency of the a.c. current is such that these atmospheres do not form, then the conductivity is significantly different from when they do not. At kHz frequencies, therefore, where most electrical conductivity meters act, the measured conductivity can be greatly different from that at 50 Hz, i.e. the operating frequency of the APV system. It is thus crucial to find the electrical conductivity under the conditions which will be employed in practice. Other electrical effects will be discussed in section 10.4.

10.3.2 Electrical conductivity of solid–liquid mixtures

The previous section demonstrated that the distribution of electrical conductivity around a particle is vital in influencing the local temperature field, and thus local hot and cold spots. The distribution of conductivity will also control the overall electrical conductivity of the material, and thus the electrical consumption of the process. Determining the electrical conductivity of mixtures is a long-standing and difficult problem dating back to Maxwell [25], who showed that the effective conductivity of a dilute dispersion of spheres of solids fraction ϕ and conductivity κ_s in a medium of κ_L was:

$$\kappa_{eff} = \kappa_L \frac{1 - 2A\phi}{1 + A\phi} \text{ where } A = \frac{\kappa_L - \kappa_S}{2\kappa_L - \kappa_S} \qquad (10.9)$$

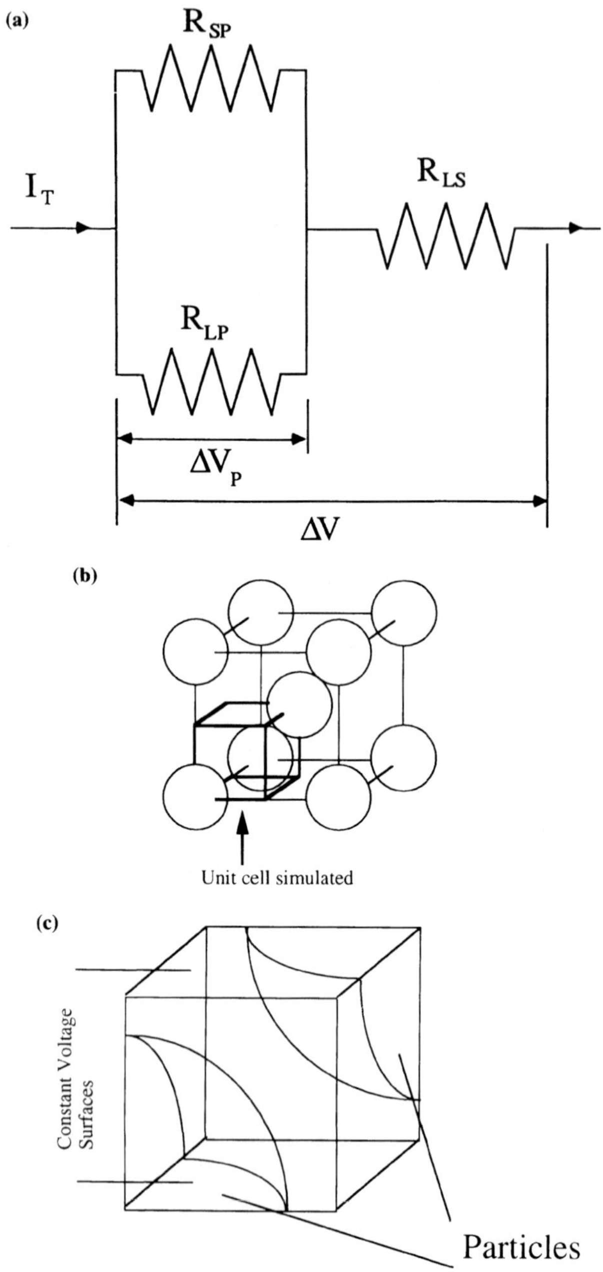

Figure 10.7 (a) Model for overall conductivity of a mixture, as a set of series and parallel resistors [27]. (b) and (c) Model for a conductivity of a mixture for a simple cube or sphere in a liquid; (b), unit cell body centred-cubic particles, showing modelled section (c); enlarged unit cell modelled using ANSYS [28].

Many modifications of this equation have been proposed and shown to give fits to data in given circumstances. It is not, however, possible to solve the general problem, as the conductivity of a food mixture will vary with the position and orientation of the particles. Although it has been shown [26] that equation (10.9) can predict the behaviour of foods at low temperature, in practice, it is important to measure the conductivity of the mixture over the full range of temperatures which will be experienced in the process.

Two models for the overall conductivity of a mixture have been proposed. Sastry [27] has modelled a mixture in terms of a circuit analysis, in which the mixture is represented as a set of series and parallel resistors, as in Figure 10.7(a). Alternatively, the conductivity of a mixture can be modelled by solving Laplace's equation for a 'unit cell' of material [28], i.e., as shown in Figure 10.7(b), i.e. for a single cube or sphere in a liquid. In concentrated suspensions or solid–liquid mixtures, it may be possible to represent the mixture in terms of circuits; here the two approaches give the same answer. The circuit approach, by assuming that all current lines are parallel, is incorrect; in practice, as shown in Figure 10.4, the field lines will be curved. This approach may be appropriate in some situations, such as high solids fractions, but will give errors in low solids–fraction cases [29].

The conductivity of a mixture may be non-linear even when the component conductivities are linear. For example, Marlow and Matthews [30] have shown that for potato cubes in CMC solutions the conductivity–temperature relationship is non-linear; this arises because current channelling around the particle creates local hot spots in the liquid which further enhance the current flow. Figure 10.8 shows this non-linearity. The effect is enhanced for large particles and for high viscosity carrier fluids; any stirring of the liquid will reduce any temperature differences, whilst for small particles the effect of thermal conduction will also reduce thermal gradients.

10.3.3 Flow and heat transfer

The above discussion shows how difficult it is to understand even the simplest case. To model the effects of flow and heat generation together in a real system, it is necessary to solve the Navier–Stokes equation together with equations (10.1), (10.2) and (10.7). This is not possible for particle–liquid mixtures at the moment, both because of the lack of information on the basic physics which governs such flows and due to the complexity of the calculations required to solve a realistic flow.

The flow and heat generation equations are coupled by the thermal dependence of the viscosity and electrical conductivity. The temperature dependence of the viscosity of foods is not easy to predict in food, but in

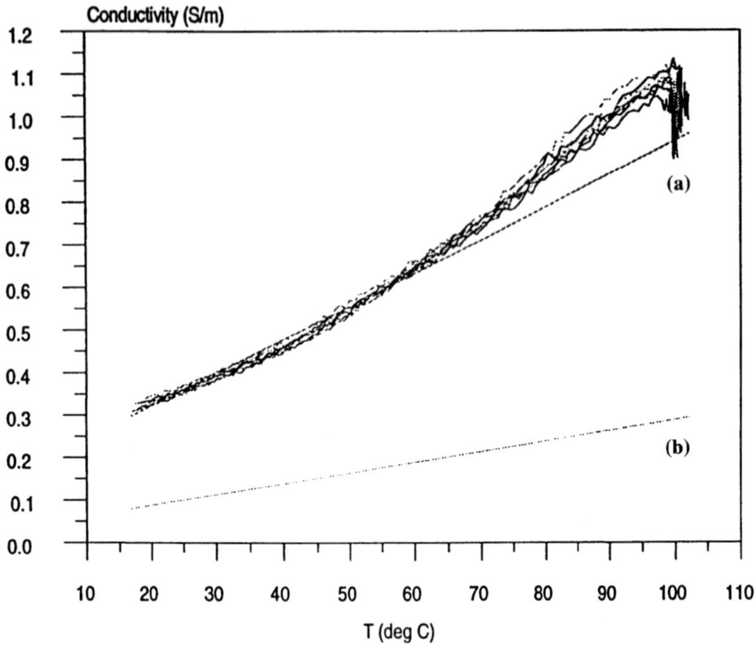

Figure 10.8 Non-linearity in conductivity in a mixture, with (a) k liquid and (b) k solid [30].

a material which does not undergo significant chemical change on heating it can be modelled as thermally activated, with an activation energy E, according to the Arrhenius equation:

$$\mu = \mu_0 \exp(-E_A/R_gT) \tag{10.10}$$

To solve the general problem of electrical heating requires simultaneous solution of the coupled equations for the electric, thermal and flow fields. This requires an unacceptable amount of computer capacity. It is however possible to model numerically the flow behaviour of single-phase fluids undergoing electrical heating. Although of limited practical interest, because the main advantage of ohmic heating is in the processing of two-phase mixtures, such models will show the range of effects found in real systems. Zhang and Fryer [31] have modelled simple laminar flows to demonstrate the type of effects which could be found during the simultaneous flow and heat transfer of real food fluids. Figure 10.9 shows the type of result predicted by the model; the effect of temperature is to decrease the viscosity near the wall where the slowest flowing liquid is hottest. This sharpens the velocity profile in comparison to that of a fluid with constant viscosity, increasing the plug flow nature of the flow. Other workers [32] have also shown that complex flow and temperature profiles can result,

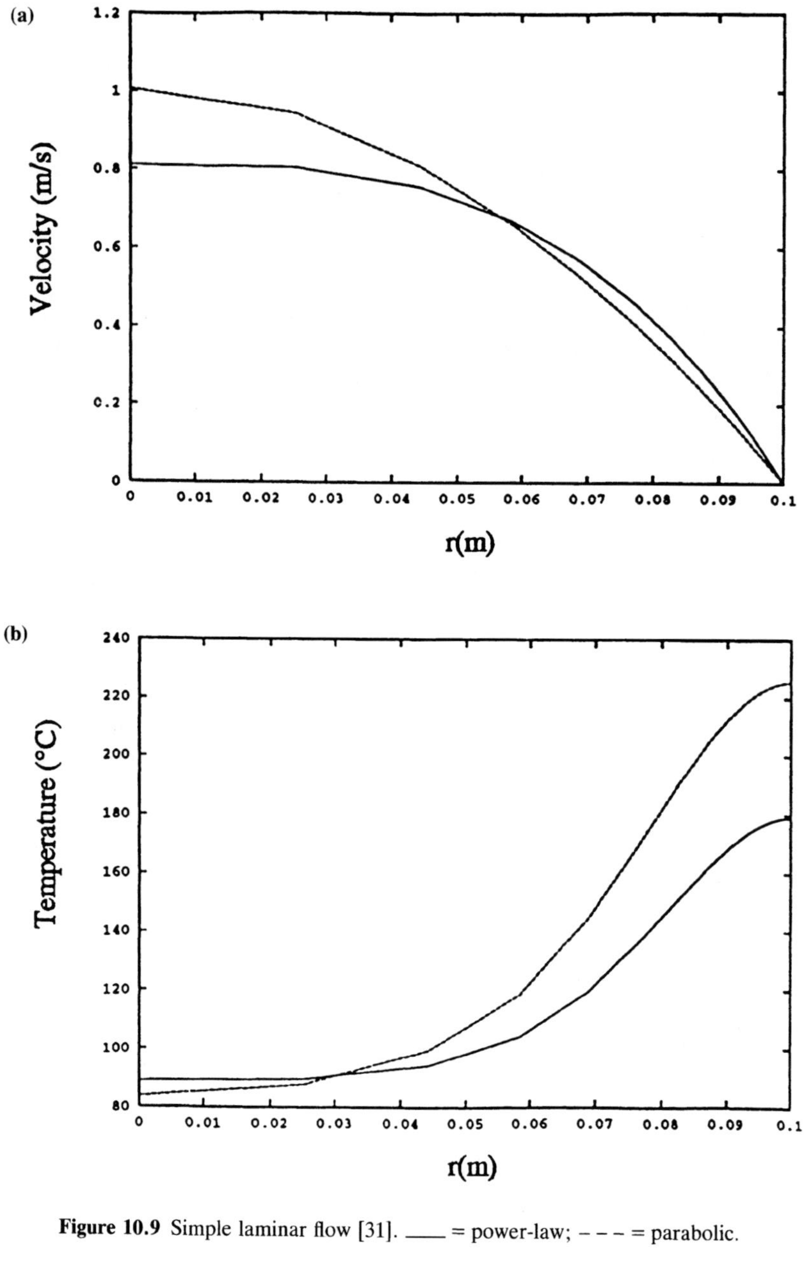

Figure 10.9 Simple laminar flow [31]. ____ = power-law; – – – = parabolic.

with high temperatures at the wall. In practice, wall cooling is sometimes applied to reduce the possibility of fouling from the slow moving wall layer of fluid; Quarini [33] has shown that in some circumstances this may lead to flow reversal in the tube.

Single phase flows may thus be very complex; two phase mixtures are even less well understood, although high solids fractions do however approximate to plug flow [34]. The rate of sterilisation of a material is highly temperature dependent, and so it is important that any model is as accurate as possible [29]. The two-phase system has been modelled using a series of approximations to simplify the problem to the point where it can be solved computationally [28, 35]. These are:

(i) That heat transfer and generation equations for the solid and liquid phases can be modelled by a one-dimensional heat balance; differences between the velocities of the phases can be accommodated.
(ii) That at each point (a) heat generation rates in the solid and liquid, (b) the temperatures of the phases, and (c) the velocities of the phases, are uniform and known. Fryer *et al.* [19] showed that uniform heating in the liquid is possible and Zhang and Fryer [28] showed that a rotating particle can heat uniformly.
(iii) Heating in the electrode housing is neglected; i.e. only heating of food during flow through the vertical tube connecting two electrodes is modelled. Boundary conditions of uniform voltage across the entrance and exit of the tube are defined for solution of the electrical field.
(iv) As the mixture flows up the tube, the temperature – and thus the electrical conductivity – of the two phases changes. The voltage gradient will thus vary up the tube, and an iterative calculation is necessary. The tube is thus modelled as a series of sections. Within each, the solid–liquid mixture has been represented as a uniform lattice of spherical particles on a regular grid, such that the 'unit cell' of the material can be modelled as representative of the whole, an approach common in symmetrical systems and already discussed in terms of the conductivity of mixtures.

Using the above assumptions, the response of the system can be modelled as a set of thermal balances. An enthalpy balance for a particle travelling at velocity v_s can be written:

$$-ha(T_S - T_L) + Q_S = v_S(\rho c_p)_S \frac{dT_S}{dx} \tag{10.11}$$

where h is the convective heat transfer coefficient between particle and liquid, a is the area of the particle per unit volume ($6/d_p$ for a sphere of diameter d_p) and Q_S is the heat generation rate per unit volume of solid.

This can be rewritten as

$$-H_S(T_S - T_L) + G_S = v_S \frac{dT_S}{dx} \tag{10.12}$$

where $H_S = ha/(\rho c_p)_S$ is a modified heat transfer term and G_s is the inherent heating rate for the solid, already defined.

Assuming no heat loss at the tube wall, the liquid phase heat balance can be written:

$$\frac{\phi}{(1 - \phi)} ha(T_S - T_L) + Q_L = v_L(\rho c_p)_L \frac{dT_L}{dx} \tag{10.13}$$

where ϕ is the fraction of the volume of the system occupied by the solid, v_L is the liquid velocity and Q_L is the heat generation rate per unit volume of liquid. This can be written

$$H_L(T_S - T_L) + G_L = v_L \frac{dT_L}{dx} \tag{10.14}$$

where

$$H_L = \frac{\phi}{(1 - \phi)} \frac{ha}{(\rho c_p)L} \text{ and } G_L = Q_L/(\rho c_p)_L.$$

To find the temperatures of the two phases, equations (10.12) and (10.14) must be solved together; data for heat generation rates, particle and fluid velocities and heat transfer coefficients must be known. If all physical properties are constant, the model can be solved as a set of two differential equations; but in general, this is not realistic. Either of the approaches for determining heating rates discussed above can be used. The source term Q has been calculated using the unit cell model for a range of R_κ and solids fraction ϕ, this data has then been fitted to polynomials which can then be incorporated into the computational scheme [35]. The equations can then predict the variation in temperature likely in a system where the physical properties vary. This includes the possibility of thermal crossover between the two phases: Figure 10.10 plots the variation in the temperature of the solid and liquid for a case where the solid is originally less conductive than the liquid but becomes more conductive with temperature.

Such a model gives an indication of how the temperature of the two phases may change during electrical processing. It does not indicate how local changes in temperature may affect the sterility pattern.

One potential problem in electrical heating is the accidental inclusion of either insulating or highly conducting material, which will not heat; food in the region round such particles may well be underprocessed. Another problem is that the unit cell model does not consider how foods of complex

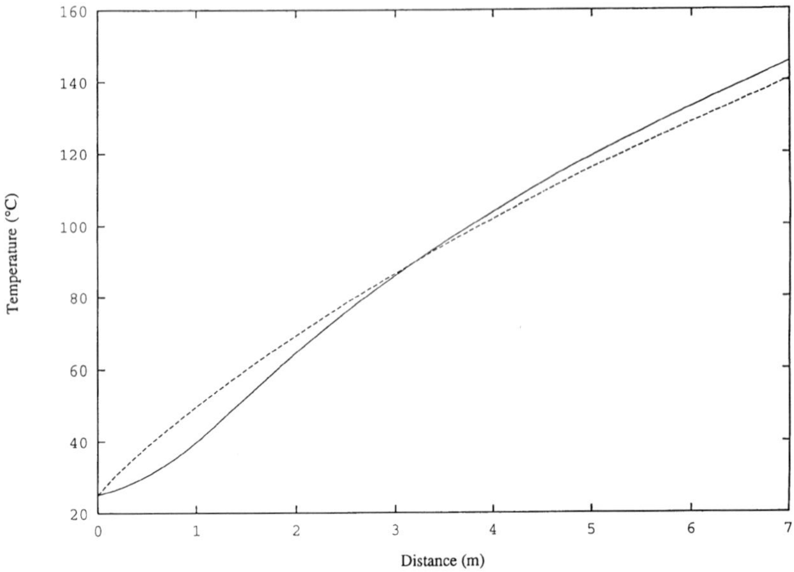

Figure 10.10 Variation in temperature of a solid (------) and a liquid (-- -- --) [35].

shapes will heat; local current concentrations can arise, for example at sharp edges or corners of particles. This type of process model is an interesting one, and shows what may happen in electrical heating, but experimental verification will be necessary before it can be used with confidence.

10.3.4 Holding and cooling sections

From the heater, material passes to the holding section and then to cooling. In a conventional process the holding section serves two purposes; to allow thermal equilibration between particle and liquid and to hold the material at high temperature long enough for the required level of sterility to be achieved [12]. In electrical heating, however, the equilibration step may well not be needed because the particles will have a temperature comparable to that of the liquid. For process control, it is useful if the particle temperature at the end of the heating section exceeds the liquid temperature. The latter temperature is straightforward to measure, and if it is known that particles are hotter then it is simple to confirm the sterility of the process. If the liquid temperature increases during thermal equilibration in the holding section, then this will show that particles have overheated.

Any overheating of particles may impair the quality of the material, because it is necessary to remove heat by thermal conduction during the cooling stage. Here, conventional models for heat transfer must be used; as it is no longer possible to assume that the particles are at a uniform temperature. Heat balances for the two phases can be written as before. For example, if it is assumed that the two phases move at uniform (but not necessarily the same) velocity in a cooling tube, then, for the liquid:

$$(wc_\mathrm{p})_\mathrm{L}\frac{\mathrm{d}T_\mathrm{L}}{\mathrm{d}x} = \pi d_\mathrm{t} h_\mathrm{w}(T_\mathrm{W} - T_\mathrm{L}) - a\,\frac{\pi d_\mathrm{t}^2}{4}\,h_\mathrm{p}(T_\mathrm{L} - T_\mathrm{S}) \qquad (10.15)$$

where h_w is the wall heat transfer coefficient, T_W the wall temperature and d_t the tube diameter and w is the mass flow. This must be solved together with unsteady-state balances for conduction within the solids, such as for spheres:

$$\frac{\partial T_\mathrm{S}}{\partial t} = \alpha_\mathrm{s}\,\frac{1}{R_\mathrm{p}^2}\frac{\partial}{\partial r}\left(r^2\frac{\partial T_\mathrm{S}}{\partial r}\right) \qquad (10.16)$$

where α_s is the thermal diffusivity. The design of two-phase systems is complex because the variation in velocities is not known. In practice, there can be a significant variation both in the velocity of the particles and in the velocity of solids and liquids [34, 36, 37]. Any variation in velocity, by changing the residence time in the system, will change the level of sterility and quality in the food.

More data will be needed before a fully accurate model is available. One use of the model is to consider and optimise the whole process, i.e. consider the effect of the temperature distribution in the heater on the subsequent process. It may in practice be advantageous for the solids to slightly underheat the liquid during processing; as it is easier to cool the liquid phase, solids, even if they are initially cooler than the liquid, will spend more time at high temperature than the liquid and thus will be sterilised during the hold and cool sections.

10.4 Electrically processed foods

10.4.1 Frequency effects in electrical processing

Foods contain complex charged species such as ions and proteins; in electrical heating, these may well behave in a different way to when conventional heating is used. Any differences between conventional and electrical processing must be quantified if the latter is to be widely used. However, the electrical properties of foods and the effects which electric fields have on foods, are not well understood. Application of direct current on foods lead to electroosmotic and electrophoretic effects; for example, dehydration of fish and tofu residues using applied electric fields has been reported in Japan [38, 39].

It might be thought that the use of alternating electrical current would preclude any ionic motion within a food. However, simple calculations can be used to suggest that this is not the case. In any ionic fluid, ions are surrounded by a set of counterions; this ionic atmosphere will take some time to form after a reversal of the direction of the electric field. The relaxation time can be estimated from the Einstein–Smoluchowski relationship [24, p. 428].

$$\tau_R = \frac{(\kappa^{-1})^2}{2D_i} \tag{10.17}$$

in which κ^{-1} is the Debye–Huckel length and D_i the diffusion coefficient for the ion. If this time is much greater than the cycle time for the electric field, then the ions will be effectively unaffected by the field.

An order-of-magnitude estimate of this time can be made as follows. For a salt solution of concentration ca. 0.025 μM conductivity is about 0.3 S/m at 20°C. For such a concentration of a 1:1 salt, the Debye–Huckel length is on the order of 2.5 nm [24, p. 199]. Diffusion coefficients for ions are also given by [24, p. 296] as; 1.3×10^{-5} cm^2/s for Na^+, and 2×10^{-5} cm^2/s for Cl^-. Substitution of these values in equation (10.18) gives a relaxation time in the region of 0.22 μs. This is several orders of magnitude less than the 0.02 s cycle time for voltage in a 50 Hz a.c. system. Ions will thus respond to the electric field. Significantly higher frequencies, with cycle times on the order of 0.2 μs, i.e. about 5 MHz, will be required to eliminate any ionic movement.

The consequences of this effect has been noted by Noguchi [38], who found: (i) that the electrical properties and heating rates of a fish protein gel were strong functions of frequency; and (ii) that the corrosion of the aluminium electrode used in his experiments decreased with increasing frequency, becoming negligible at 50 kHz. In the APV process, specially designed electrodes are used to eliminate corrosion.

This section describes recent experiments to investigate differences in electrical and conventional processing. These are only preliminary; much more work will be needed to understand fully the effects of electrical fields on foods.

10.4.2 Enhanced diffusion in electrical processing

There is evidence that the electrical conductivity of a food heated electrically is different to the same material cooked under conventional heating [21]. Some experiments have been carried out to study differences in the behaviour of foods due to the two techniques. The diffusion of betanin

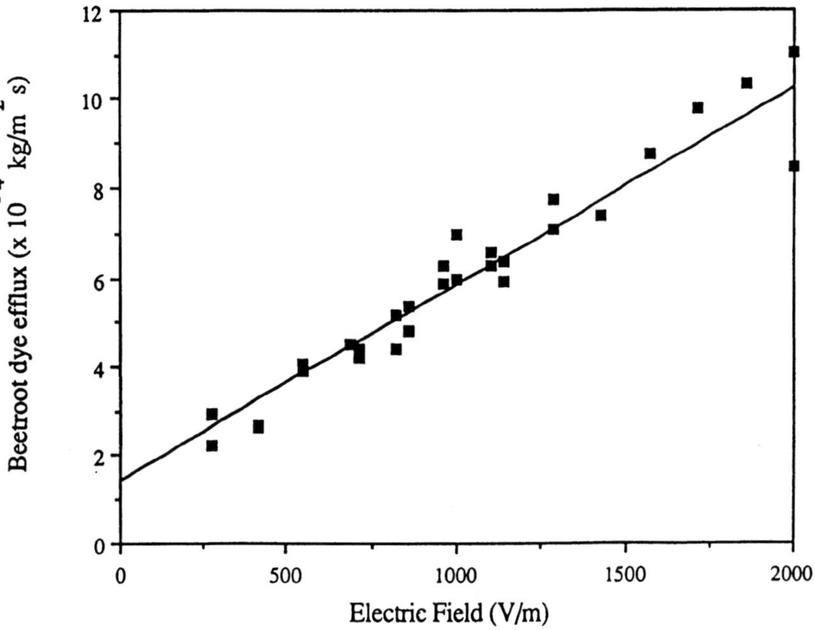

Figure 10.11 Diffusion of betanin dye between solid beetroot and surrounding fluid as a function of surrounding electrical field [37].

dye between solid beetroot and the surrounding fluid has been studied as a function of the applied electric field strength in a 50 Hz environment. Results are shown in Figure 10.11; the amount of mass transfer increases linearly with the electric field strength [40].

Similar effects have been found for other dyes, and demonstrate clearly that the motion of charged species is significantly enhanced in electrical heating. The amount of mass transfer is proportional to the surface area available, showing that the effect obeys the standard laws of mass transport. The presence of an electric field enhances the motion of ions [41]; this, for example, is the basis of electrophoresis. In a direct current, the force on an ion of charge number z_i is equal to

$$-Fz_i \text{ grad } V \tag{10.18}$$

where F is the Faraday constant; this will enhance the velocity of the ion. The rapid formation of the ionic atmosphere noted above suggests that significant ionic motion will occur in each cycle. Although the reversal of the voltage gradient will reverse the motion of the ion, the effect of diffusion will ensure that at the end of the cycle the ion has been displaced from its original point.

Other effects are also possible; electroosmosis has been proposed by Sielaff and Holt [20] as an explanation for the differences in electrical conductivity between foods processed electrically and conventionally. It is worth nothing that the data fit the equation shown by Pikal and Shah [42] to describe electroosmosis in a biological system:

$$N = K_N \text{ grad } V \tag{10.19}$$

Effects may also arise due to the presence of cell membranes. Although food dyes are charged, it is unlikely that they will carry much of the total charge in the system. The transport number of an ionic species is defined as the fraction of the total current carried by the particular species:

$$t_i = \frac{I_i}{\sum_i I_i} \tag{10.20}$$

The current I_i carried by species i is a function of the concentration and the ionic mobility, which is a function of particle size. The radius of the dye molecules will be significantly greater than the sodium and chloride ions, so that the mobility of the dye molecules will be less. In addition, the concentration of dye in the bulk will be significantly less than that of the salt. As well as the effects of electroosmosis, Barry and Hope [43] describe the effects of such transport number differences between a membrane-bound system and the surrounding bulk solution. This transport number difference leads to enhancement or depletion of concentrations at the membrane-solution interface:

$$\Delta c = \frac{\alpha I}{F} \tag{10.21}$$

where α is the transport number difference between the bulk solution and the membrane, I the current and F the Faraday constant. The concentration difference is a direct result of the passage of current through the system and leads to an enhancement of diffusion flux across the membrane. The effect may well be to enhance diffusion flux in addition to the effects of electroosmosis. Barry and Hope [43] point out that, in many cases, the transport number effect is larger than any electroosmotic effect present in a system, though the two effects may act together.

10.4.3 Differences between diffusion in conventional and electrically processed foods

The diffusion of species within food solids can be followed using pulsed field gradient spin echo (PFGSE) NMR [such as in 44]. A detailed account of this technique is given by Karger *et al.* [45]. In a PFG experiment the NMR signal is reduced by the movement of the nucleii in a sample during a defined time interval Δ, the time between two gradient pulses of duration δ which respectively follow 90° and 180° rf pulses. If the probability $P(r, \Delta)$ for a molecule to diffuse a distance r in the direction of the magnetic field gradient in the time D is known, then the signal attenuation can be calculated. For unrestricted Brownian motion in 3D, the probability distribution can be used to find the signal attenuation:

$$\ln \Psi = D(\gamma g \delta)^2\left(\Delta - \frac{\delta}{3}\right) = Dk \tag{10.22}$$

where the quantity k contains a number of known factors, including the gyromagnetic ratio g, the magnetic field gradient γ, the times Δ and δ, and D is a self-diffusion coefficient. This equation applies to unrestricted diffusion; any solid structure, such as a cell wall, will restrict diffusion, an effect which will be greater for longer diffusion times. Various models for restricted diffusion can be fitted to the attenuation data, such as for molecules confined between parallel plates [46] and in spherical shells [47]. In the former case, the attenuation between two parallel impermeable plates a distance L apart is given by:

$$\Psi = \frac{2(1 - \cos \gamma g \delta L)}{(\gamma g \delta L)^2} + 4(\gamma g \delta L)^2 \sum_{n=1}^{\infty} \exp\left\{\frac{-n^2\pi^2 D\Delta}{L^2}\right\} \left\{\frac{1 - (-1)^n \cos(\gamma g \delta L)}{(\gamma g \delta L)^2 - (n\pi)^2}\right\} \tag{10.23}$$

To a given set of experimental data, D and L can be adjusted to find both the plate spacing and the self-diffusion coefficient. The length L can be associated with a cell wall spacing, although the physical model is not really appropriate.

The self-diffusion coefficient and the cell spacing L were studied as a function of the time of cooking for both conventional and electrical heating of a test material, carrots. Figure 10.12 shows the type of results for cooked and uncooked material; in part (a), for uncooked material, significant differences are seen for different values of the diffusion time, and the Figure also shows that equation (10.23) can be fitted well to the data. Figure 10.12(b) shows results for material which has been electrically cooked for one minute; there is much less separation of the curves, reflecting the breakdown in structure which results from the cooking process. Values for D and L are given in Table 10.1 for different parts of the carrot and different cooking times and techniques [48]; as cooking time increases, L increases, as the structure of the material breaks down. Table 10.1 also gives differences between the self-diffusion coefficient for the two techniques. The differences were statistically significant, but much less than that between cooked and uncooked material [48]. This suggests that

(a)

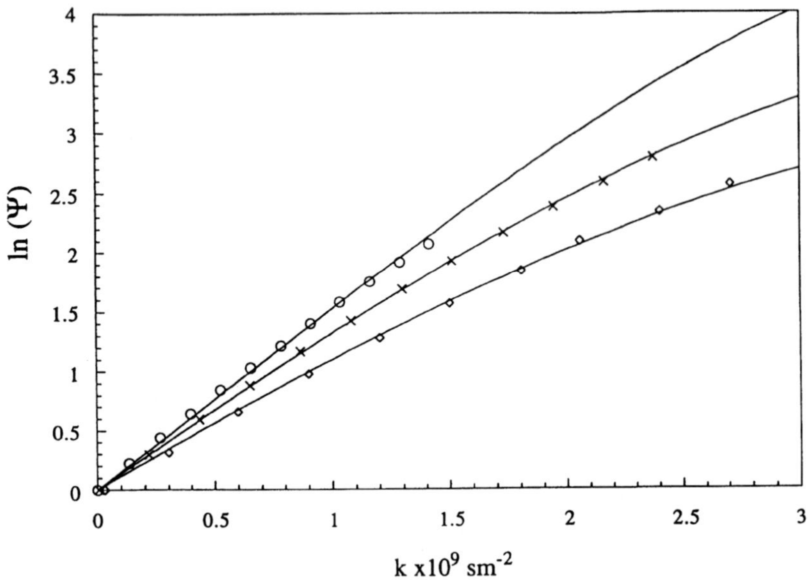

(b)

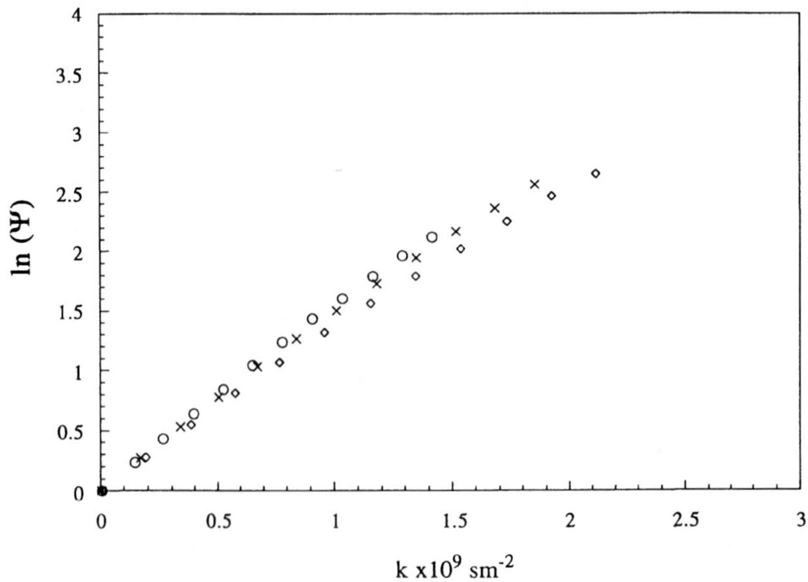

Figure 10.12 NMR studies of electrical and conventional cooking. (a) Effect of time on PFG response of uncooked carrot. (b) PFG response of electronically cooked carrot [48]. ○ = 20 ms; X = 100 ms; ◇ = 400 ms; ------ = model.

Table 10.1 Results of least square fit to parallel plate model: data for axial diffusion coefficients in carrot

Diffusion coefficient (m^2/s): Barrier separation (μm):	Raw	1 minute processing		10 minutes processing	
		Conventional	Electrical	Conventional	Electrical
Exterior	1.58 58	1.54 550	1.52 640	1.68 920	1.78 1750
Interior	1.60 139	1.77 910	1.53 410	1.70 1280	1.68 690

there may be differences in the structure of materials cooked electrically and conventionally. The reason for this is unclear, but if motion of food components is significantly different in the two cooking methods, as suggested above, then cell walls may be disrupted in different ways during processing.

These observations suggest that complex effects may arise in the electrical processing of foods, which must be understood in order to exploit the technology fully. More work is thus required.

10.5 Conclusions

In electrical processing, heating occurs volumetrically throughout the material as a result of passing an electric current through the food. Electrical heating is an effective way of processing solid–liquid mixtures, as it overcomes the limitations of slow thermal conduction between solids and liquids. Although the technique is not novel, interest has been revived by the development of a commercial process for the sterilisation and pasteurisation of solid–liquid mixtures by APV Baker Ltd. The process is well suited to high solids fractions foods containing large particulates, that are difficult to sterilise in other ways.

This chapter has reviewed progress in the understanding of the basic principles of electrical heating. At the low frequencies of the APV process, the heat generation mechanism is given by conventional I^2R; it is necessary to know both the distribution of current and resistance (or electrical conductivity) in the system. Complicated heating effects can result when particles of different aspect ratios and significant differences in electrical conductivity to the liquid are processed. In general, it is not possible to solve Laplace's equation for the electric field distribution analytically; numerical techniques are needed to predict heating rates.

Two types of model have been used to predict the behaviour of the mixtures; one based on a circuit analogy, and one which uses a solution to Laplace's equation in a 'unit cell' of the mixture. The two models give similar results for high solids fractions, but the unit cell model, being more physically realistic, is more appropriate for many cases. A flow model based on solving Laplace's equation for a unit cell of the mixture, which is assumed to be uniform, has been used to model a two-phase system and demonstrate the range of effects which may occur in commercial practice. Further experimental data on flow patterns and heat transfer rates are needed to improve the applicability of any model; it is also important that any model produced can be validated.

A number of differences between conventionally and electrically processed foods have been found. At 50 Hz, foods essentially behave as if processed in direct current; much higher frequencies are needed to eliminate any d.c. effects. Experiments on test food materials have demonstrated the existence of enhanced diffusion in electrical processing; the rate of mass transport is linear in the applied electric field strength. This may well cause differences between food processed by electrical and conventional heating, especially in foods such as meat where texture depends on water content. Further experiments will be necessary to establish both its cause and its effects, if any, on the properties of electrically processed food materials. Preliminary experiments on diffusion in foods measured by NMR have suggested slight differences between electrical and conventionally processed foods, but these differences are small in comparison with those between cooked and uncooked foods. These effects remain largely unexplained; one useful series of experiments would be to establish the effect of variations in the frequency of the applied electric field on electrically induced diffusion enhancement. At frequencies significantly higher than 50 Hz, of cycle times comparable to τ_r, it is likely that the enhancement would be significantly reduced. This may offer new opportunities for electrical processing.

Acknowledgements

Work at Cambridge on electrical heating described here was carried out by Dr A.A.P. de Alwis, sponsored by APV Baker Ltd and the Cambridge Commonwealth Trust, and Dr L. Zhang, sponsored by AFRC, together with a number of final year students.

References

1. Hallstrom, B., Skjoldebrand, C. and Tragardh, C. *Heat Transfer and Food Products*, Elsevier, London, 1988.
2. Tucker, G.S. and Holdsworth, S.D. *Trans. IChemE C*, **69**, 5–12, 1991.
3. Holdsworth, S.D. *Aseptic Processing and Packaging of Food Products*, Elsevier, London, 1993.
4. Ayeppa, K.G., Davis, H.T., Crapiste, G., Davis, E.A. and Gordon, J. *Chem. Eng. Sci.*, **46**, 1005, 1991.

5. de Alwis, A.A.P. and Fryer, P.J. *J. Food Eng.*, **11**, 3–27, 1990.
6. He, H. and Hoseney, R.C. *Cereal Chem.*, **68**, 151, 1991.
7. Skudder, P.J. and Biss, C.H. *The Chemical Engineer*, **433**, 26–8, 1987.
8. Biss, C.H., Coombes, S.A. and Skudder, P.J. In *Process Engineering for the Food Industry*, eds Field, R.W. and Howell, J.A., pp. 17–27, 1987.
9. Stirling, R. *Power Engineering*, **6**, 137–58, 1990.
10. Parrott, D.L. *Food Technol.*, **46**, 68–72, 1992.
11. Gaze, J.E. and Brown, G.D. *J. Soc. Dairy Technol.*, **43**, 49–52, 1990.
12. Sastry, S.K. *J. Fd Proc. Eng.*, **15**, 163–178, 1992.
13. Skoldebrand, C. and Ohlsson, T. *J. Fd Eng.*, **20**, 149–166, 1993.
14. Skoldebrand, C. and Ohlsson, T. *J. Fd Eng.*, **20**, 167–182, 1993.
15. de Alwis, A.A.P., Halden, K. and Fryer, P.J. *Chem. Eng. Res. Des.*, **67**, 159–168, 1989.
16. de Alwis, A.A.P. and Fryer, P.J. *J. Fd Eng.*, **15**, 21–48, 1992.
17. Sastry, S.K. and Palaniappan, S. *J. Food Proc. Eng.*, **15**, 213–227, 1992.
18. de Alwis, A.A.P. and Fryer, P.J. *Chem. Eng. Sci.*, **45**(6), 1547–1560, 1990.
19. Fryer, P.J., de Alwis, A.A.P., Koury, E., Stapley, A.G.F. and Zhang, L. *J. Fd Eng.*, **18**, 101–125, 1993.
20. Sielaff, H. and Holt, K.H. *Fleisch*, **33**, 164–167 and 196–198, 1979.
21. Halden, K., de Alwis, A.A.P. and Fryer, P.J. *Int. J. Fd Sci. Technol.*, **25**, 9–25, 1990.
22. Palaniappan, S. and Sastry, S.K. *J. Food Proc. Eng.*, **14**, 221–236, 1991.
23. Palaniappan, S. and Sastry, S.K. *J. Food Proc. Eng.*, **14**, 247–260, 1991.
24. Bockris, J.O'M. and Reddy, A.K.N. *Modern Electrochemistry*, Plenum, New York, 1970.
25. Maxwell, J.C. *A Treatise on Electricity and Magnetism*, Vol. 1, Clarendon Press, Oxford, 1886.
26. Palaniappan, S. and Sastry, S.K. *Trans. IChemE C*, **69**, 167–174, 1991.
27. Sastry, S. *J. Food Proc. Eng.*, **15**, 163–178, 1992.
28. Zhang Li and Fryer, P.J. *Chem. Eng. Sci.*, **48**, 633–643, 1993.
29. Zhang Li and Fryer, P.J. *J. Food Proc. Eng.*, 1994 (in press).
30. Marlow, S. and Matthews, G. *Part II Tripos Report*, University of Cambridge.
31. Zhang Li and Fryer, P.J. presented at ICEF6, Tokyo, May 1993, and pp. 760–762 in *Developments in Food Engineering*, eds T. Yano, R. Matsuno and K. Nakamura, Blackie, Glasgow, 1994.
32. Muller, F.L., Pain, J.P., Villon, P. and Fryer, P.J. Presented at Colloque Annuel de la societe francaise des thermiciens, 1993.
33. Quarini, J. *J. Food Eng.*, 1994(in press).
34. Proctor, J.M, de Alwis, A.A.P., Fryer, P.J., Liu, S. and Pain, J.-P. *Chem. Eng. Commun*, **124**, 97–114, 1993..
35. Zhang Li and Fryer, P.J. *AIChEJ*, **40**, 888–898, 1994.
36. Mankad, S., Branch, C.A. and Fryer, P.J. *Chem. Eng. Sci.*, 1994 (in press).
37. Lareo, C., Branch, C.A. and Fryer, P.J. presented at "Food Process Engineering", Bath, 1994.
38. Noguchi, A., presented at Nestec New Technologies Workshop, Salle Louis Pasteur, Switzerland, 1994; and Uemura, K., Noguchi, A., Park, S.J. and Kim, D.U., presented at ICEF6, Tokyo, May 1994, and pp. 310–312 in *Developments of Food Engineering*, eds T.Yano, R. Matsuno and K. Nakamura, Blackie Glasgow, 1994.
39. Suzuki, Y., Konno, M., Sato, T. and Shishido, I. *Kaguku Kogaku Ronbunshu*, **16**, 1133 –1137, 1990.
40. Schreier, P.J.R., Reid, D.G. and Fryer, P.J. *Int. J. Fd Sci. Tech.*, **28**, 249–260, 1993.
41. Wesselingh, J.A. and Krishna, R. *Mass Transfer*, Chichester, Ellis Horwood, 1988.
42. Pikal, M.J. and Shah, S. *Pharmaceutical Research*, **7**, 213–221, 1990.
43. Barry, P.H. and Hope, A.B. *Biophysical Journal*, **9**, 700–728, 1969.
44. Jenner, C.F., Xia, Y., Eccles, C.D. and Callaghan, P.T. *Nature*, **336**, 199–402, 1988.
45. Karger, J., Pfiefer, H. and Heink, W. In *Advances in Magnetic Resonance*, ed. W.S. Warren, Academic Press, New York, 1988.
46. Stejska, E.O. *J. Chem. Phys.*, **43**, 1768–1777, 1965.
47. Murday, J.S. and Cotts, R.M. *J. Chem. Phys.*, **48**, 4938–4945, 1968.
48. Stapley, A.G.F., Sousa Goncalves, J.A., Hollewand, M., Gladden, L.F. and Fryer, P.J. *Int. J. Fd Sci. Tech.*, 1994 (in press).

Nomenclature

A: Constant defined in equation (10.9)
a: Area of the particle per unit volume
c_p: Specific heat
D: Self-diffusion coefficient
D_i Diffusion coefficient for an ion
d_p Spherical particle diameter
d_t Tube diameter
E: Field strength
E_A: Activation energy
E_o Field strength a distance from the particle
F: Faraday constant;
G: Inherent heating rate of the material $= \dfrac{Q}{pc_p}$
g: Gyromagnetic ratio
h: Convective heat transfer coefficient between particle and liquid
H_L: Modified heat transfer term for liquid $= \dfrac{\phi}{(1-\phi)} \dfrac{ha}{(\rho c_p)_L}$
H_S: Modified heat transfer term for solid $= \dfrac{ha}{(\rho c_p)_S}$
h_w Wall heat transfer coefficient
I: Electric current
κ: Electrical conductivity
k: Variable in equation (10.22)
k_N Constant in equation (10.19)
L: Spacing of barrier
N Flux
n Number
Q: Heat generation rate
R: Electrical resistance
r: Radius
R_g: Gas constant
R_κ Ratio of solid to liuquid conductivity κ_S/κ_L
R_p: Particle radius
T_0: Base temperature
T: Temperature
t_i: Transport number
T_W: Wall temperature
V: Voltage
v: Velocity
V_o: Voltage a distance from the particle

w: Mass flowrate
z_i: Charge number

α: Transport number difference between the bulk solution and the membrane
α_s: Solid thermal diffusivity
Δ: Defined time interval between pulses
δ: Pulse duration
Δc: Difference in concentration
$\Delta \kappa$: Temperature coefficient of electrical conductivity
ϕ: Solid fraction
γ: Magnetic field gradient
κ^{-1}: Debye Huckel length
κ: Electrical conductivity
κ_0: Base temperature electrical conductivity
κ_{eff}: Mean system electrical conductivity
λ: Thermal conductivity
Θ: Angle
ρ: Density
τ_R: Ionic relaxation time
Ψ: Signal attenuation
μ: Viscosity
μ_o: Pre-exponential factor for viscosity

Subscript:
L: Liquid phase
S: Solid phase

11 High-voltage pulse techniques for food preservation

W. SITZMANN

11.1 Introduction

The large-scale mass production of foodstuffs, nowadays quite usual – at least in industrial nations – is only reasonable if at the same time it can be ensured that the products are storable for a comparably long time. To achieve this goal, the total microbial cell count in foodstuffs at the beginning of their storage time must not exceed a justifiable maximum. At the same time, the elimination of pathogenic cells must be ensured in order to exclude any danger to consumer's health.

Depending on the specific requirements, various technological measures are available for separating or killing microorganisms and thus reaching the required product quality. Cell count reduction in foodstuffs is achieved either by filtration, pasteurization, sterilization, chemical treatment or irradiation, the latter being a method which, to date, is allowed in some, but not all countries, but little exploited. Usually sterilization occurs as a result of heating, to be precise by thermal damage to cell macromolecules, including coagulation of proteins and failure of membrane function.

Whereas filtration sterilization can be used for clear liquids only, e.g. beer or wine, the other procedures mentioned have certain disadvantages. For instance, they initiate chemical reactions in the product that lead to a loss in quality, or they destroy vitamins, antioxidants and lead to changes in colour and odour as a result of the thermally induced Maillard and other reactions. This becomes quite clear when we take milk as an example. There are considerable differences in the taste of untreated certified milk, fresh milk that has been pasteurized, and sterilized and homogenized milk. Also with some heating procedures, there are undesired effects when substances contained in the product, such as salts or proteins, are deposited on heat exchanger surfaces. This leads to a reduction of the heat transfer coefficient so that periodic cleaning of the systems is absolutely necessary.

These disadvantages, characteristic of conventional pasteurization and sterilization procedures, contribute much to the fact that non-thermal procedures gain more and more significance.

11.2 Cell count reduction by using electricity: a historical review

With regard to commercialization, two out of a series of various novel methods of foodstuff conservation [1] have been developed to a high standard. The one is a method already applied in Japan, at least in partial sectors, and using high hydrostatic pressures for the elimination of vegetative cells of bacteria, yeasts and moulds, e.g. in jams and fruit juices. The other uses high voltage pulses. Since the middle of the last century it is a well-known fact that electric currents have an influence on the metabolism and viability of microorganisms [2]. Gillissen and Carlson [3] succeeded, for instance, in showing that the metabolism of mould fungi, when under the influence of electric currents, could be significantly increased. However, the opposite effect, i.e. the killing of microorganisms, has been investigated much more comprehensively. The principal precondition for this to happen is that the treated products must be pumpable and conductive, and that the electric energy is supplied via at least one pair of electrodes which are in contact with the product.

The strength of the either bacteriostatic or a bacteriocidal effect of three principal phenomena, varies with the way electric energy is supplied into the product to be treated. The three phenomena in question are:

(i) Heat, which is produced by transformation of induced electric energy.
(ii) Electrolysis products or free radicals, that may occur depending on the electrode material used and the chemical composition of the treated products.
(iii) A mechanical destruction of the bacterial cell membrane under sufficiently high energy densities, which occurs when using extremely short high-voltage pulses.

The fact that these phenomena, which occur when treating biological systems with electric pulses, cannot be completely isolated from one another, makes it clear why different groups of authors often mention different causes for the cell count reductions observed.

In the 1920s and 1930s a process for electric pasteurization of milk was introduced in the USA from Europe, the so-called 'Electropure Process' which, contrary to the discontinuous process usual at that time, was described as product-saving treatment [4, 5]. In the United States about 50 plants [6] of this type were in operation until the 1960s. The biggest of these plants had a processing capacity of somewhat more than 100 cu.m. per day. Through carbon electrodes the milk was subjected to 220 V d.c. This caused it to be heated up to and held at a temperature of about 70°C for a period of around 15 seconds [7]. Although with this process it was possible to make inactivate pathogenic bacteria and other microorganisms, not one of these plants was in operation after the 1950s [8]. Maybe the reasons were the rising energy costs and/or competitive heating technologies with more and more sophisticated plate-type heat exchangers and optimized heat recovery. From today's point of view the Electropure Process was purely thermic, as in Ohmic Heating.

Already in 1925, Beattie and Lewis [9] had established that the Electropure Process, due to the relatively low voltages applied, had only thermic, but no electric effects. Tests carried out using higher voltages, in a range between 3000 and 4000 V, showed that, in addition to thermic effects, electrically caused killing became more and more effective. This assertion was supported by investigations carried out in parallel with heat sterilization.

Predominantly in the 1950s and 1960s various groups of researchers worked on a process called 'Electrohydraulic Treatment'. For the purpose of killing microorganism cultures in watery solutions, very high electric field strengths were applied pulsatingly using capacitor discharges which created electric arcs and a quickly expanding plasma. Ultraviolet light, electrochemical reactions and, mainly, a shock wave of up to 5000 bar were identified as responsible for the bacteriocidal effects.

Früngel [10] for instance, who in 1957 made an application to patent a method and an apparatus for electric sterilization and cleaning of dairy machinery, attributed the bacteriocidal effect of the discharges to the shock waves that were produced. Gossling [11] described in his patent application an apparatus with a killing and mutation effect of microorganisms under the influence of electric high-voltage pulses. Unfortunately he did not deal with design details. From his observations (which largely deviate from later works by other authors) he concluded that the lower the treatment temperature, the higher the killing rates of *Escherichia coli*. The results were virtually independent of the suspension medium in which the microorganisms were suspended (a statement likewise in contradiction to the work of other authors).

Brandt *et al.* [12] carried out many killing tests involving a great variety of bacteria, e.g. *E. coli*, *Staphylocococcus aureus*, *Streptococcus faecalis*, as well as spores of *Bacillus subtilis*. The repeated discharge of a 0.6 μF capacitor with a discharge voltage of 47 kV made it possible to reduce the initial cell count by up to five powers of ten. Even spores were killed in this process at a rate of nearly 99%. Contrary to Gossling, the authors observed a strong connection between the suspension medium used and the killing results attainable. When using protein-rich media, the killing results were clearly lower. Furthermore the authors established that there was a cell count reduction when suspending, immediately *after* the electric treatment, microorganisms in water (with or without salt) that had been treated with electric shocks. This phenomenon lead to the assumption that, besides the effect of the shock waves, the measured lethal effects were to be attributed mainly to short-lived, highly reactive, free radicals that were possibly created by the spark discharge produced in the suspension medium.

The apparatus described in [12] was used in the mid-1960s by Edebo and Selin [13] as well as by Edebo *et al.* [14] for further investigations into the influence of the electrode materials on cell destruction. When using copper and silver electrodes, the bacteriocidal effects were clearly stronger than with brass or tungsten. Allen and Soike [15, 16] conducted experiments with distinctly lower voltages, of a maximum of 14 kV, but with much higher capacitances of up to 95 μF. In a discharge chamber of 1.2 l watery suspensions of *E. coli*, *Saccharomyces cerevisiae*, bacteriophages T-2 and spores of *B. subtilis* were treated electrohydraulically and practically complete sterilization was achieved. The bacteriocidal effect of the shock treatment was attributed to similar causes as those proposed by Brandt *et al.* [12].

Gilliland and Speck [17, 18], too, established in their tests with capacitor discharges (10 kV, 24 μF), that the presence of proteins in the treated bacterial suspensions had a somewhat protective effect. In systematic investigations they proved that the kind of energy supply had a decisive influence on the lethal effects attainable. If a predetermined quantity of energy was supplied in few pulses of high voltage and with high capacitance into the system, the killing rates were lower than with a correspondingly higher number of pulses of lower voltage and capacitance. Furthermore, the authors observed a clear reduction in the activity of various enzymes as a result of the electric shock treatment.

However, the electrohydraulic treatment has never been developed up to the point of practical application. The publications mentioned give information about interesting effects when subjecting suspensions of microorganisms to electric high-voltage pulses. It is, however, practically impossible to directly compare the test results of different authors with one another and to extrapolate them in order to make them available for the design of large-scale plants. To achieve the energy densities that are necessary for generating spark discharges, needle-shaped electrodes are needed. These electrodes are arranged only a few millimetres apart in discharge chambers of different sizes. (Brandt *et al.* [12] used chambers of up to some hundreds of litres.) Due to the fact that the local bacteriocidal effect of electric shocks in the suspension volumes treated is not known and its intensity is difficult to assess, the lethal effects measured in the above-mentioned tests depend quite specifically on the apparatus used. Hence, they cannot be compared with one another.

At the beginning of the 1960s, Doevenspeck [19, 20] succeeded in making decisive progress in pulse technology development and in achieving substantial cell count reductions. Contrary to electrohydraulic shock treatment, in which disruptive discharges are produced in suspension medium for the purpose of killing the microorganisms contained in it, he used defined homogenous electric fields in his experiments. Doevenspeck found that homogenous fields of different field strengths have different effects on biological systems. With so-called 'hard' pulses, cells of vegetable and animal origin can be disrupted, 'medium strong' pulses can kill microorganisms, while 'soft' pulses of low field strengths can be used to increase the metabolism of microorganisms. Even weak electric fields can effect the flocculation of solids dissolved in certain sewages [21].

The first reports on systematic investigations in the field of microorganism killing under the influence of homogeneous electric high-voltage fields, were made by Sale and Hamilton [22–24]. These authors treated suspensions of bacteria and yeast cells as well as spores of *Bacillus cereus* and *Bacillus polymyxa* in sodium chloride solutions and, whilst observing the inactivation of vegetative microorganisms, found that even under the influence of a field strength of 30 kV/cm, spores were resistant. A reduction in the activity of various enzymes under the influence of electric pulses could not be observed, any more than bacteriocidal influences of thermic or electrolytic effects.

The results of their work proved that the lethal effects of high-voltage pulses only depend on the applied electric field strength, the treatment time (= the product of the duration of one single pulse times the number of pulses), and the cell size. They observed that radical cell count reductions begin at the moment when transmembrane potentials of about 1 Volt are reached or exceeded as a result of the applied electric fields. The resulting changes of the membrane structure and, consequently, the permeability of the membrane were identified as the causes of the lethal effects.

Sakurauchi and Kondo [25] investigated in their experiments the killing possibilities for spores of *Bacillus subtilis* and vegetative cells of *E. coli*. It was observed that under very long treatment periods (in the range of several seconds) and under high capacitances, even spores could be killed. A conceivable cause for these results could be the bacteriocidal effect of electrolysis products which are expected to be formed under the test parameters chosen. Here the authors set up, for the first time, an inactivation equation which allowed mathematic definition of the killing rates measured.

When treating yeast cells, Jacob *et al.* [26] found similar correlations between the killing rate and the electric field strength, as well as the number of pulses. The observation that microorganisms in their logarithmic phase of growth are much more sensitive to electric fields than during their stationary phase of growth, is an interesting result of their work (see Figure 11.1). This was confirmed by the work of Hülsheger *et al.* [27–30]. In very comprehensive and systematic studies a mathematical model was formulated for the description of the killing kinetics. This model consi-

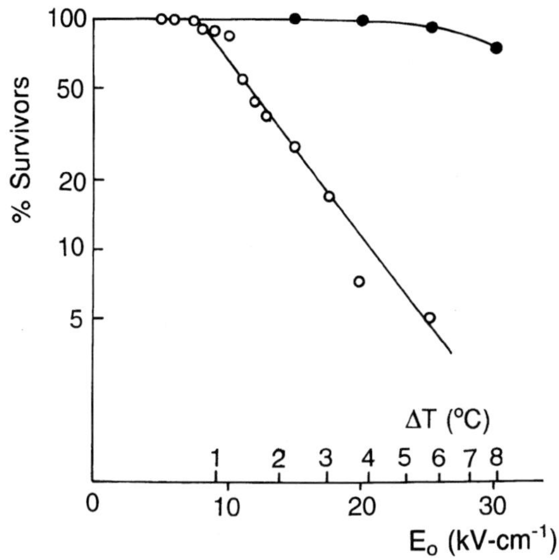

Figure 11.1 Electrical field effect and growth phase. Killing of *S. cerevisiae* caused by electrical discharges. (○–○) Logarithmic growth phase; (●–●) stationary growth phase. Redrawn from [26].

dered the influences of field strength and pulsing duration on the cell count reduction.

Mizuno and Hori [31] tested four different discharge chamber types for the purpose of optimizing the energy utilization when killing *S. cerevisiae* in watery solutions. However, one has to be sceptical about whether the killing effects measured in the different systems can be directly compared with one another. In some of the tests temperature increases in the media subjected to electric pulses were taken into account. With certainty these temperature increases have led to a purely thermic cell reduction and consequently to a falsification of the test results. Mizuno and Hayamizu [32] also investigated the destruction of yeast under the influence of electric high-voltage pulses using *S. cerevisiae* cells suspended in salt solution. At a maximum field strength of up to 17 kV/cm the cell count reduction amounted to two or three powers of ten. Examination of the treated cells under an electron microscope showed modification of the cell surface with little 'craters' visible. Sato *et al.* [33, 34] discovered in their experiments, aimed at killing microorganisms that had been immobilized on ceramic and calcium alginate carriers, that, contrary to freely suspended microorganisms, the killing rate achievable was distinctly lower. Under the electron microscope it was clear that yeast cells treated with electric pulses looked quite different from untreated cells. The high-voltage pulses had an obvious physically disruptive effect on the cells.

Dunn and Pearlman [35], as well as Bushnell *et al.* [36], discovered a strong synergistic effect between the initial (45–55°C) temperature of the media to be treated and the subsequent application of high-voltage pulses. The lethal effects observed during such treatment were remarkably higher than those from purely thermic pasteurization carried out in a parallel test. Jayaram *et al.* [37] also discovered a distinct synergistic effect between high-voltage pulse treatment and the treatment temperature when killing *Lactobacillus brevis* in phosphate buffer. They reached cell count reductions of up to nine powers of ten. The lethal effects were attributed to the destruction of the cell membrane, which could be observed under the electron microscope. The question of how far electric pulse treatment carried out under increased temperatures [35–37] results in thermic damages to the treated, liquid, pumpable foodstuff was not investigated by the authors.

11.3 The Elsteril Process

In the 1980s, based on the work of Doevenspeck [20, 21], Krupp Maschinentechnik GmbH in Hamburg, FRG, investigated the electric pulse process and its application to two possible fields.

On the one hand they developed the so-called ELCRACK process for electric cracking of vegetable and animal cells, on the other hand the ELSTERIL process for electric sterilization (pasteurization) of pumpable, electrically conductive media. The research work on the ELSTERIL process has been supported by the Department of Biochemical Engineering of the Technical University Hamburg-Harburg, FRG [42–44].

Both processes – which differ mainly in their applied field strength levels – are based on the same principle outlined in Figure 11.2. The principal components of an ELSTERIL plant are the pulse generator (3) for producing high electric charging voltages, the capacitor (5), and the discharge chamber (1). The material to be treated (A) is continuously fed through the discharge chamber which consists of two plane-parallel carbon electrodes mounted inside a pipe. The high voltage produced by the generator charges the capacitor when the switch is set to L.

Once the charging process has ended, the switch (4) is moved into position E. Within microseconds, the capacitor is discharged through the electrode pair (6), i.e. through the material to be treated. Subsequently the switch is again set to position L and a new charging cycle commences. The electric field strength which is important for cell count reduction is the quotient calculated by dividing the max. discharge voltage applied to the electrodes (2) by the electrode distance.

The discharge voltage time curve measured on the electrodes (2) is shown in Figure 11.3. One can see that the voltage drops following an

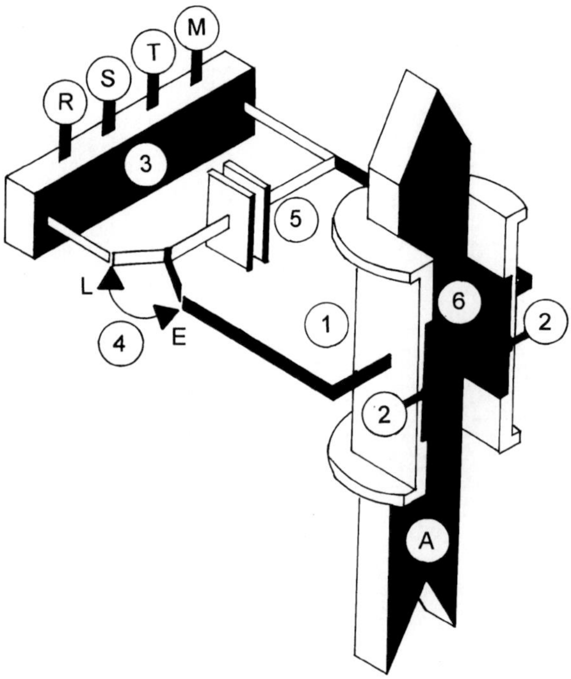

Figure 11.2 Principal components of an ELSTERIL plant. 1 = discharge chamber; 2 = electrodes; 3 = pulse generator; 4 = capacitor switch; 5 = capacitor; 6 = electrode pair.

exponential function and that the curve can be characterized by the maximum value of the voltage U_{max} and the pulse duration $\tau = 1/e \times U_{max}$. Both parameters are mainly influenced by the electric conductivity of the material to be treated, the field strength of the pulses applied and the capacitance of the capacitor.

For tests at the laboratory scale, discharge vessels were used, examples of which are shown on Figure 11.4. The left discharge chamber, with a filling volume of 25 ml and an electrode distance of 0.5 cm, is used for batchwise preliminary tests. The chamber on the right side has approximately the same volume and the medium to be treated can flow continuously through it.

Unlike most authors who usually carry out their tests with suspensions of microorganisms in artificial buffer systems, Grahl [45] investigated natural systems. He made his cell count reduction tests with pumpable foodstuffs and the spoilage generating microorganism contained therein. He examined the energy consumption that is necessary for killing microorganisms and spores in the ELSTERIL process, the effects of high-

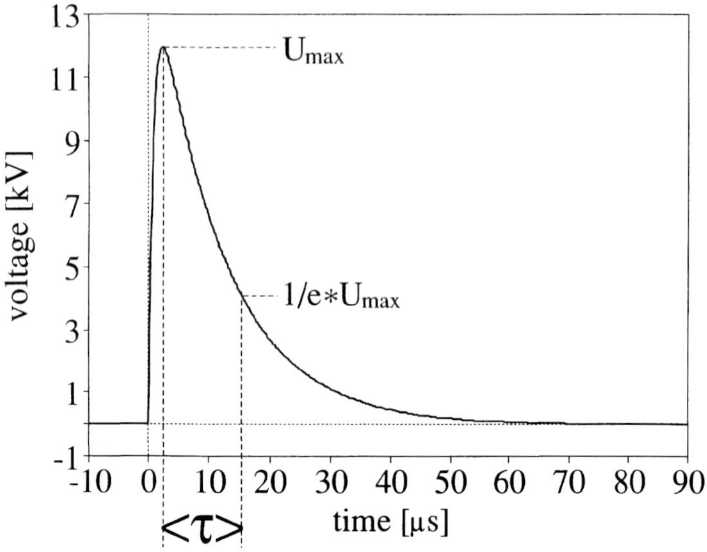

Figure 11.3 High voltage pulse measured on electrodes of ELSTERIL plant. $U_0 = 15\,\text{kV}$; $C = 5\,\mu\text{F}$; $A = 50\,\text{cm}^2$; $l = 5$ mm. Milk, 1.5% fat.

voltage pulses on the ingredients of foodstuff, and the question as to whether it is possible to destroy microorganisms *in situ* by means of high-voltage pulses.

11.4 Influence of high-voltage pulses on microorganisms

Sale and Hamilton [22] had already established that microorganisms are killed when electric high-voltage fields are applied for a short time. They determined, by means of calculations, the critical membrane potential induced by the exterior electric field which has a cell count reduction effect. It is about 1 Volt. This value has been confirmed by other authors [30, 46, 47].

According to Zimmermann *et al.* [48] the membrane potential V which is generated under the influence of an exterior electric field E can be mathematically calculated with the following equation:

$$V = F \times a \times E \times \cos \alpha$$

F being a factor characterizing the shape of the respective cell, a being the cell radius and α representing the angle between a point on the cell surface and the field lines of the exterior homogenous field.

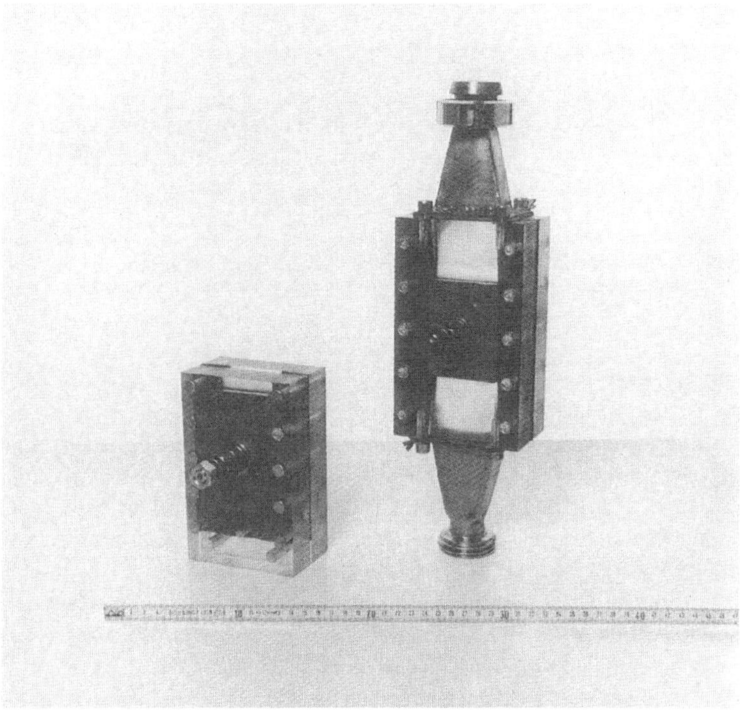

Figure 11.4 Examples of discharge vessels for ELESTRIL processes.

Notwithstanding the fact that it has not yet been possible to explain the mechanism leading to the lethal effect after the critical membrane potential of 1 V has been exceeded, it is a matter of fact that the cell membrane's permeability drastically increases as soon as this value is approached. With the 'electric breakdown' model as per Zimmermann [49] these processes can be elucidated (Figure 11.5). Reversible pore formation in cell membranes is made use of in the process of 'electroporation' where substances are injected into the cell and in the process of 'electrofusion' [50, 51]. Further investigations concerning these applications are described in the literature [52–66]. The application of electric field strengths which are clearly higher than the critical field strength results in irreversible pore formation and, inevitably, in the killing of the microorganisms treated.

A semi-logarithmic illustration (Figure 11.6) shows that there is a linear correlation between the electric field strength E and the fraction of surviving cells s [44]. One can see, for example, that for killing *S. cerevisiae* in orange juice, only $n = 5$ pulses of comparably low strength are necessary, whilst for killing *E. coli* in milk, with $n = 20$ pulses, consider-

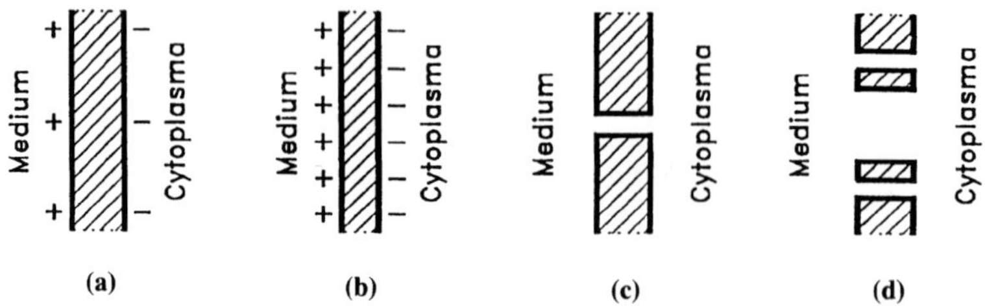

Figure 11.5 Electrical breakdown model [48]. E = electrical field strength; E_k = critical electrical field strength. (a) $E = 0$; (b) $E < E_k$; (c) $E = E_k$; (d) $E \gg E_k$.

ably higher field strengths are necessary. One can also see that for each microorganism a critical minimum field strength (for reaching the critical membrane potential) E_c has to be exceeded. A similar connection becomes evident when entering the fraction of surviving cells as a function of the treatment time (= the sum of the duration of the individual pulses). Here, too, it proves necessary that a critical minimum treatment time depending on the microorganism and the medium is exceeded [44].

The influence of electric high-voltage pulses on microorganisms as examined within the scope of Grahl's work [45] with milk, orange juice,

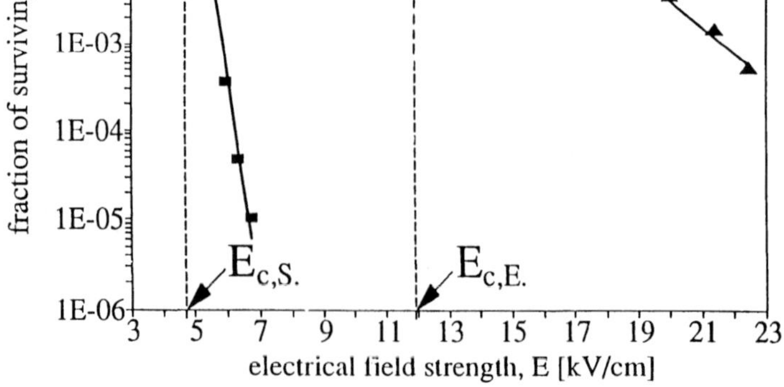

Figure 11.6 Effect of electrical field strength on cell survival. ■–■ = *S. cervisiae* in orange juice, $n = 5$; $N_0 = 5 \times 10^6$ CFU/ml. ▲–▲ = *E. coli* in milk (3.5% fat), $n = 20$, $N_0 = 4 \times 10^7$ CFU/ml.

and alginate solutions can be summarized as follows:

- Efficient killing of microorganisms is possible with field strengths exceeding 20 kV/cm.
- The killing of yeasts is much easier than the killing of bacteria.
- It is virtually impossible, even at very high field strengths, to kill endospores and ascospores.
- The killing of microorganisms in orange juice is significantly easier than in milk.

11.5 The influence of electric high-voltage pulses on food ingredients

In their tests with inhomogenous fields of high energy density (electro-hydraulic treatment), Gilliland and Speck [17, 18] observed that enzymes were inactivated. Grahl [45] investigated the influence of homogenous electric high-voltage fields on food ingredients in various media (see Table 11.1).

Figure 11.7 shows that noticeable inactivation of enzymes was only observed with lipase and only with relatively high energy supplies, from about 200 kJ/l upwards. Generally we can note that vitamins and enzymes are only inactivated to a small degree when subjected to the influence of electric high-voltage pulses.

Investigations as to the taste of the products treated, as in the case of Grahl's [45] example of orange juice, showed that the application of

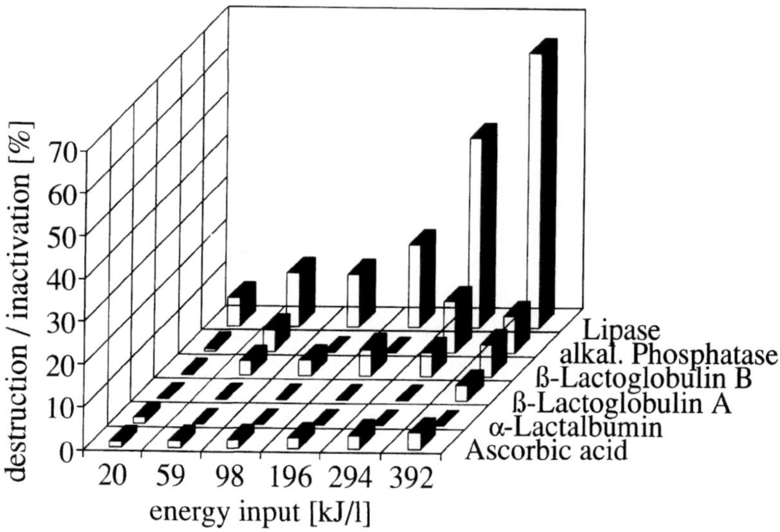

Figure 11.7 Effect of high-voltage pulses on food ingredients. $T_{max} = 40°C$; $E = 21.5$ kV/cm.

Table 11.1 Influence of homogeneous electric high-voltage fields on food ingredients [45]

Food ingredients	Buffer	Milk	Orange juice
Vitamin A		×	
Vitamin B_2 (riboflavin)	×		
Vitamin C (ascorbic acid)	×	×	×
Whey proteins		×	
Alkaline phosphatase		×	
Lactoperoxidase		×	
Lipase		×	
Pectinesterase			×

electric fields with an energy supply which was low, but still high enough to ensure a sufficient cell count reduction, did not change the taste in a noticeable manner. Even with a supply of energy much higher than necessary to guarantee a successful cell count reduction negative influences on the taste could not be observed.

11.6 Mathematical modelling of cell count reduction

Once the dependence of cell count reduction on electric field strength and treatment period was determined on a laboratory scale with relatively slight experimental and analytical cost, it was possible to deduce mathematical relationships from these results to describe cell count reductions in various systems:

$$\ell g\, s = B_{\mathrm{E}}(E - E_{\mathrm{c}}) \tag{13.1}$$

$$\ell g s = B_{\mathrm{t}} \ell g(t/t_{\mathrm{c}}) \tag{13.2}$$

Equation (13.1) describes the dependence of the fraction of surviving cells s as a function of electric field strength E, equation (13.2) as a function of the treatment period t, which is defined as the product of pulse quantity and pulse duration.

By combining equations (13.1) and (13.2) one obtains a third parameter, k, specific to the system involved in the particular case.

$$\frac{E - E_{\mathrm{c}}}{B_{\mathrm{t}}} = \frac{Ig(t/t_{\mathrm{c}})}{B_{\mathrm{E}}} = K \tag{13.3}$$

Equation (13.4), which was already formulated in 1981 by Hülsheger of the University of Hannover, FRG, describes the fraction of surviving cells s for a certain microorganism in a defined medium, as a function of the empirically determined parameters E_c, t_c, and k, whereby the value of k is always less than zero.

$$s = \left(\frac{t}{t_c}\right)^{\frac{E - E_c}{k}} \tag{13.4}$$

Equation (13.5) describes the calculation of the specific quantity of energy Q, consumed by the application of n pulses with a capacity C defined by the apparatus (V = Volume) and with a charge voltage U_0.

$$Q = \frac{U_0^2 C n}{2V} \tag{13.5}$$

By combining equations (13.4) and (13.5), a diagram can be compiled for the process designing procedures, as illustrated for the systems *E. coli* in milk and *S. cerevisiae* in orange juice (Figure 11.8). In this diagram, the energy required Q is entered as a function of the electric field strength E for the lethal effects specifically required, expressed here by the parameter s, the fraction of surviving cells. It can be seen that it is much easier to kill yeast in orange juice than *E. coli* in milk, i.e. smaller field strengths and less energy are required. It can also be seen that there is a minimum level

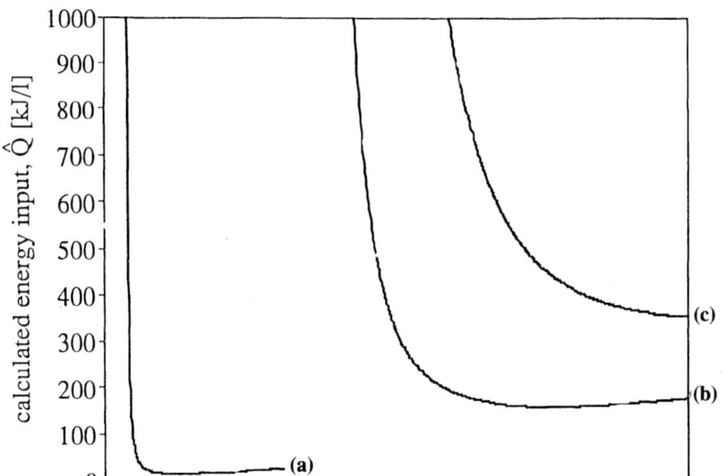

Figure 11.8 Relationship between calculated energy input and electrical field strength. The lethal effects are expressed by the parameter s. (a) *S. cerevisiae* in orange juice, $E_c = 4.69$ kV/cm; $t_c = 76.1\,\mu s$; $k = -0.196$ kV/cm; $s = 1E - 4$. (b) *E. coli* and (c) in milk (3.5% fat), $E_c = 11.92$ kV/cm; $t_c = 29.98\,\mu s$; $k = -2.625$ kV/cm. For (c), $s = 1E - 4$; for (b), $s = 1E - 2$.

relative to the energy demand. Given a prescribed fraction of surviving cells, it is therefore possible to select the most favourable procedure parameters with regard to the energy, by means of this kind of diagram.

11.7 Conclusions

The application of homogenous and inhomogenous, pulsating, electrical fields with high field strengths for killing microorganisms can be, in competition with the conventional thermic process, a method for pasteurizing pumpable media.

In particular the application of homogenous fields at low treatment temperatures (e.g. ELSTERIL process) has proved to be a product-saving sterilization process in various natural systems, mainly juices.

However, when designing plants for large scale production it is essential that preliminary batch tests are carried out on a small scale. This requirement is based on the experience that the mortality of a certain species of microorganisms is very dependent on the nature of the suspension medium.

References

1. B. Mertens and D. Knorr, *Food Technology* **5** (1992) 124.
2. S. Carlson, *Zeitsch. f. Hygiene* **135** (1952) 346.
3. Gillissen and S. Carlson, *C.r. Acad. Sci. Paris* **234** (1952) 1494.
4. A.K. Anderson and R. Finkelstein, *Journal of Dairy Sci.* **2** (1919) 374.
5. J.C. Fetterman, *Agricultural Engineering* **9** (1928) 107.
6. B.D. Moses, *Agricultural Engineering* **19** (1938) 525.
7. B.E. Getchell, *Agricultural Engineering* **16** (1935).
8. S. Palaniappan, S.K. Sastry and E.R. Richter, *Journal of Food Proc. and Pres.* **14** (1990) 393.
9. F.C. Lewis and J.M. Beattie, *Journal of Hygiene* **24** (1925) 123.
10. F. Früngel, *United States Patent* 2,931,947, April 5, 1960.
11. B.S. Gossling, *British Patent* 845,743, Oct. 5, 1955.
12. B. Brandt, L. Edebo, C.-G. Heden, B. Hjotzberg-Nordlund, I. Selin and M. Tigerschiöld, *Tekniskvetenskapligforskining (TVF)* **33**, (1962) 222.
13. L. Edebo and I. Selin, *Journal of General Microbiology* **50** (1968) 253.
14. L. Edebo, T. Holme and I. Selin, *Journal of General Microbiology* **53** (1968) 1.
15. M. Allen and K. Soike, *Science* **154** (1967) 155.
16. M. Allen and K. Soike, *Science* **154** (1967) 524.
17. S.E. Gilliland and M.L. Speck, *Applied Microbiology* **15** (1967) 1031.
18. S.E. Gilliland and M.L. Speck, *Applied Microbiology* **15** (1967) 1038.
19. H. Doevenspeck, *Die Fleischwirtschaft*, **13** (1961) 986.
20. H. Doevenspeck, *German Patent* 1 237 541, Feb. 8, 1960.
21. H. Doevenspeck, *Archiv für Lebensmittelhygiene* **13** (1962) 3.
22. A.J.H. Sale and W.A. Hamilton, *Biochimica et Biophysica Acta* **148** (1967) 781.
23. W.A. Hamilton and A.J.H. Sale, *Biochimica et Biophysica Acta* **148** (1967) 789.
24. A.J.H. Sale and W.A. Hamilton, *Biochimica et Biophysica Acta* **163** (1968) 37.

25. Y. Sakurauchi and E. Kondo, *Nippon Nogeikagaku Kaishi* **54** (1980) 837.
26. H.-E. Jacob, W. Förster and H. Berg, *Zeitsch. f. Allg. Mikrobiol.* **21** (1981) 225.
27. H. Hülsheger and E.-G. Niemann, *Radiation and Environmental Biophysics* **18** (1980) 281.
28. H. Hülsheger, J. Potel and E.-G. Niemann, *Radiation and Environmental Biophysics* **20** (1981) 53.
29. H. Hülsheger, J. Potel and E.-G. Niemann, *Radiation and Environmental Biophysics* **22** (1983) 149.
30. H. Hülsheger, *Medizinische Hochschule* Hannover, Germany, Dissertation, 1984.
31. A. Mizuno and Y. Hori, *IEEE Transactions on Industry Applications* **24** (1988) 387.
32. A. Mizuno and M. Hayamizu, Sixth Intern. Symp. of High Voltage Engineering, Aug. 28–Sept. 1, 1989, New Orleans, LA, USA.
33. M. Sato, K. Tokita, M. Sadakata, T. Sakai and K. Nakanishi, *Kagaku Kougaku Ronbunshu* **14** (1988) 556.
34. M. Sato, *Kagaku Kougaku* **53** (1989) 818.
35. J.E. Dunn and J.S. Pearlman, *United States Patent*, 4,695,472, Sept. 22, 1987.
36. A.H. Bushnell, J.E. Dunn and R.W. Clark, *United States Patent*, 5,048,404, Sept. 17, 1991.
37. S. Jayaram, G.S.P. Castle and A. Maragaritis, *Biotechnology and Bioengineering* **40** (1992) 1412.
38. W. Sitzmann and E.-W. Münch, *Fat Sci. Technol.* **89** (1987) 368.
39. W. Sitzmann and E.-W. Münch, *Die Fleischmehlindustrie* **40** (1988) 22.
40. W. Sitzmann and E.-W. Münch, lecture, Gemeinsame Sitzung der Fachausschüsse "Grenzflächen" und "Lebensmittelverfahrenstechnik", April 11–12, 1988, Brussels, Belgium.
41. J. Schröder, F. Schneider and W. Sitzmann, lecture, IAFMM-Conference, Nov. 19–24, 1989, Hong Kong.
42. W. Sitzmann and E.-W. Münch, *Die Ernährungsindustrie* **6** (1989) 54.
43. W. Sitzmann and E.-W. Münch, lecture, Conference "Keimarmes und aseptisches Verarbeiten von Lebensmitteln", Nov. 27–28, 1990, Cologne, Germany.
44. W. Sitzmann, Th. Grahl and H. Märkl, lecture, IFTEC Intern Food Technol. Exposition & Conference, Nov. 15–18, 1992, The Hague, Netherlands.
45. Th. Grahl, Dissertation, Technical University Hamburg-Hamburg, 1994, in preparation.
46. U. Zimmermann, J. Vienken and G. Pilwat, *Biochemistry and Bioenergetics* **7** (1980) 553.
47. U. Zimmermann, *Biochimica and Biophysica Acta* **694** (1982) 227.
48. U. Zimmermann, G. Pilwat and F. Riemann, *Biophysical Journal* **14** (1974) 881.
49. U. Zimmermann, *Biochemistry and Pharmacology* **105** (1986) 175.
50. U. Zimmermann, W.M. Arnold and W. Mehrle, *Journal of Electrostatics* **21** (1988) 309.
51. U. Zimmermann, P. Gessner, M. Wander and S.K.H. Foung, in: *Electromanipulation in Hybridoma Technology – A Laboratory Manual*. C.A.K Borrebaeck and I. Hagen (eds), Stockton Press, New York, 1989, 1.
52. H.G.L. Coster and U. Zimmermann, *Journal of Membrane Biology* **22** (1975) 73.
53. U. Zimmermann, G. Pilwat, F. Beckers and F. Riemann, *Bioelectrochemistry and Bioenergetics* **3** (1976) 58.
54. E. Jeltsch and U. Zimmermann, *Bioelectrochemistry and Bioenergetics* **6** (1979) 349.
55. U. Zimmermann and R. Benz, *The Journal of Membrane Biology* **53** (1980) 33.
56. E. Neumann, workshop, ZiF, University of Bielefeld, April 3–4, 1986, Bielefeld, Germany.
57. J.S. Clements, M. Sato and R.H. Davis, *IEEE Transactions on Industry Applications* **Ia-23** (1987) 224.
58. K. Kinosata Jr, I. Ashikawa, N. Saita, H. Yoshimura, H. Itoh, K. Nagayama and A. Ikegami, *Biophysical Journal* **53** (1988) 1015.
59. R.W. Glaser, S.L. Leikin, L.V. Chernomordik, V.F. Pasushenko and A.I. Sokirko, *Biochimica et Biophysica Acta* **940** (1988) 275.
60. L.V. Cernomordik, Y.A. Chizmadzhev, in: *Electroporation and Electrofusion in Cell Biology*. E. Neumann, A.E. Sowers and C.A. Jordan (eds), Plenum Press, New York, 1989, 83.

61. E. Neumann, in: *Electroporation and Electrofusion in Cell Biology*. E. Neumann, A.E. Sowers and C.A. Jordan (eds), Plenum Press, New York, 1989, 61.
62. J.C. Weaver and K.T. Powell, in: *Electroporation and Electrofusion in Cell Biology*. E. Neumann, A.E. Sowers and C.A. Jordan (eds), Plenum Press, New York, 1989, 111.
63. M. Joersbo and J. Brunstedt, *Plant Cell Reports* **8** (1990) 701.
64. A. Barnett and J.C. Weaver, *Bioelectrochemistry and Bioenergetics* **25** (1991) 163.
65. S. Elvin and A.H.A. Bingham, *Letters in Applied Microbiology* **12** (1991) 39.
66. T.Y. Tsong, *Biophysical Journal* **60** (1991) 297.

12 Preservation by microbial decontamination; the surface treatment of meats by organic acids

F.J.M. SMULDERS

12.1 Introduction

Over the past few years, increasing attention has been focused on the safety of our food supply. As a major food item, and one produced by a complex series of events, each with an associated risk of contamination, meat is an object of special scrutiny. While the interior of muscle from living, healthy animals is generally sterile, organs like liver and kidney may contain microorganisms even before slaughter. In addition, certain ante-mortem handling factors such as stress on the animal during transport or lairing, and feed availability in the period preceding slaughter interact with inherent genetic and physiological traits. Post-mortem processing procedures enable the prevailing microflora to colonise the surface of carcasses and cuts and hence determine the safety and shelf life of meats. Integrated hygiene control along the meat production line will therefore be the processor's most effective approach to increasing the storage life of his products.

The modern approach to hygiene control relies on the Hazard Analysis Critical Control Point system [1]. A 'hazard analysis group' jointly studies all steps of production until they have a clear understanding of the implications of deviations of standard procedures. Risks resulting from these deviations are ranked according to severity and likelihood of occurrence. If such deviations could pose a realistic threat to safety, those processing steps are recorded as 'Critical Control Points' (CCPs). This procedure leads to a listing of hazards and the risk of their occurrence. Subsequently, the group considers how control is best attained and suggests 'Good Manufacturing Practices' (GMPs) to achieve this goal. In addition, extra sanitation measures may be considered to further reduce public health risks. Before elaborating on the usefulness of acid decontamination for this purpose, the major contamination mechanisms are discussed to stress that such extra measures can never substitute for the strict adherence to GMPs.

12.2 Critical control points in carcass contamination

Table 12.1, including the flow of material through a fresh meat production line, presents a general overview of hazards and control points. To systematically address the underlying mechanisms, the approach of Ishikawa [2] is followed, by which four major factors – 'material', 'machine', 'method' and 'man' – are distinguished.

12.2.1 Material: the animal

The principal sources of microbial contamination are microorganisms present in the intestines or on the hide or fleece of the live animal, which will vary depending on the husbandry conditions. Intensive agriculture involves the rearing of very young animals having immature gut flora and lacking resistance to colonisation by pathogens. The farm environment permits a recycling of excreted pathogens, as a result of which livestock are trapped in a vicious circle of infection, excretion and re-infection, and environmental pressure is so heavy that sanitation and regimentation are insufficient preventive measures. The stress of moving the animals [3], mixing them at markets [4], insufficient disinfection of transport vehicles and extended time in lairage [5] will allow pathogens to be excreted

Table 12.1 General overview of hazards and control points

Species				
Ruminants/horses	Pigs	Poultry	Site of contamination	Type of CCP*
On-farm production	On-farm production	On-farm production	MAJOR	2
Transport	Transport	Transport	minor	
Lairage	Lairage		minor	
Slaughter	Slaughter	Slaughter		2
	Scalding	Scalding		
Dehiding			MAJOR	2
	Dehairing	Plucking	MAJOR	
		Washing		2
	Singeing			2
	Polishing		MAJOR	
Evisceration	Evisceration	Evisceration	MAJOR	2
Washing		Washing		
Chilling	Chilling	Chilling		1
(Transport)	(Transport)		minor	2
Packaging	Packaging	Packaging		
Transport	Transport	Transport		

*1. Effective CCP (= critical control point), i.e. control of this operation can completely eliminate one or more microbial hazards.
2. non-absolute CCP, i.e. control will minimize, but not completely prevent, a hazard.

and cross-infect the entire shipment. Also, bacteria may be released from hepatic and mesenteric lymph nodes [6]. Showering upon arrival at the abattoir reduces the contamination of pigs (and consequently the scalding tank), but in sheep and cattle it will do little more than disperse bacteria over hide and fleece [7]. Washing poultry is generally considered impractical as it is extremely difficult to get the birds dry again.

12.2.2 Machine: equipment and utensils

Slaughtering and processing of meat animals is only allowed in facilities complying with certain regulations. However, it must be conceded that many international hygiene requirements are sometimes rather inconsequent. For instance, whereas the walls of slaughter facilities should be 'smooth' to allow proper cleaning and disinfection, such requirements do not apply for machines. Consequently, many such specifications have been characterised as being based on preconceived ideas rather than on science [8].

Captive bolts and sticking knives have been demonstrated to be a source of internal contamination of organs as well as muscle [9–12]. Yet only very high levels of contamination will cause viable bacteria to remain in the meat interior, as low numbers will be eliminated by the anti-bacterial action of blood [13]. Knives, steels and aprons of slaughter personnel who handle beef carcasses before dehiding may be an important source of contamination [14]. The polishing equipment is probably the most important factor in contaminating pig carcasses [15]; yet surprisingly little effort has been put into re-designing this equipment to allow a more effective cleaning and disinfection [16].

Protective gloves, often used in the dressing and boning line, have also been reported as sources of cross-contamination [17], as have the surfaces of cutting tables for further breakdown of carcasses into subprimals [18, 19].

12.2.3 Method: slaughter and fresh meat processing

Immersion of carcasses in scalding water of about 60°C will remove hairs from the pig's epidermis and may reduce bacterial contamination [15]. However, given that dirt and faeces are constantly being introduced into the scalding tank, contaminated water can enter lungs and stick wounds. Scalding carcasses by steam has been proposed as an alternative [20]. Unless the flaying procedure is performed with care, dehiding and defleecing can be a major source of contamination in ruminants, regardless if this process is performed manually or automated [21]. Stolle [22] identified the following as the most hazardous sites in the beef dressing line: removal of the horns and the freeing of the skin round the lower parts of the legs and the sternal region.

The process of evisceration is critical for all meat species, particularly the freeing of the rectum and the opening of the abdominal cavity [23, 24]. Chilling carcasses is generally regarded as essential to extend the shelf life of meats. The number of colony forming units (CFU) of Enterobacteriaceae will remain relatively constant, or can even slightly decrease at refrigeration temperatures around 0°C [7]. The desiccation of carcass surfaces is particularly important to secure elimination of *Campylobacter jejuni* from chicken and pig carcasses [25]. Spray-chilling, however, has become commonplace in beef processing abattoirs in North America.

12.2.4 Man: the slaughter personnel

The most important human sources of bacteria include the oral and nasal cavity, digestive tract and the skin. The so-called 'transient' skin flora, prevailing on the skin surface, can be more or less removed by washing. It has been demonstrated that the skin of workers in the meat industry have more coliform bacteria and salmonellae than the hands of workers in other food sectors, obviously a result from extensive contact with products of animal origin [26–28]. Therefore, regular cleaning and disinfection may help to prevent contamination.

It has been shown that unhygienic working conditions have little effect on the bacteriological condition of meats when the initial levels of carcass contamination are high [29, 30]. At low levels of contamination, however, strict hygienic behaviour has proven to contribute to low bacterial loads on beef [31, 32], veal [33] and pig carcasses [34, 35]. Mackey and Roberts [36] showed that even under relatively primitive conditions low levels of bacterial contamination may be achieved, provided the slaughter personnel are skilled and careful. Conversely, the workers' inclination to practise hygiene was demonstrated to largely depend on the management's commitment to the control of quality and safety [37, 38].

The preceding concise overview serves to show that the potential exposure of meat to bacteria is determined by a variety of interrelated factors. Mackey and Roberts [36] demonstrated the cumulative effects of these factors on the bacterial load of carcasses of various meat animal species from commercial British abattoirs. Locations most consistently contaminated (in order of importance) were:

- Pigs: hind limb (distal and lateral), abdomen (lateral side), mid-dorsal region, abdomen (medial surface)
- Beef: brisket, fore-rib, flank, groin, round (lateral surface)
- Lamb: abdomen (flank), thorax (lateral surface), crutch, breast (lateral surface).

Johanson *et al.* [39] reported similar results from studies in Norway. The nature of the bird's skin and the specific processing procedures, result in a special contamination pattern in poultry carcasses, described in detail by Thomas *et al.* [40]. In essence, all areas where connective tissue is exposed are major sites of contamination, the neck flap area being the most heavily contaminated carcass site.

12.3 Organic acids as meat decontaminants

Smulders and Woolthuis [33] demonstrated that, under model processing conditions (i.e. adherence to strict measures of GMP during slaughter, consequent cleaning and disinfection of utensils and wearing a fresh pair of surgical gloves at the start of the boning process), it was possible to produce veal cuts with extremely low bacterial loads after two weeks of refrigerated storage in vacuum. In these circumstances, application of a 1% lactic acid spray afforded no additional reduction of the microbial load. Unfortunately, in practice, despite the use of increasingly sophisticated hygiene measures, carcasses and cuts are still found contaminated with pathogens such as salmonellae and campylobacters, and also a number of other important groups, i.e. *E. coli*, *Yersinia*, *Listeria* and spoilage types. Consequently, over the past two decades, it has often been suggested that meat processing should include a terminal decontamination step, not in lieu of, but in addition to GMPs along the slaughter-line. As observed with pasteurisation in the dairy industry, such a terminal treatment would not only reduce, or even eliminate pathogens, but also further extend the keeping quality. Apart from the more controversial issue of meat irradiation (addressed elsewhere in this volume), surface treatment with organic acids stands out as a more realistic option to achieve this goal without arousing major consumer reactions. The objective of the following section is to review the state-of-the-art in research and industrial application of microbial decontamination of meats with organic acids.

12.3.1 The antimicrobial properties of organic acids

The antimicrobial effects of organic acids have been known since times immemorial. For instance, bacteria have been used for over 2000 years to convert lactose into lactic acid, and a variety of products have been generated varying in flavour, composition and consistency. Smulders *et al.* [41] cited examples of foods, such as rye and wheat breads, cheese, sauerkraut, and various types of fermented sausages, containing lactic acid. Post-rigor skeletal muscle generally contains approximately 0.9% lactic acid, which contributes markedly to its flavour and can affect the keeping quality.

The bactericidal and bacteriostatic effects of organic acids are well-known. Their antimicrobial action depends on three factors: (1) the effect solely of pH, (2) the extent of dissociation of the acid, and (3) a specific effect related to the acid molecule [42]. The relative importance of each of these contributions has been a point of discussion. Gill and Newton [43] suggest that the pH-fall is the most important of the three. Other investigators [42, 44–49] consider that the nature of acid is also of significance, such that at a given pH the antimicrobial activity of an acid solution also depends on the nature of the anion.

Microorganisms multiply only within certain species-dependent pH-ranges, and for each group of bacteria these are rather stable [50]. In particular, acid-forming bacteria, yeasts and moulds are more tolerant of acids as they do not interfere with the organisms' energy-yielding biochemical activities, whilst some may even be metabolised [51]. An acid which decreases the pH of a food will extend the lag phase of acid-sensitive microorganisms [33, 52–54] and will eventually result in the death of the organisms [55, 56]. The pH drop attained and the period during which it is maintained depend on the amount of acid and the buffering capacity of a food. After spraying of hot calf carcasses with 1.25% (v/v) lactic acid, Woolthuis and Smulders [57] found a surface pH fall of more than three units. However, after 72 h the pH had returned to its initial value. Immersion of pig livers in 0.2% (v/v) lactic acid effected a surface pH fall of about 0.6 units, whereupon it took 5 h to restore the initial surface pH value [58]. Successive treatments of broiler carcasses with a 2% lactic acid spray did not further reduce the surface pH nor enhanced the bacteriostatic and bactericidal effects [59].

A shift in pH within the species-specific growth range will usually have little effect upon microorganisms. However, when using weak acids and salts as preservatives, such pH shifts may cause considerable differences in growth inhibition and death [42]. This is explained by the fact that the pH affects the extent of dissociation [60] and that undissociated weak acids are 10 to 600 times as effective as the undissociated form [47]. Undissociated acids penetrate the cell by means of diffusion and then dissociate and acidify the 'cell interior', thus interfering with cellular metabolism or decreasing the biological activity as a result of pH changes of the cell's environment [61–63]. Effective growth inhibition by an acid only occurs when an appropriate amount of the undissociated molecule is present [42, 64–66]. This amount may be obtained by either applying more acid or by lowering the pH. Most organic acids are therefore effective only at low pH values (i.e. below about pH 5.5). However, at pH levels above 6, dissociated sorbic acid is reported to cause more than 50% growth inhibition of *Bacillus subtilis*, *Bacillus cereus*, *Escherichia coli*, *Pseudomonas aeruginosa*, *Staphylococcus aureus* and *Candida albicans*; apparently the anion exerts some influence under these circumstances [47]. Under the same conditions of pH and acid dissociation there are still differences in antimicrobial activities of the various organic acids. Ingram *et al.* [42] suggested that this 'specific effect' is related to (1) the potency to penetrate a cell, (2) the part of the cell which is attacked, (and 3) the chemical nature of that attack. As regards (1) and (2) particularly the heteropolar nature of the molecule is important. Heteropolar molecules are surface active, i.e. will concentrate at the bacterial cell surface, interfere with its permeability and eventually result in acidification of the 'cell interior'. The moderately lipophilic part partly determines the penetration of the cell membranes. Wyss [67] mentioned as possible points of impact of acids in the cell: competition with co-enzymes, inactivation of S-S enzymes and the suppression of fumarate oxidation in catalase-positive organisms.

12.3.2 Factors influencing the efficacy of meat decontamination by acids

(a) Nature of the meat surface and its contamination with organic material. Jarvis and Burke [68] indicated that the accessibility and fat content of the meat substrate affected the efficacy of acid decontamination, an observation also made in various tests in The Netherlands in the early 1980s [41]. Recently, Prasai *et al.* [69] failed to demonstrate a decontaminating effect of 1% lactic acid applied at 55°C on unskinned pork carcasses and suggested that acids may not be able to contact properly with firmly attached bacteria in crevices of pork skin. Dickson [70, 71] demonstrated that acid treatment was more efficient to reduce *Salmonella typhimurium*, *Listeria monocytogenes* and *Escherichia coli* 0157:H7 on fat than on lean beef, which was also found to be true for spoilage bacteria (*Pseudomonas fragi* and *Brochothrix thermosphacta*) on pork fat and lean tissue [72]. The latter study substantiates the fact that the buffering capacity of lean is many times greater than that of fat, which therefore allows a much faster initial drop in surface pH.

Probably, the degree of contamination of the meat surface with organic material and the nature of the contaminating material also plays a significant role. Van Netten *et al.* (unpublished results) observed that with increasing levels of organic material in scalding tanks for pig carcasses, higher concentrations of lactic acid were necessary to produce the desired bactericidal effects on pork. Dickson [71] compared the decontaminating potential of a 2% acetic acid rinse of beef lean and fat tissue, contaminated with 'low moisture' menstrua (sand and soil) *vs.* a phosphate buffer. *Salmonella typhimurium* counts on the samples treated with the low-moisture menstrua were not found to be reduced in higher numbers than the buffer-treated controls. However, in another study, Dickson [73], reported that on fatty surfaces of samples, immersed in menstrua containing manure and subsequently treated with 2% (v/v) acetic acid, *Salmonella typhimurium* survived in higher numbers than on samples treated with menstrua not containing manure. Lean surfaces were not affected. These observations once more stress that, unless care is taken to keep carcass surfaces as clean as possible by adhering to strict GMPs (see above), sanitising under commercial conditions will not be very effective.

(b) The initial bacterial load. At high initial levels of contamination 1–3% concentrations of lactic acid may not effect marked microbial lethality as theorised by Baird Parker [60]. In Dutch experiments, the lethality of lactic acid on aerobic colony counts, though significant, did not exceed 2 or exceptionally 3 $\log_{10}$ cycles on beef [74], veal [33, 54] and pork [56, 75]. Dickson [73] found that the reduction in the *Salmonella typhimurium* population on artificially contaminated beef lean and fat by 2% (v/v) acetic acid was consistently proportional to the initial inoculum level. Greer and Dilts [76] found that the reduction in the number of pathogens (*Listeria monocytogenes*, *Yersinia enterocolitica*, *Salmonella typhimurium*, *Escherichia coli*, *Campylobacter jejuni*) or spoilage bacteria (*Pseudomonas fragi*, *Brochothrix thermosphacta*) on acid-treated lean beef discs was significantly affected by the initial levels of bacterial contamination, but that this effect followed no consistent pattern and varied between bacterial species (see below).

(c) Choice of acids. Many studies have been dedicated to assessing which of the food-grade acids might be the most suitable for purposes of decontamination. Mountney and O'Malley [52] compared 11 acids for their suitability for decontaminating poultry and, although all acids were found to inhibit growth, recommended acetic, adipic and succinic acid rather than hydrochloric, lactic or sorbic acid as most effective in extending shelf life. Chung and Goepfert [50] examined 13 acids as inhibitors of salmonellae in laboratory media and recommended acetic and propionic acid as the most effective agents. Similarly, Carpenter [77], determined the inhibitory effects against *Salmonella enteritidis* of various compounds, amongst which citric, phosphoric and acetic acid were compared. Whilst all compounds produced zones of inhibition, acetic acid at pH 2 was regarded the most acceptable option to reduce *Salmonella* and aerobic counts on hog carcasses. Ockerman *et al.* [78] compared various concentrations of acetic and lactic acid for their decontaminating effects on lamb carcasses and indicated acetic acid (18%) to be most effective. However, discoloration seemed to appear more readily than with lactic acid. For this reason the attention of many researchers has focussed on lactic acid as the preferable agent [41].

More recent observations provide more arguments for the separate or combined use of acetic and lactic acid, although they do not afford consistent recommendations.

Acetic acid was found to have superior decontaminating qualities in a study on sliced beef [43] and on vacuum packaged pork [79]. Ahamad and Marth [66] found higher anti-listerial activity of acetic as compared to lactic or citric acids in culture media and indicated the larger dissociation constants of the latter two acids as the main cause. Recently, El-Katheib *et al.* [80] demonstrated that on aseptically obtained, lactic acid-treated beef, subsequently inoculated with *Listeria monocytogenes*, lactic acid was unable to block *Listeria* attachment. Acetic acid was found to be more effective than lactic acid in raising the minimum pH for growth of *Yersinia enterocolitica* in a culture-media study by Brocklehurst and Lund [81].

Slightly better microbiocidal effects of a single spray treatment of beef carcasses with lactic- as opposed to acetic acid were found by Hamby *et al.* [82]. Anderson *et al.* [83], compared the effects of 3% acetic acid, 3% lactic acid, and two mixtures of these acids with citric and ascorbic acids in an inoculation study on beef. It was shown that 3% lactic acid was the most effective against *Salmonella typhimurium* with a reduction of 2 $\log_{10}$ at 70°C. Brackett *et al.* [84] found lactic acid to be most effective against *E. coli* 0157:H7 in a comparison with acetic acid, albeit that this effect was only noticeable immediately after acid treatment. In a recent study, Nettles Cutter and Siragusa [85] found no significant differences between the efficacy of various concentrations of acetic, lactic or citric acid at 24°C for reducing *E. coli* 0:157 H:7 or *Pseudomonas fluorescens* on inoculated beef and fat tissues.

Acetic and lactic acid were found to have similar effects in a culture-media study on growth inhibition and survival of *Listeria monocytogenes* by Conner *et al.* [49], whilst Greer and Dilts [76] observed effects of both these acids on lean beef, inoculated with *Listeria monocytogenes*, *Yersinia enterocolitica*, *Salmonella typhimurium*, *Escherichia coli*, *Campylobacter jejuni*, *Staphylococcus aureus*, *Pseudomonas fragi* and *Brochothrix thermosphacta* to be similar.

Arguments in favour of the combined use of acetic and lactic acid were provided by Adams and Hall [65], who reported a potentiation by lactic acid of the inhibiting effect of acetic acid on *Salmonella enteritidis* and *Escherichia coli* in a culture media study (an observation also made for *Salmonella typhimurium* by Rubin [86], and by Mendonca *et al.* [87], who found that a combination of 1% acetic and 1% lactic acid effected a similar surface pH drop as did a 3% acetic acid solution.

Several authors have relied on mixtures containing various other acids in an attempt to make use of possible synergistic effects of the anions. Although Quartey-Papafio *et al.* [88] failed to demonstrate enhanced antimicrobial effects of 1% formic/1% acetic acid mixtures over 2% acetic acid, Bell *et al.* [89] found a mixture of 0.6% acetic and 0.046% formic acid to produce similar bactericidal effects as did 1.2% acetic acid. Osthold *et al.* [90], using a mixture of acetic, lactic, citric and ascorbic acid, reported selective inhibition of Enterobacteriaceae and argued that mixtures containing acetic acid will produce a lower pH on the meat surface. Similar mixtures have been put on the market [91]. However, more recently, Anderson and Marshall [92] demonstrated that the overall microbiological effects of such mixtures were similar to those of acetic or lactic acid under the same conditions. Surve *et al.* [93] compared 1 to 4% mixtures of acetic/lactic or acetic/propionic acid and found similar microbiological effects; an ecological analysis of the effects of these mixtures on counts of aerobic organisms, Enterobacteriaceae, staphylococci, *Bacillus* spp. and clostridia revealed a slightly higher susceptibility of the latter group to the acetic/lactic acid mixture.

In summary, both acetic and lactic acid seem to produce the desired microbial reduction, whilst mixtures of the two may possibly enhance the antimicrobial effects as a result of potentiation.

(d) Concentration and temperature of the acid. Organic acids are generally found to exert a more marked antimicrobial effect in high than in low concentrations [57, 74, 76, 78, 83, 85, 94]. Surve *et al.* [93] observed that the bactericidal and bacteriostatic effects of 1 to 4% acetic/lactic and acetic/propionic acid mixtures increased with increasing concentration, but that the effect was reduced as time progressed. Some researchers report negligible differences between solutions of lactic or acetic acid in low concentrations. Van der Marel *et al.* [59] observed that immersion in lactic acid at 15°C reduced the bacterial contamination per gram chicken skin by about 1 log unit, irrespective of the concentration of the solution (1 or 2%). Similarly, Anderson *et al.* [95], found no significant differences in aerobic-, psychrotrophic-, Gram-negative and coliform counts, nor in counts of *Brochothrix thermosphacta* and lactobacilli on lamb carcasses treated with 1.5 and 3% acetic acid.

Dickson [96] demonstrated that 2% acetic acid combined with carcass desiccation reduced the numbers of *Salmonella typhimurium* and *Listeria monocytogenes* to a greater degree than did acid treatment alone. Chilling procedures which include spraying with water obviously dilute the acids and may negate any potential residual antibacterial effects during extended storage [70] or even result in some recovery of injured *Salmonella* [97].

Various studies [49, 66, 81, 98] have substantiated the higher sensitivity of several microorganisms to acids at higher temperatures. Gill and Newton [43] demonstrated that the ability of spoilage bacteria to initiate growth on nutrient agar acidified to pH 5.5 with lactic acid was markedly lower at 2 than at 25°C. Anderson *et al.* [95] found numbers of Gram-negative bacteria and lactobacilli on lamb to be affected significantly by the temperature of the acid solutions, 55°C acetic acid rendering superior results over 25°C. Anderson and Marshall [92, 99], Anderson *et al.* [83] and Greer and Dilts [76] comparing acid treatment at various temperatures reported similar results when treating beef. In contrast, Brackett *et al.* [84] found warm acid sprays (20°C) to reduce *Escherichia coli* 0157:H7 on raw beef slightly better than hot acids (55°C), although differences in reduction were so small as to be inconsequential.

In summary, higher acid concentrations and higher temperatures generally yield the best antimicrobial effects.

(e) Acid sensitivity of various organisms. The variable sensitivities of different taxa of microorganisms to acids has been established in several experiments. The relative tolerance of yeasts and mounds to acids is known [100]. This was confirmed in decontamination experiments with carcasses [101–103]. However, the possibility that this selective inhibition by acid might result in a flora shift towards pathogenic organisms at the cost of microbial antagonists was refuted [58, 74, 90, 104, 105, 106]. A flora-shift after acid treatment from Gram-negative to Gram-positive bacteria has been reported [91, 93, 105], but the latter may serve as antagonists of the enteropathogenic Gram-negative bacteria [107–109].

Pathogenic bacteria. The effects of acid decontamination upon the microbial ecology of meat have been the subject of various studies, both under laboratory and industry conditions.

Relying on *in vitro* trials, Van Netten *et al.* [56, 105] demonstrated that a 2 min treatment with 0.2 mol/l lactic acid at pH 2.5 at room temperature resulted in the following lethalities (= $\log_{10}$ initial − $\log_{10}$ final CFUs): *Campylobacter jejuni*: >5.3, *Salmonella spp*. 4.3–5.2; *Yersinia enterocolitica*: 3.8–5.1; enteropathogenic *E. coli* 2.2–4.1 and mesophilic *Enterobacteriaceae*: 2.2–2.8. Bell *et al.* [89] immersed 1 cm cubes of beef, inoculated with 5.2×10^6 CFU of various microbial genera, in 1.2% (v/v) acetic acid or in distilled water; expressed as percent of bacterial reduction the delta's (treatment minus control) were *Pseudomonas aeruginosa*: 33.6, *Salmonella typhimurium*: 31.2, *Shigella sonnei*: 28.7, *Yersinia enterocolitica*: 26.6, *Streptococcus faecalis*: 17.2 and *Escherichia coli*: 5.1. Anderson and Marshall [92] determined that 15 s immersion of cylindrical beef samples in various concentrations of acetic acid at 4 different temperatures reduced bacterial numbers in the following order: total aerobic counts, Enterobacteriaceae, *Salmonella typhimurium*, *Escherichia coli*. These investigators questioned if the reductions found (0.5–1.5 log cycles) were very meaningful with regard to meat safety.

Dixon *et al.* [102], studied the effects of a combination of strict hygiene and spraying with 1% (v/v) lactic acid at 55°C on the bacterial association of beef carcasses in industrial practice. They found *Micrococcus* and yeasts to be the predominant (>50%) microbial types immediately upon acid treatment, whereas in control carcasses, *Micrococcus*, and to a lesser extent *Streptococcus* and coagulase-negative *Staphylococcus* and/or coryneform bacteria, were dominant. However, this flora shift was no longer noticeable after 20 d of refrigerated storage.

Cudjoe and Kapperud [110] reported that *Campylobacter jejuni* was removed rather efficiently from broiler carcasses by 1% lactic acid sprays, reflecting the organism's low resistance to acids, which had already been demonstrated in culture and inoculation experiments by Stern *et al.* [111]. Cudjoe [112] found 1% lactic acid to have limited bacteriostatic effects on *Yersinia enterocolitica* counts (0.6 to 1.5 log cycles reduction) on pig skin. Greer and Dilts [76], comparing the susceptibilities of various pathogens and spoilage bacteria, reported that of the pathogens studied *S. aureus* was the most (1.4 log cycle reduction) and *S. typhimurium* the least (0.4 log cycle reduction) susceptible to treatment with 1–3% (v/v) acetic or lactic acid at 20 or 55°C. Van Netten *et al.* [113] recently reported that a 2% lactic acid treatment of pork bellies at 21°C effected a reduction of *Campylobacter jejuni*, *Salmonella typhimurium* and *Listeria monocytogenes* of 3.1, 0.8 and 0.3 log cycles, respectively. The limited susceptibility of *Listeria* to lactic acid was also confirmed by El-Khateib *et al.* [80], who found a 1 min incubation of 1 cm^3 cubical pieces of beef in 2% lactic acid (prior to immersion in a *Listeria* suspension) to reduce the *Listeria* count 2 d after treatment by only 0.2 log cycles per cm^2. The same authors also showed that *Listeria* significantly increased in the presence of lactic acid, suggesting that lactic acid is unable to block attachment of listeriae [80]. Data of Greer and Dilts [114] suggest a bacteriostatic (delayed) effect of 3% (v/v) lactic acid at 55°C on *Listeria monocytogenes* for up to 15 days of storage at 4°C and these authors clearly demonstrated that acid treatment is much more effective in inhibiting *Listeria* for fatty than lean tissue.

The susceptibility of *Escherichia coli* 0157:H7 to acid treatment is of special interest, given the recent outbreaks of this organism in the northwest of the United States. It has been suggested that acid rinses might change the bacteria's ability to attach to the carcass and that acid sprays might be useful to combat the prevalence of *E. coli* on carcasses [115]. However, Dickson and Frank [116], studying amongst other factors the sensitivity of starved and unstarved cultures of *Salmonella typhimurium*, *Listeria monocytogenes* and *E. coli* 0:157 H:7 to acid, demonstrated that, although starved cultures of the former two bacteria were slightly more sensitive to a 2% acetic acid rinse than were unstarved controls, even the starved *E. coli* cultures maintained their relative insensitivity. The latest information on the efficiency of up to 1.5% (v/v) acetic, citric and lactic acid sprays for eliminating *E. coli* 0157:H7 from meat is that none of these acids appreciably affect this organism, even at 55°C [84]. Nettles Cutter and Siragusa [85], applied the same acids at 24°C and, although significant reductions in three strains of *E. coli* 0157:H7 of 1.4 (1% acid) to 1.8 log cycles (5% acid) were observed, complete inactivation of the pathogen (even at 5% acid level) was not possible. It was also found (Dr G. Gordon

Greer, personal communication) that a 3% (v/v) lactic acid solution at 20°C had no effect upon the number of *E. coli* 0157:H7 recovered from inoculated lean beef and that only a 1 log reduction was achieved at 55°C.

Spoilage bacteria. As meat is generally kept in chilled storage rooms, the surface microflora will be dominated by organisms capable of growing at refrigeration temperatures. Many of these psychrotrophic organisms will have contaminated the carcass during various dressing procedures. Yet it is only under certain storage conditions that the various fractions of this initial population will colonise the meat surface. Major spoilage organisms are *Pseudomonas* and *Acinetobacter/Moraxella* (strictly aerobic), *Alteromonas putrefaciens* and *Brochothrix thermosphacta* (both facultative anaerobes), *Lactobacillus* spp. (facultatively anaerobic, aerotolerant) and the Enterobacteriaceae, a group including a wide range of facultative anaerobes, amongst which are pathogens such as *Salmonella* (see above).

Provided they are applied to the carcass immediately upon slaughter, single 1.25% lactic acid sprays extended the shelf life of veal loins stored 14 d at 3 ± 1°C both under aerobic and anaerobic conditions, with hardly any growth of lactobacilli, whilst Enterobacteriaceae counts after 14 d of refrigerated storage remained as low as 2.2 $\log_{10}$ CFU [54]. Shelf life extension was also found after treatment with acid mixtures containing lactic and/or acetic acid by Osthold *et al.* (beef and mutton stored for 5–8 d at 7 and 10°C, respectively) [90], Ozari and Kotter (beef and pork stored for 9 d at 7°C) [91], Surve *et al.* (buffalo stored for 7 d at 7 ± 1°C) [93], Zeitoun and Debevere (chicken legs stored at 6°C in modified atmosphere packages for up to 17 d) [94], Matthieu and van Hoof (hot and cold boned beef stored frozen and subsequently thawed during 16 h) [103] and Hamby *et al.* (beef stored for 28 d at 2°C) [82]. In the latter study only lactic acid afforded shelf life extension and only in some of the subprimals.

Cacciarelli *et al.* [117] treated pork loins with a 12.8°C 2% acetic acid spray before vacuum packaging and observed a significant reduction in aerobic and anaerobic flora up to 21 d of storage at 4°C. After an additional week of storage only the aerobic counts remained significantly lower than untreated controls, whilst Lactobacillaceae counts remained largely unaffected throughout the entire storage period. Shay *et al.* [79] studied the effects of total immersion of skinned pork loins with relatively high pH (6.2–6.5) in 1.5% or 2% acetic or lactic acid at 55°C, and found an extension of the shelf-life of vacuum packed sections from 3 weeks to 5 (lactic acid) or 6 (acetic acid).

In contrast to the previous findings, Greer and Jones [118] found that, although 1% lactic acid sprays produced a 0.7 log cycle reduction in bacteria on beef carcasses, this effect was not reflected in an extended case life of loins and rib-eye steaks prepared from these carcasses. Similar results were obtained by Prasai *et al.* [119].

Gill and Newton [43] were among the few who specifically studied the sensitivity of Gram-negative psychrotrophic spoilage organisms to lactic acid in meat. They established that 80% of the Enterobacteriaceae and 98% of the *Pseudomonas* strains (isolated from carcasses, meat and work surfaces) initiated growth on culture media at pH 5.5 at 2°C. The other spoilage bacteria investigated (*Pseudomonas*, *Enterobacteriaceae*, *Moraxella*, *Acinetobacter* and *Flavobacterium*) were hardly affected by the higher meat pH. The effects of surface pH values lowered by acid treatment were studied recently [76, 114]. Greer and Dilts [76] showed that *B. thermosphacta* (3.5 log cycle reduction) and *Ps. fragi* (2.3 log cycle reduction) were more sensitive to the lethal effects of acid treatment of beef than were pathogenic bacteria. More recently, Greer and Dilts [114] further demonstrated that a 3% lactic acid treatment at 55°C had a more significant bactericidal effect on pork fat tissue, where spoilage bacteria were reduced by 3 (*Ps. fragi*) or 4 (*B. thermosphacta*) log cycles, as opposed to the 1 and 2 log cycle reductions on pork lean.

In summary, at generally used concentrations (1–3%), acids are very effective against pathogenic bacteria such as *Campylobacter jejuni*, *Yersinia enterocolitica* and reasonably effective against salmonellae. *Listeria monocytogenes* and *E. coli* are rather insensitive to acids at these concentrations. The growth of potent spoilage bacteria is well controlled by acid decontamination, retaining populations well below thresholds necessary for deleterious sensory changes.

(f) Time of application, contact time of the acid with the meat surface. Bacterial attachment is the result of each of the following processes: retention (bacteria remain in a liquid surface film), entrapment (maintenance of the bacteria by the specific microtopography of the surface), adsorption (bacteria are adsorbed to a solid–liquid interface by short-range attractive forces) and adhesion (establishment of intimate contact between bacteria and surface, sometimes mediated by polymer-bridging or by fimbriae and holdfasts). The reader is referred to papers by Thomas *et al.* [40] and Dickson and Anderson [120] for more detailed information on the mechanisms involved.

Contaminating bacteria, unless quickly removed from the liquid film covering the meat surface, may attach after relatively short periods of exposure, followed by colonisation of the meat surface [121, 122]. Consequently, carcass and meat surfaces should be treated as soon after dressing/boning as possible [123]. Following this procedure Smulders and Woolthuis [54] observed both immediate effects of a 1.25% lactic acid treatment on veal carcasses [54, 57], and delayed bacteriostatic effects on veal subprimals stored aerobically for 14 d [54]. With few exceptions [114, 117] the late application of acids was found to render poor results. Mulder and Krol [124] treated beef steaks with 0.5 to 5% lactic acid at 1 and 2 days post-mortem and failed to detect marked microbiological effects. Similarly, when spraying beef subprimals with various acid mixtures after grading [125], or as late as 96 h post-mortem [102], only an immediate bactericidal – but negligible delayed effects of acid decontamination were found after refrigerated vacuum storage. The authors ascribed this inactivity to application of the acids in a too late stage of processing. Also, Mendonca *et al.* [87] used chops from pork loins obtained from a commercial source (and hence presumably at least at 1–2 days post-mortem (FS); and established that concentrations lower than 3% acetic acid failed to exhibit Enterobacteriaceae after 6 week chilled vacuum storage. Sirami *et al.* [126], using a mixture of acetic, lactic, citric and tartaric acid as decontaminants for diced pork and raw ham, showed that this treatment was of limited efficacy and impaired sensory and processing quality. Gauthier and Jacquet [127] compared the effects of 1.50, 1.45 and 1.70 g/kg acetic, lactic and citric acid, respectively, taken up by cubes intended for further processing and prepared at a late stage post mortem. Only when combined with nitrite curing salts a 1 g/kg acid uptake gave a shelf life up to 8 days at 4°C.

The possibility of improving efficacy using a series of acid treatments was studied by various investigators. When subprimals from acid-treated veal carcasses were sprayed a second time with 2% lactic acid immediately after hot boning and before vacuum packaging, Smulders and Woolthuis [54] found that, although the combined acid treatments resulted in aerobic counts at 14 d post mortem similar to the initial levels of carcass contamination (around 3 $\log_{10}$ CFU/cm^2), the Enterobacteriaceae count remained below the limit of detection. Single treatment with 1.25% lactic acid resulted in Enterobacteriaceae counts around 2 $\log_{10}$ CFU. Lactic acid sprays, 1% at 55°C, during various steps (after dehiding/dehairing, after evisceration, or at both locations) in the processing line, were studied by Prasai *et al.* for beef [119] and pork carcasses [69]. In the pork study neither of the acid treatments were effective. The double acid treatment yielded lower bacterial counts in the beef study. Hamby *et al.* [82] applied 1% acetic and/or lactic acid to beef sides, as a single 30 s spray, or intermittently for 12 consecutive hours (2 × 30 s per hour, i.e. a total of 12 min of spraying), whereafter subprimal cuts from these sides were vacuum packaged and stored for 28 d at 2°C. Only the intermittent spray treatment resulted in significant reductions throughout the storage period. In contrast, Van der Marel *et al.* [59] failed to confirm the cumulative effects of repetitive lactic acid treatment of broiler carcasses. This finding is possibly related to the rather specific microtopography of poultry carcasses [40].

Siragusa and Dickson [128, 129] investigated the effects of acetic and lactic acid immobilised in alginate gels and determined that such gels have increased bactericidal activity on lean beef (not on fatty surfaces) inoculated with *Listeria monocytogenes*, *Salmonella typhimurium* and *E. coli*

0157:H7. The effects on the latter two organisms were less pronounced than on *Listeria*. Probably the contact time that the inhibiting acid maintains with the meat surface before evaporation or neutralisation is increased and the moisture loss slowed down. Since alginate can be sprayed on carcass surfaces and has a GRAS status it would be suitable to currently used meat processing operations.

In summary, acids are to be applied to meat before contaminating organisms have colonised the surfaces. In addition, the meat surface should be allowed sufficient contact time with the acid solution.

12.4 Effects of acid treatment on sensory properties

12.4.1 Effects on colour

(a) Ruminants. Dezeure-Wallays and Van Hoof [101] treated beef carcasses with an acid mixture containing lactic acid, sugars and antioxidants and found an irreversible brown discoloration. After having studied several concentrations of lactic acid on beef carcasses, Snijders *et al.* [74] concluded that the application of 1% lactic acid spray resulted in a slight but reversible discoloration of beef carcasses. Similarly, Prasai *et al.* [119] did not observe any adverse colour effects of beef loins from carcasses that had been treated with hot (55°C) 1% lactic acid sprays. Osthold *et al.* [90] treated beef and sheep carcasses with an acid spray of 2% acetic acid, 1% lactic acid and 0.25% citric acid and found negligible differences in carcass colour in beef, whereas the lean (not the fat) of sheep carcasses was found to be slightly darker than controls. On veal carcasses Woolthuis and Smulders [54] observed no off-colours of fat or lean treated with up to 1.25% acid. When deboned, in the course of which fat is trimmed off, concentrations up to 2% lactic acid were used without noticeable bleaching of meat; higher concentrations resulted in an unacceptable bleaching of the lean and in considerable off-colours of the fat [54]. When many blood spots are present acid treatment, particularly at high concentrations, may result in blood coagulation and consequently in the appearance of rusty brown–black spots, as was confirmed in experiments with beef [74], veal [57] and buffalo carcasses [93].

Mulder and Krol [124] immersed beef steaks in 0.5, 1, 2, 3, 5 and 8% lactic acid at 6°C for 2 s and established that, after 1 day of refrigerated storage, all treatments resulted in brown discoloration and lower panel colour scores. Spraying the steaks after treatment with 0.05% ascorbic, 0.01% citric, or 0.015% nicotinic acid was found useful to counteract these colour changes, an observation also reported by Quartey-Papafio *et al.* [88]. Steaks prepared from subprimals treated with 1% lactic acid, 1%

acetic acid or a mixture of 1% lactic, 2% acetic, 0.25% citric and 0.1% ascorbic acid, applied at ambient temperature, had no consistent effect on panel scores for lean and fat colour, surface discolorations and overall appearance [130]. In another test by the same group of investigators [102], these panel scores were significantly higher than controls when a 1% lactic acid spray at 55°C was used. Hamby *et al.* (1% lactic, 1% acetic or a 1% lactic/1% acetic acid mixture at 2–3°C) [82] and Greer and Dilts (1% lactic acid at 20°C) [76] found no colour effects on beef cuts from vacuum-packaged and stored subprimals from acid-treated carcasses. Quartey-Papafio *et al.* [88] used formic, acetic and propionic acid as sanitisers for beef, and found that the microbiologically most effective solutions, 2% formic acid or 1% formic plus 1% acetic acid, produced brown off-colours within seconds after spraying, a phenomenon seen at formic acid concentrations of 0.5% and higher. Bell *et al.* [89] comparing the colour effects of immersion of diced beef in various acetic and acetic/formic acid solutions, established that immersion in an acetic acid solution at higher concentrations than 0.6% resulted in discoloration.

(b) Pork. Reynolds and Carpenter [131] sprayed pork carcasses with various concentrations of an acetic–propionic acid solution mixed in 60:40 ratio (w/w) and found excessive darkening and discoloration on all lean cut surfaces when applying the 3.0 M solution, which was attributed to the propionic acid. Labots *et al.* [132] reported an irreversible discoloration of pig carcasses after treatment with 2.40% lactic acid, especially in the thorax, cavities and locations in which lactic acid was collected. These authors therefore recommend 1% lactic acid, as this concentration would not result in discoloration of the carcass, which was confirmed by Prasai *et al.* [69] even for hot (55°C) sprays. Snijders *et al.* [75] determined that the liquid film covering pig carcasses will slightly dilute the acid applied to the surface, allowing concentrations of lactic acid to be raised up to 1.5% with only a minimal effect on colour.

Cacciarelli *et al.* [117] observed that pork loins, spray-washed and sanitised with 2% acetic acid, were significantly less pinkish-red and had a less desirable colour than dry controls. Mendonca *et al.* [87] dipped fresh pork chops in various acid solutions (1% acetic acid, 1% acetic–1% lactic acid, 1.5% acetic acid/1.5% sodium acetate, 3% acetic acid/3% sodium ascorbate, 3% acetic acid/2% sodium chloride); all acid treatments, but particularly the ascorbate-containing mixture produced higher Hunter L* values than water-treated controls. As exception was the sodium chloride-containing solution, which had lower Hunter L*- and a*-values, possibly resulting from oxidation of the myoglobin by salt. Gauthier and Jacquet [127] reported that 1.5 g/kg acetic acid and 1.45 g/kg lactic acid, taken up by diced pork after immersion, caused only minor colour changes, in contrast to a 1.70 g/kg citric acid uptake which caused brown discoloration.

(c) Poultry. Mountney and O'Malley [52] observed that broilers treated with acetic-, adipic- and succinic- (but not with lactic) acid of various concentrations caused the carcasses to whiten slightly. This effect was more pronounced with acetic- than with succinic acid. Dresel and Leistner [133] reported a slight discoloration after treating chicken carcasses with 2.8% acetic, 1.8% lactic, 0.25% citric and 0.1% ascorbic acid solutions. Stern *et al.* [111] reported carcass discoloration when broilers were treated with 1% lactic acid at 5°C, also observed by Van der Marel *et al.* [134] at 2% lactic acid and higher. Although Mulder *et al.* [135] (using 1% lactic acid) and Cudjoe and Kapperud [110] (using 2% lactic acid) did notice bleaching (suggested by Mulder *et al.* to possibly be the result of the natural catalase activity of skin and blood, releasing oxygen after treatment) or 'graying' of broiler carcasses, this discoloration reverted to normal after 24 h.

(d) Variety meats. When decontaminating variety meats by immersion in acid solutions the surface may discolour, depending on the nature of the tissue. High concentrations are possible, for instance, with tongues, where surfaces will be peeled off in the course of preparation [136]. However, livers discolour after immersion in lactic acid concentrations higher than 0.20% [58]. Calf brain exhibits off-colours at 1.25%, whereas lower concentrations do not provide adequate decontamination [137].

12.4.2 Effects on flavour and odour

(a) Ruminants. Woolthuis and Smulders [57] assessed the effects of 2 and 4% lactic acid sprays (followed by 2 h of incubation) of fresh veal longissimus chops on the flavour after 3 min of frying in butter. Only the 4% lactic acid spray resulted in panel flavour scores different from the untreated controls. Gill and Penney [106] found even 5% lactic acid had no significant effect on flavour and odour and vacuum-packaged lamb after storage periods of 3, 6, 9 and 12 weeks. In a study by Bell *et al.* [89], panelists identified flavour differences between baked ground beef prepared from diced beef, that had been immersed either in 1.2% acetic acid or distilled water. Treatment with 0.6% acetic acid did not lead to flavour differences. Jeremiah *et al.* [138] sprayed beef carcasses with 1% lactic acid or water and found acid treatment had little influence (although a slight metallic aftertaste was noticed) on the flavour scores of roasts prepared from acid-treated loins aged in vacuum for 7 d. In a second trial, where acid-treated subprimals were stored up to 28 days before evaluation, lactic acid significantly increased flavour scores (intensity of the beefy and fatty character notes). Three days after treatment of buffalo meat with 1 and 2% acetic–/lactic- or acetic–/propionic acid mixtures, Surve *et al.* [93] noticed a slightly sweetish odour. A 3% solution of the acetic–/lactic acid

mixture effected no off-odours, but a 3% acetic–/propionic acid as well as 4% solutions of both mixtures resulted in acidic odours.

(b) Pork. Reynolds and Carpenter [131] observed slight residual off-odours of pork carcasses treated with 2.32 M acetic acid or 1.67 M acetic/1.36 M propionic acid solutions. Sirami *et al.* [126] found impaired flavour in meat products prepared from diced pork and raw ham treated with an acid mixture of acetic-, lactic-, citric- and tartaric acid. Gauthier and Jacquet [127] reported vinegar-like off-flavours in garlic sausage prepared from diced pork that had taken up 1.5 k/kg acetic acid; when pork was processed into paté such differences were not noticed.

(c) Poultry. Mountney and O'Malley [52] noticed that the strong odour of broiler carcasses treated with acetic acid was offensive, whilst the odours associated with adipic- and succinic acid, although detectable, were not unpleasant. No adverse effects of lactic acid on flavour and odour were reported. Mulder *et al.* [135] found no off-odour of broiler carcasses after 1% lactic acid treatment. Van der Marel *et al.* [134] established that lactic acid accumulated in the skin of broilers with no increase in acid in the underlying muscle tissue. Concentrations up to 5% lactic acid produced no flavour differences in grilled legs. Dorn *et al.* [139] compared various concentrations and combinations of solutions for their efficacy as decontaminants against salmonellae on broiler carcasses and reported that at effective microbicidal doses none of the acids tested (benzoic, *p*-hydroxybenzoic, succinic, peracetic and lactic acid, and a citric acid + potassium sorbate mixture) was without adverse effects on appearance, odour and flavour.

(d) Variety meats. Visser *et al.* [138] did not observe off-odours of veal tongues after immersion in 3% lactic acid.

12.4.3 Effects on drip loss

Offer and Trinick [140] stated that lowered intracellular pH may cause myofibrillar shrinkage and may thus result in drip loss. Mendonca *et al.* [87] measured the percent exudate from pork chops in the course of 4 weeks of refrigerated vacuum storage and observed increases in drip loss in all chops immersed in acid solutions, particularly those solutions (3% acetic acid/3% sodium ascorbate, 3% acetic acid/1.5% sodium chloride and 1% acetic acid/1% lactic acid) that caused lower pH values. Jeremiah *et al.* [138] cut loin roasts from beef carcasses treated with 1% lactic acid sprays and assessed thaw drip- and cooking losses after 90 d of frozen vacuum storage were not affected by the acid-treatment. Gauthier and Jacquet [127] observed that diced pork that had taken up 3 g/kg citric acid after immersion treatment released higher amounts of exsudate after 10 d of refrigerated storage, while 1.5 g/kg acetic or 1.45 g/kg lactic acid uptake did not result in increased exsudation.

In summary, decontamination with acetic or lactic acids at effective concentrations (1–2%) hardly impairs the sensory quality of meat surfaces. When using acetic acid, sensory scores are affected more readily than with lactic acid. Mixtures containing acids with a marked antioxidant effect may help to alleviate colour problems. Slight discolorations visible immediately after application at low concentrations usually disappear upon diffusion of the acid.

12.5 Mode of application of acids; technologies available

12.5.1 Spraying and spray cabinets

Spraying is the most common way of applying acids to carcasses. For experimental purposes, hand-held spray devices have been used with some degree of success. Yet, it must be conceded that such procedures are too time-consuming and surely not suitable for high-speed industrial operations. They may also be rather ineffective, depending upon the influence of various physical parameters. Essential information on these parameters has been collected in laboratory studies on spraying with water and various added compounds, e.g. those in England in studies by Bailey [141], in Australia by Smith and Graham [142], Graham *et al.* [143], Davey [144] and Davey and Smith [145], in Ireland by Kelly *et al.* [146], and a large series of experiments in the United States by Anderson and co-workers [120].

Of the physical parameters, spray pressure and flow rate, nozzle type and configuration and angle of the spray seem to be particularly relevant. In an experiment with beef carcasses Kotula *et al.* [147] increased the water pressure from 85 kPa to 498.5 kPa and determined a markedly higher reduction in surface flora to levels often less than 2 CFU/cm^2; these investigators therefore concluded that the effectiveness of high pressures overshadowed the effects of lowered pH or wash water temperature. In a model-experiment meat was inoculated with *Rhodatorula rubra*, whereupon various spray angles, pressures, flow rates and conveyor speeds were assessed [148]. It was established that the effect of spray angle was only important at low pressures (e.g. at 70.9 kPa a 90° *vs.* 30° spray angle resulted in about 79 *vs.* 90% reduction), that, irrespective of line pressure (nozzle), yeasts were removed in higher numbers at higher flow rates, that a 141.8 kPa line pressure removed 89% of this yeast and that a four-fold increase in pressure only resulted in a 5% extra reduction. Finally, yeast removal was decreased at higher conveyor speeds. Figure 12.1 presents

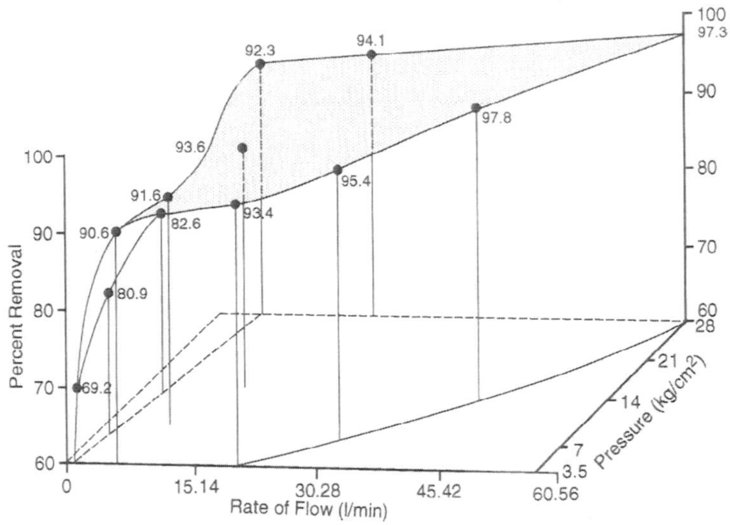

Figure 12.1 Relationship between removal, rate of flow and pressure of spray.

a three-dimensional plot of percent removal against rate of flow and pressure. The concern of Smith *et al.* [149] that higher pressures would force microorganisms into the carcass tissue was investigated by De Zuniga *et al.* [150], who concluded that in beef carcasses sprayed at pressures beyond 2070 kPa this phenomenon became important.

In many countries the use of organic acids for purposes of carcass decontamination has not, or only recently, been approved (see below). Consequently, the design options for (acid) spray equipment and the amount of practical experience are limited. A beef carcass wash chamber has been operating since 1980 in a commercial plant in the USA to a fair degree of satisfaction [151]. Recently, similar units have been designed and USDA-approved for the use in pig processing lines.

In a theoretical analysis Davey [144] predicted that a water-flood cabinet would yield better results than would conventional spraying with nozzles. A laboratory evaluation [145] indicated that the 'deluge'-cabinet for hot water decontamination of beef carcasses would reduce the running cost for this unit to two-third of a spray operation when compared at the maximum reduction of $\log_{10}$ 1.3 (95%) achieved by the spray cabinet.

In the Netherlands a prototype carcass spray unit was designed for a veal slaughter plant (Smulders *et al.*, in preparation). Sanitising relies on the combination of mechanical removal by water sprays of blood and dirt and decontamination by hot water or acid. Figure 12.2 shows the major features of this unit.

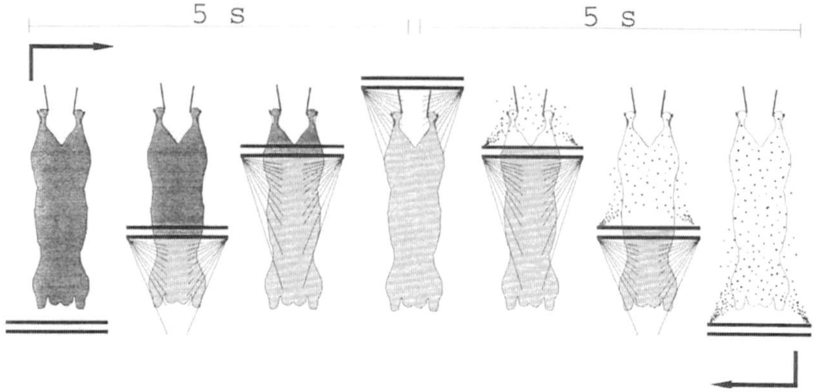

Figure 12.2 Features of prototype spray unit.

In a sequence of one upward and one downward movement the unit, consisting of two rings, each equipped with 8 fan type nozzles spraying at 202.5 kPa, is moved around the carcass. During the 5 s upward movement the lower ring sprays water at a 30° downward angle, detaching dirt, hair and blood from the surface. This leaves some of these particles suspended in a liquid film on the surface. Subsequently, during the 5 s downward movement the lower ring 'shaves off' this suspension leaving a fresh water film, whilst the upper ring sprays a lactic acid mist which solubilises (and hence is diluted) in the liquid film. This unit was used for an in-plant evaluation of the microbiological effects of 10° and 80°C water (lower ring alone), and those of a combination of cold water and 1.25% or 2% lactic acid (lower and upper ring combined). Whilst all treatments rendered a visually clean carcass with no discoloration, cold water failed to significantly reduce microbial counts. Only moderate reductions were achieved with hot water and 1.25% lactic acid (as a result of dilution of the acid in the water film); 2% lactic acid yielded the desired reduction. Figure 12.3 illustrates this.

12.5.2 *Immersion*

Variety meats and (sub)primals may be decontaminated by immersion for a certain period. In this procedure organic acids very effectively reaches all locations, possibly even including the interior of organs such as the liver via the bile ducts [58]. On the other hand, acid solutions used in this way were found to gradually lose part of their decontaminating potential, as the anions are easily bound by peptides and proteins which are progressively released into the immersion tanks (Van Netten, unpublished

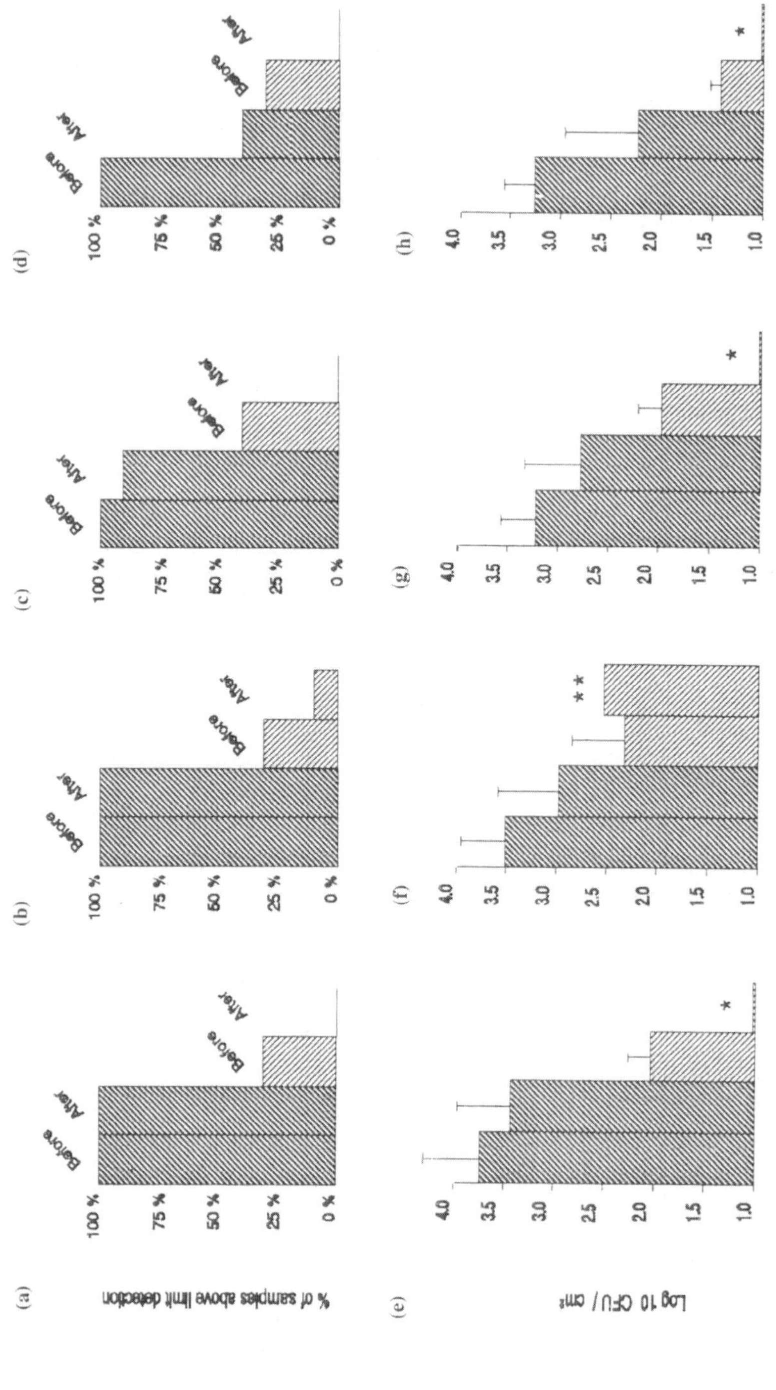

Figure 12.3 Microbiological effects of water and lactic acid sprays on CFU. (a) 10°C water; (b) 80°C water; (c) 1.25% lactic acid; (d) 2% lactic acid; (e) $p < .10$; (f) $p < .001$; (g) $p < .025$; (h) $p < 0.001$.

results). For similar reasons Visser *et al.* [136] found it necessary to clean veal tongues thoroughly before acid decontamination to dispose of the mucus covering the numerous papillae. To this end they successfully applied centrifugation in the presence of 2% lactic acid. Delicate tissues such as brain require more gentle treatment, but, unless previously liberated from hair, skin and bone fragments, these particles will substantially interfere with the decontaminating potency of lactic acid [137]. Thus, immersion may not prove to be very cost-efficient. For instance, Dorn *et al.* [139] found that immersion of large numbers of *Salmonella*-contaminated broiler carcasses in solutions containing various acid decontaminants was not very successful. Whereas under experimental conditions the immersion of pork loins in acid solutions has been reported to be effective for pork loins [79], it seems unlikely that such a procedure will prove cost-efficient under commercial circumstances.

12.5.3 Other methods

Logtenberg [152] experimentally applied lactic acid to pig carcasses by electrostatic dispersion, in an attempt to reduce the amount of acid needed for effective decontamination; obviously, the washing effect is practically absent in such procedures.

Lactobacilli produce antibacterial agents (bacteriocins) as by-products of their metabolism. Consequently, starter cultures of lactic acid bacteria have been applied as *in situ* lactic acid producers, particularly incomminuted meats [153]. Murali *et al.* [154] achieved suppression of the spoilage flora on fresh mutton by applying cultures of *Lactobacillus plantarum*, *Lactobacillus bulgaricus* and *Streptococcus lactis* onto the meat surfaces. All cultures proved inhibitory to psychrotrophic Gram-negative rods, bacteria of the coli-aerogenes group, staphylococci and proteolytic and lipolytic spoilers. This effect was primarily attributed to lactic acid production. More has to be learned about the effects of this mode of lactic acid decontamination on sensory traits. Gill and Penney [106] reported that vacuum-packaged lamb treated with a *Lactobacillus* culture and stored for 9 weeks at −0.5°C had lower odour and flavour scores. Extended storage (12 weeks) resulted in the development of off-colours and unacceptable 'dairy' flavours.

Recently, attempts have been made to apply isolated bacteriocins to contaminated beef carcass surface tissue using a pilot scale spray washer. The results of this work look very promising (Dr Gregory R. Siragusa, 1994, personal communication).

12.6 Acceptability of acid treatment

12.6.1 Toxicological considerations

Clearly, substances suitable for chemical decontamination of carcass surfaces must be toxicologically acceptable if they are to be used on food because of the potential of residues in the final product. Elias [155] provided a toxicological evaluation of various organic acids (and their potassium and sodium salts) for purposes of decontamination. Acetic, propionic, lactic and formic acid were considered to be acceptable from a toxicological perspective.

12.6.2 Legislation and regulations

Legislation in the meat-producing countries as to the authorisation of decontaminating agents in general is rather unclear. All meat inspection laws contain some paragraph indicating that it is prohibited to treat raw meat with foreign substances or with additives, unless these are specifically listed. Essentially, in Europe little has changed since 1987 when the interpretation of the legislation on the use of organic acids was monitored [156]. The most recent information on legislative aspects of the use of additives within Europe [157] contains the essentials on the use of acids in meat products, but states no specifics on the use of organic acids on carcass surfaces. In the United States, pre-evisceration carcass sprays with organic acids have been allowed, subject to several conditions the most important of which is that only those operations will be considered, that maintain sanitary dressing practices (established by a written and accepted 'Partial Quality Control' programme, covering the CCPs in, and inspected through monitoring of, the spray system as well as the quality control programme) [158]. Also, in Canada the use of (at the present time only acetic) acid has recently been approved [159].

12.7 Conclusions and actions needed

The use of organic acids as decontaminating agents should only be allowed as an integrated part of extensive hygiene programs at the slaughter line. The data presented here indicate that, provided meat surfaces contain low amounts of organic material and/or bacteria as a result of adherence to GMP and provided the acid is applied as soon after slaughter as possible, treatment with the suitable food-grade acids (particularly at high temperatures) may result in a significant improvement in the bacteriological condition. At concentrations not impairing the sensory quality, microbiological effects may persist after extended refrigerated storage. Whereas the effects on some pathogenic bacteria such as *Listeria monocytogenes* and *E. coli* 0157:H7 appear to be limited, acid decontamination has proven largely to control growth of pathogens such as *Campylobacter*, *Listeria* and *Salmonella*, and particularly that of potent spoilage organisms. Currently several spray units are commercially available, which perform to a fair degree of satisfaction.

Decontamination of carcasses by organic acids has recently been allowed in North America. However, most other meat-producing countries have not made any commitments on this issue. Since the GATT (General Agreements on Tariff and Trade) negotiations have recently led to a formal agreement, the aforementioned change of policy will lead to distortion of competition between the European Union and the North America's. It seems inevitable that, in Europe, a change in strategy must ensue.

Acknowledgements

Thanks are due to the Institute of Food Technologists, Chicago (IL) for their kind permission to reproduce Fig 5 from the paper by Anderson *et al.* [149]. The author is also thankful to G. Gordon Greer PhD (Agriculture and Agri-Food Canada, Lacombe Research Station, Lacombe, Alberta, Canada) and Gregory R. Siragusa PhD (USDA, Agricultural Research Service, Clay Center, Nebraska, USA) for sharing their latest, unpublished, findings. In addition, Dr Greer is gratefully acknowledged for reading the final script and suggesting necessary changes.

References

1. ICMSF, *Microorganisms in Foods, 4. Application of the Hazard Analysis Critical Control Point (HACCP) System to ensure Microbiological Safety and Quality*, Blackwell Scientific Publications, Oxford, UK, (1988), 1.
2. K. Ishikawa, *Guide to Quality Control*, Asian Productivity Organisation, Tokyo, Japan, (1976), 1.
3. A.H. Linton and M.H. Hinton, in *Elimination of Pathogenic Organisms from Meat and Poultry*, ed. F.J.M. Smulders, Elsevier, Amsterdam, (1987), 9.
4. F.G. van Zijderveld, J.F. Frik and A.M. van Zijderveld-van Bemmel, *Tijdschr. Diergeneesk.* **107** (1982), 123.
5. F.H. Grau and M.G. Smith, *J. Appl. Bacteriol.* **37**, (1974) 111.
6. H. Grønstol, A.D. Osborne and S. Pethiyagoda, *J. Hyg. Camb.* **72** (1974), 155.
7. T.A. Roberts, *Royal Soc. Health J.* **100** (1980), 3.
8. T.A. Roberts and W.R. Hudson, in *Elimination of Pathogenic Organisms from Meat and Poultry*, ed. F.J.M. Smulders, Elsevier, Amsterdam, (1987), 235.
9. J. Labadie, Ph. Gouet and J. Fournaud, *Zentralbl. f. Bakteriol. Hyg.* **164** (1977), 390.
10. J.M. de Kruijf, *Bakteriologische kwaliteit van varkenslever*, PhD Thesis, Utrecht University, (1979), 1.
11. B.M. Mackey and C.M. Derrick, *J. Appl. Bacteriol.* **46** (1979), 355.

12. F.J.M. Smulders, F. Korteknie and C.H.J. Woolthuis, *Int. J. Food Microbiol.* **2** (1985), 169.
13. C.O. Gill, *J. Appl. Bacteriol.* **47** (1979), 367.
14. F.H. Grau, in *Elimination of pathogenic Organisms from Meat and Poultry*, ed. F.J.M. Smulders, Elsevier, Amsterdam, (1987), 221.
15. J.M.A. Snijders, *Hygiene bij het slachten van varkens*, PhD Thesis, Utrecht University, (1976), 1.
16. G.E. Gerats, in *Elimination of Pathogenic Organisms from Meat and Poultry*, ed. F.J.M. Smulders, Elsevier, Amsterdam, (1987), 269.
17. E.G.M. van Klink and F.J.M. Smulders, *J. Food Prot.* **9** (1989), 660.
18. M.E. Stiles and L.K. Ng, *Appl. Environm. Microbiol.* **41** (1981), 867.
19. F.J.M. Smulders and G. Eikelenboom, in *Accelerated Processing of Meat,* ed. A. Romita, C. Valin, A.A. Taylor, Elsevier Applied Science Publ., London, (1987), 79.
20. H. Grätz, *Schlachten und Vermarkten* **76** (1976), 111.
21. O.P. Whelehan, W.R. Hudson and T.A. Roberts, *J. Hyg. Camb.* **96** (1986), 205.
22. F. Stolle, *J. Appl. Bacteriol.* **50** (1981), 235.
23. F.H. Grau, *Arch. Lebensmittelhyg.* **30** (1979), 87.
24. G.E. Gerats, J.M.A. Snijders and J.G. van Logtestijn, *Proc. 27th EMMRW, Vienna*, Egermann Drückereigesellschaft **1** (1981), 198.
25. J. Oosterom, *Studies on the epidemiology of* Campylobacter jejuni, PhD Thesis, Erasmas University Rotterdam, (1985), 1.
26. J.C. de Wit and E.H. Kampelmacher, *Zbl. Bakt. Hug., I Abt. Orig. B* **176** (1981), 390.
27. J.C. de Wit and E.H. Kampelmacher, *Zbl. Bakt. Hyg. I. Abt. Orig. B.* **176** (1982), 553.
28. M.E. Stiles and A.Z. Sheena, *J. Food Prot.* **41** (1987), 867.
29. G.G. Greer and L.E. Jeremiah, *J. Food Prot.* **43** (1980), 277.
30. C.J. Wyatt and V. Guy, *J. Food Prot.* **43** (1980), 385.
31. E. Hess and G. Lott, *Fleischwirtsch.* **50** (1970), 47.
32. M. Catsaras, A.W. Gulistani and D.A.A. Mossel, *Receuil de Medicine Vétérinaire* **150** (1974), 287.
33. F.J.M. Smulders and C.H.J. Woolthuis, *J. Food Prot.* **46** (1983), 1032.
34. A.B. Childers, E.E. Keahey and A.W. Kotula, *J. Am. Vet. Ass.* **171** (1977), 1161.
35. J. Oosterom and S. Notermans, *J. Hyg. Camb.* **91** (1985), 59.
36. B.M. Mackey and T.A. Roberts, in *The Scientific Basis for Harmonising Trade in Red Meat*, ed. J. Hannan, J.D. Collins, Publ. University College Dublin, (1990), 3.
37. F. Tazelaar, in *Elimination of Pathogenic Organisms from Meat and Poultry*, ed. F.J.M. Smulders, Elsevier, Amsterdam, (1987), 251.
38. G.E.C. Gerats, *Werken aan Kwaliteit*, PhD Thesis, Utrecht University, (1990), 1.
39. L. Johanson, B. Underal, K. Grosland, O.P. Whelehan and T.A. Roberts, *Acta Vet. Scand.* **24** (1983), 1.
40. C.J. Thomas, T.A. McMeekin and J.T. Patterson, in *Elimination of Pathogenic Organisms from Meat and Poultry*, ed. F.J.M. Smulders, Elsevier, Amsterdam, (1987), 163.
41. F.J.M. Smulders, P. Barendsen, J.G. van Logtestijn, D.A.A. Mossel and G.M. van der Marel, *J. Food Technol.* **21** (1986), 419.
42. M. Ingram, F.J.H. Ottoway and J.B.M. Coppock, *Chem. Ind.* **42** (1956), 1154.
43. C.O. Gill and K.G. Newton, *Appl. Environm. Microbiol.* **43** (1982), 284.
44. D.D.A. Mossel and A.S. de Bruin, *Ann. Inst. pasteur, Lille* **11** (1960), 65.
45. F.H. Grau, *Appl. Environm. Microbiol.* **42** (1981), 1043.
46. H.E. Rubin, T. Nerad and F. Vaughan, *J. Dairy Sci.* **65** (1982), 197.
47. T. Eklund, *J. Appl. Bacteriol.* **54** (1983), 383.
48. D.J. Lynch and N.N. Potter, *J. Food Prot.* **51** (1988), 475.
49. D.E. Conner, V.N. Scott and D.T. Bernard, *J. Food Prot.* **53** (1990), 652.
50. K.C. CHung and J.M. Goepfert, *J. Food Sci.* **35** (1970), 326.
51. D.A. Corlett and M.H. Brown, in *Microbial Ecology of Foods. I. Factors Affecting Life and Death of Microorganisms*, ed. J.H. Silliker *et al.*, Academic Press, New York, (1980), 92.
52. G.J. Mountney and J. O'Malley, *Poultry Sci.* **44** (1965), 582.
53. G.G. Greer, *J. Food Prot.* **45** (1982), 1043.

54. F.J.M. Smulders and C.H.J. Woolthuis, J. *Food Prot.* **48** (1985), 838.
55. A.S. Levine and C.R. Fellers, *J. Bacteriol.* **39** (1940), 449.
56. P. van Netten, H. van der Zee and D.A.A. Mossel, *J. Appl. Bacteriol.* **57** (1984), 169.
57. C.H.J. Woolthuis and F.J.M. Smulders, *J. Food Prot.* **48** (1985), 832.
58. C.H.J. Woolthuis, D.A.A. Mossel, J.G. van Logtestijn, J.M. de Kruijf and F.J.M. Smulders, *J. Food Prot.* **47** (1984), 220.
59. G.M. van der Marel, J.G. van Logtestijn and D.A.A. Mossel, *Int. J. Food Microbiol.* **6** (1988), 31.
60. A.C. Baird-Parker, in *Microbial Ecology of Foods. I. Factors Affecting Life and Death of Microorganisms*, ed. J.H. Silliker *et al.*, Academic Press, New York, (1980), 126.
61. C.A. Cherrington, M. Hinton, G.R. Pearson and I. Chopra, *J. Appl. Bacteriol.* **70** (1991), 156.
62. S. Doores, in *Antimicrobials in Foods*, ed. A.L. Branen, P.M. Davidson, Marcel Dekker Inc., New York, (1983), 1184.
63. D.R. Hunter and I.H. Segel, *J. Bacteriol.* **113** (1973), 1184.
64. B.J. Maeris, *Appl. Microbiol.* **30** (1975), 503.
65. M.R. Adams and C.J. Hall, *Int. J. Food Microbiol.* **23** (1988), 287.
66. N. Ahamad and E.H. Marth, *J. Food Prot.* **52** (1989), 688.
67. O. Wyss, *Advances in Food Research*, Academic Press, New York, (1948), 373.
68. B. Jarvis and C.S. Burke, *Inhibition and Inactivation of Vegetative Microbes*, ed. F.A. Skinner, W.B. Hugo, Academic Press, new York, (1977), 345.
69. R.K. Prasai, G.R. Acuff, L.M. Lucia, J.B. Morgan, S.G. May and J.W. Savell, *Meat Sci.* 32 (1992), 413.
70. J.S. Dickson, *J. Food Sci.* **56** (1991), 191.
71. J.S. Dickson, *J. Food Sci.* **57** (1992), 297.
72. G.G. Greer, *Meat Focus Int.* May (1993), 207.
73. J.S. Dickson, *J. Food Safety* **12** (1992), 181.
74. J.M.A. Snijders, M.J.G. Schoenmakers, G.E. Gerats and T.W. de Pijper, *Fleischwirtsch.* **59** (1979), 656.
75. J.M.A. Snijders, J.G. van Logtestijn, D.A.A. Mossel and F.J.M. Smulders, *Vet. Quart.* **7** (1985), 277.
76. G.G. Greer and B.D. Dilts, *Food Res. Int.* **25** (1992), 355.
77. J.A. Carpenter, *Proc. Meat Ind. Res. Conf.*, Univ. of Georgia, Athens, Georgia, (1973), 35.
78. H.W. Ockerman, R.J. Borton, V.R. Cahill, N.A. Parett and H.D. Hoffman, *J. Milk Food Technol.* **37** (1974), 203.
79. B.J. Shay, A.F. Egan, D. Miller and T. Ai-Jia, *Proc. ICoMST, Brisbane, Australia*, (1989), 489.
80. T. El-Khateib, A.E. Yousef and H.W. Ockerman, *J. Food Prot.* **56** (1993), 29.
81. T.F. Brocklehurst and B.M. Lund, *J. Appl. Bacteriol.* **69** (1990), 390.
82. P.L. Hamby, J.W. Savell, G.R. Acuff, C. Vanderzant and H.R. Cross, *Meat Sci.* **21** (1987), 1.
83. M.E. Anderson, R.T. Marshall and J.S. Dickson, *J. Food Safety* **12** (1992), 139.
84. R.E. Brackett, Y.-Y. Hao and M.P. Doyle, *J. Food Prot.* **57** (1994), 198.
85. C. Nettles Cutter and G.R. Siragusa, *J. Food Prot.* **57** (1994), 97.
86. H.E. Rubin, *Appl. Environm. Microbiol.* **36** (1978), 623.
87. A.F. Mendonca, R.A. Molins, A.A. Kraft and H.W. Walker, *J. Food Sci.* **54** (1989), 18.
88. E.A. Quartey-Papafio, R.T. Marshall and M.E. Anderson, *J. Food Prot.* **43** (1980), 168.
89. M.F. Bell, R.T. Marshall and M.E. Anderson, *J. Food Prot.* **49** (1986), 207.
90. W. Osthold, H.K. Shin, J. Dresel and L. Leistner, *Fleischwirtsch.* **63** (1983), 603.
91. R. Ozari and H.C.L. Kotter, *Fleischerei* **36** (1985), 887.
92. M.E. Anderson and R.T. Marshall, *J. Food Sci.* **55** (1990), 903.
93. A.N. Surve, A.T. Sherikar, K.N. Bhilegaonkar and U.D. Karkare, *Meat Sci.* **29** (1991), 309.
94. A.A.M. Zeitoun and J.M. Debevere, *Int. J. Food Microbiol.* **11** (1990), 305.
95. M.E. Anderson, H.E. Huff, H.D. Naumann and R.T. Marshall, *J. Food Prot.* **51** (1988), 874.

96. J.S. Dickson, *J. Food Prot.* **53** (1990), 674.
97. J.S. Dickson and M.E. Anderson, *J. Food Prot.* **54** (1991), 514.
98. H.S. Park and E.H. Marth, *J. Milk Food Technol.* **35** (1972), 532.
99. M.E. Anderson and R.T. Marshall, *J. Food Prot.* **52** (1989), 312.
100. E. Lueck, *Antimicrobial Food Additives*, Springer Verlag, Berlin, (1980), 265.
101. B. Dezeure-Wallays and J. van Hoof, *Proc. 27th EMMRW, Vienna*, Egermann Drückereigesellschaft, Vienna, (1980), 316.
102. Z.R. Dixon, G.R. Acuff, L.M. Lucia, Z.R. Vanderzant, J.B. Morgan, S.G. May and J.W. Savell, *J. Food Prot.* **54** (1991), 200.
103. A.M. Matthieu and J. van Hoof, *Fleischwirtsch.* **72** (1992), 912.
104. P. van Netten and D.A.A. Mossel, *Arch. Lebensmittelhyg.* **31** (1980), 190.
105. P. van Netten, M.E.B. Werdler and D.A.A. Mossel, *J. Appl. Bacteriol.* **53** (1982), XII.
106. C.O. Gill and N. Penney, *Meat Sci.* **14** (1985), 43.
107. S.E. Gilliland and M.L. Speck, *J. Milk Food Technol.* **35** (1972), 307.
108. S.E. Gilliland and M.L. Speck, *J. Food Prot.* **40** (1977), 820.
109. C. Daley, M. Lachance, W.E. Sandine and P.R. Elliker, *J. Food Sci.* **38** (1973), 426.
110. K.S. Cudjoe and G. Kapperud, *Act. Vet. Scand.* **32** (1991), 491.
111. N.J. Stern, P.J. Rothenberg and J.M. Stone, *J. Food Prot.* **48** (1985), 606.
112. K.S. Cudjoe, *Aspects of elimination of pathogenic bacteria from meat and poultry by the use of lactic acid with special reference to* Campylobacter jejuni *and* Yersinia enterocolitica, PhD Thesis, Norwegian College of Veterinary Medicine, Oslo, (1990), 1.
113. P. van Netten, J. Huis in 't Veld and D.A.A. Mossel, *J. Appl. Bacteriol.* **76** (1994), 49.
114. G.G. Greer and B.D. Dilts, *Int. J. Food Microbiol.* (1994), (in press).
115. Anonymous, *Meat Focus Int.* November (1993).
116. J.S. Dickson and J.F. Frank, *Food Microbiol.* **10** (1993), 215.
117. M.A. Cacciarelli, W.C. Stringer, M.E. Anderson and H.D. Naumann, *J. Food Prot.* **46** (1983), 231.
118. G.G. Greer and S.D.M. Jones, *Can. Inst. Sci. Technol.* **24** (1991), 161.
119. R.K. Prasai, G.R. Acuff, L.M. Lucia, D.S. Hale, J.W. Savell and J.B. Morgan, *J. Food Prot.* **54** (1991), 868.
120. J.S. Dickson and M.E. Anderson, *J. Food Prot.* **55** (1992), 133.
121. S. Notermans and E.H. Kampelmacher, *Br. Poultry Sci.* **16** (1975), 351.
122. R. Firstenberg-Eden, *J. Food Prot.* **44** (1981), 602.
123. S. Notermans and M. van Schothorst, *Fleischwirtsch.* **2** (1977), 249.
124. S.J. Mulder and B. Krol, *Fleischwirsch.* **9** (1975), 1255.
125. G.R. Acuff, C. Vanderzant, J.W. Savell, D.K. Jones, D.B. Griffin and J.G. Ehlers, *Meat Sci.* **19** (1987), 217.
126. J. Sirami, B. Jacquet, D. Thuault and L. Tissot, *Viandes et Produits Carnés* **9** (1989), 57.
127. M. Gauthier and B. Jacquet, *Viandes et Produits Carnés* **12** (1992), 131.
128. G.R. Siragusa and J.S. Dickson, *J. Food Sci.* **57** (1992), 293.
129. G.R. Siragusa and J.S. Dickson, *J. Food Safety* **13** (1993), 147.
130. Z.R. Dixon, C. Vanderzant, G.R. Acuff, J.W. Savell and D.K. Jones, *Int. J. Food Microbiol.* **5** (1987), 181.
131. A.E. Reynolds and J.A. Carpenter, *J. Anim. Sci.* **38** (1974), 515.
132. H. Labots, H. Logtenberg, F.K. Stekelenburg and J.M.A. Snijders, *Central Institute for Nutrition and Food Research*, Zeist, The Netherlands, Report T83-183/110264, (1983), 1.
133. J. Dresel and L. Leistner, *Mitteilungsblatt der Bundesanstalt für Fleischforschung*, Kulmbach, Number 85, (1984), 6040.
134. G.M. van der Marel, A.W. de Vries, J.G. van Logtestijn and D.A.A. Mossel, *Int. J. Food Sci. Technol.* **24** (1989), 11.
135. R.W.A.W. Mulder, M.C. van der Hulst and N.M. Bolder, *Poultry Sci.* **66** (1987), 1555.
136. I.J.R. Visser, P.A. Koolmees and P.G.H. Bijker, *J. Food Prot.* **51** (1988), 208.
137. F.J.M. Smulders and F. Korteknie, *Int. J. Food Microbiol.* **2** (1985), 293.
138. L.E. Jeremiah, G.G. Greer and L.L. Gibson, *J. Muscle Foods* **2** (1991), 119.
139. P. Dorn, P. Krabisch and H. Gehra, *Arch. f. Geflügelk.* **53** (1990), 123.
140. G. Offer and J. Trinick, *Meat Sci.* **8** (1983), 245.
141. C. Bailey, *Inst. Meat Bull.*, Number 75, (1972), 3.
142. M.G. Smith and A. Graham, *Meat Sci.* **2** (1978) 119.

143. A. Graham, B.P. Cain and I.G. Eustace (1978), *CSIRO Meat Research Report*, Number 11/78, CSIRO Meat Laboratory, Division of Food Research, (1978), 1.
144. K.R. Davey, *Int. J. Food Sci. Technol.* **24** (1989), 291.
145. K.R. Davey and M.G. Smith, *Int. J. Food Sci. Technol.* **24** (1989), 305.
146. C.A. Kelly, J.F. Dempster and A.J. McLoughlin, *J. Appl. Bacteriol.* **51** (1981), 415.
147. A.W. Kotula, W.R. Lusby, J.D. Crouse and B. de Vries, *J. Anim. Sci.* **39** (1974), 674.
148. M.E. Anderson, R.T. Marshall, H.D. Naumann and W.C. Stringer, *J. Food Sci.* **40** (1975), 1232.
149. G.C. Smith, W.L. Varnadore, Z.L. Carpenter and M.C. Calhoun, *J. Anim. Sci.* **42** (1976), 1167.
150. A.G. Dezuniga, M.E. Anderson, R.T. Marshall and E.L. Ianotti, *J. Food Prot.* **54** (1991), 256.
151. M.E. Anderson, R.T. Marshall, W.C. Stringer and H.D. Naumann, *J. Food Prot.* **44** (1981), 35.
152. H. Logtenberg, *Central Institute for Nutrition and Food Research*, Zeist, The Netherlands, Report T *2.002/110264, (1981), 1.
153. F-K. Lücke, in *New Technologies for Meat and Meat Products*, ed. F.J.M. Smulders *et al.*, ECCEAMST, Audet Tijdschriften, Nijmegen, The Netherlands, (1992), 37.
154. H.S. Murali, R.K. Leela, R. Sankaran and T.R. Sharma, *Chem. Mikrobiol. Technol. Lebensmittelhyg.* **9** (1985), 19.
155. P.S. Elias, in *Elimination of Pathogenic Organisms from Meat and Poultry*, ed. F.J.M. Smulders, Elsevier, Amsterdam, (1987), 345.
156. F.J.M. Smulders, in *Elimination of Pathogenic Organisms from Meat and Poultry*, ed. F.J.M. Smulders, Elsevier, Amsterdam, (1987), 319.
157. ECCEAMST (European Consortium for Continuing Education in Advanced Meat Science and Technology), *The use of additives in Meat Products throughout Europe*, tentative Report, Bundesanstalt für Fleischforschung, Kulmbach, Germany, (1994), 1.
158. FSIS, *Directive Number 6340.1-11/24/92*, United States Department of Agriculture, Food Safety and Inspection Service, Washington, DC, (1992), 1.
159. Agriculture Canada, *Guidelines for submitting partial quality control programs for organic acid used on meat and poultry during slaughter operations*, Food Production and Inspection Branch, Ottawa, Ontario, (1993), 1.

13 Advances and potential for aseptic processing

D. ROSE

13.1 Aseptic technology

Business frameworks based on aseptic technology continue to be of interest to the food processor for the production of ambient shelf-stable foods. UHT milk, soon followed by pasteurised fruit juices, were among the first commercially successful products to be sold in modern flexible and semi-rigid packaging whose barrier materials may comprise plastics, paperboard and aluminium foil. It is in the continued use of these and their new applications for other foods to which the main advantages relate in terms of economy, convenience and lightness. This is especially so when compared with the packaging alternatives such as metal cans or glass jars.

More recent additions to the range of low-acid products (i.e. pH 4.5 or more for any component of the food) which are or have been commercially available include homogeneous and particle-containing desserts, and several brands of soups, sauces and ready meals whose varieties include some with particles of meat, fish or vegetables.

There are in existence around 38 types of commercially available aseptic filling and packaging machines for retail pack sizes and perhaps twelve for bulk systems such as bag-in-box or bag-in-drum. In retail pack sizes the containers are commonly metal or laminate cans, plastic pots, laminate cartons, sachets and plastic or glass bottles or jars. The mode of production operations may, however, differ widely in the formation of what otherwise seem to be similar aseptic packages. Cursory examinations by consumers could fail to differentiate in some cases between ambient shelf-stable aseptic packages and microbiologically limited shelf life products distributed through chill chains.

The basic principles of aseptic transfer of sterilised materials into sterile containers have been understood and used by microbiologists since the era of Louis Pasteur. Developments with foods, however, have almost exclusively concerned those which are pumpable, although a few batch systems such as the APV 'Jupiter' have been developed for the processing of discrete particles in steam [1] as have other unexploited systems such as Steriglen [2] or sterilisation by deep frying [3].

Characteristically the food is subjected to rapid heating in a heat exchanger, held for sufficient time to achieve commercial sterility and then cooled before filling into a decontaminated container within an aseptic filling zone. It is critical that all food contact surfaces are sterilised before production starts and that any other surfaces in proximity to the food should be sterilised too, in order to avoid reinfection. Avoidance of product reinfection is usually achieved by maintaining an over-pressure within the processing, storage and filling system. According to Bean [4] microbiologists' concern with aseptic packaging has principally been with spore heat resistance at ultra high temperatures and with the methods used to sterilise empty containers.

The first commercially acceptable aseptic canning system produced by Martin and Dole in the 1940s used heat as the decontaminating method. This provided the means to aseptically fill pre-sterilised, cooled food firstly into metal and then later into composite cans. Some energy advantages from the use of the James Dole system were presented as recently as 1979 [5]. Burton [6] referred to vast amounts of literature on ultra-high-temperature (UHT) processes and cited UHT milk processing plant mentioned by Orla-Jensen in 1913, and a report by Procter in 1951 that aseptically canned milk was shipped from South Africa to London in 1923. Burton specifically described the European meaning of UHT as defining continuous flow heat treatment of at least 130°C with a holding time of 1–2 s or more to achieve a state of commercial sterility. This he contrasted strongly with the North American view that UHT denoted any heat treatment which was at a temperature higher than conventional pasteurisation, although he observed that there was an increasing tendency in the United States and Canada to adopt the European view.

13.2 Regulatory effects

Outstanding differences are apparent between the activities of European food processors (including those in ownership by United States companies) and food processors within the United States as the direct result of impositions by regulatory authorities. UK and European Union laws require that foods shall be safe and wholesome for the consumer, but leave the safety issues to self-regulation by the processors. This contrasts sharply with the heavily legislated procedures insisted upon by the United States Department of Agriculture and the Food and Drug Administration: an example of their attitude towards canners of low-acid foods was expressed in the mandatory provisions for heat process filing requirements [7]. As a result of those requirements, 22 000 forms were received representing 80 000 filed processes each of which had to be recorded and inspected. Only by 1984 had the National Food Processors Association achieved important changes in the USDA's guidelines document for meat and poultry processors concerning the addition of aseptic processing systems to their plants. This amounted to USDA recognition that 'shelf stable' and 'commercially sterile' were synonymous [8]. Johnston and Terry [9] revealed that chemical sterilant packaging systems could not be used at one period for commercial processing and aseptic packaging of foods. Furthermore there had been an arousal of major concern by FDA resulting from the perceived ignorance shown by some processors of basic principles of thermal processing, and the abuse of mechanical systems that were the original reasons for the formulation of the Good Manufacturing Practice regulation published in 1973. By 1981, Davis and Dignan [10] reported a final rule published in the Federal Register which then permitted the safe use of hydrogen peroxide as a sterilising agent for polyethylene surfaces of containers used in contact with food. This paper introduced the appropriate documentation for processors to complete and points to extant difficulties on how to distinguish between commonly refrigerated products where confusion could have arisen as the result of unclear terminology. Between 1978 and 1990, the numbers of USA companies packaging low-acid foods increased from 13 to 28. All of these products, however, were homogeneous dairy based foods. Regulatory avoidance was recognised by Johnson [11] who observed that many single service creamers purveyed through the hotel, restaurant and institutional ('HRI') trade used Class I milk and were products intended to be sold refrigerated, but that commercial kitchens were aware of their long shelf life and tended to ignore the instructions to store under refrigeration.

Bernard [12] confirmed how the general failure by industry to obtain regulatory clearance for aseptically packaged low-acid foods containing particles of significant size had exclusively relegated the application of the technology in the USA to homogeneous products or those containing relatively small particles, such as rice pudding, tapioca pudding and some dips and sauces. He pointed out in contrast the opportunities for marketing soups and stews in user-friendly packaging through the refrigerated case, with relative ease from the regulatory standpoint. Here, thermal processes could be designed to inactivate pathogens whilst not achieving commercial sterility and less rigorous standards need apply to the control of microorganisms on the packaging material and in the filling zone as well.

13.3 Aspects of good manufacturing practice

13.3.1 Scheduled processes

The process flow diagram (Figure 13.1) shows some generic processing and aseptic packaging alternatives. It depicts separate sterilisation of food and decontamination of packaging materials, and the stages of aseptic fill and seal. Whichever equipment is used, the requirement of production is to make commercially sterile products through the achievement of scheduled

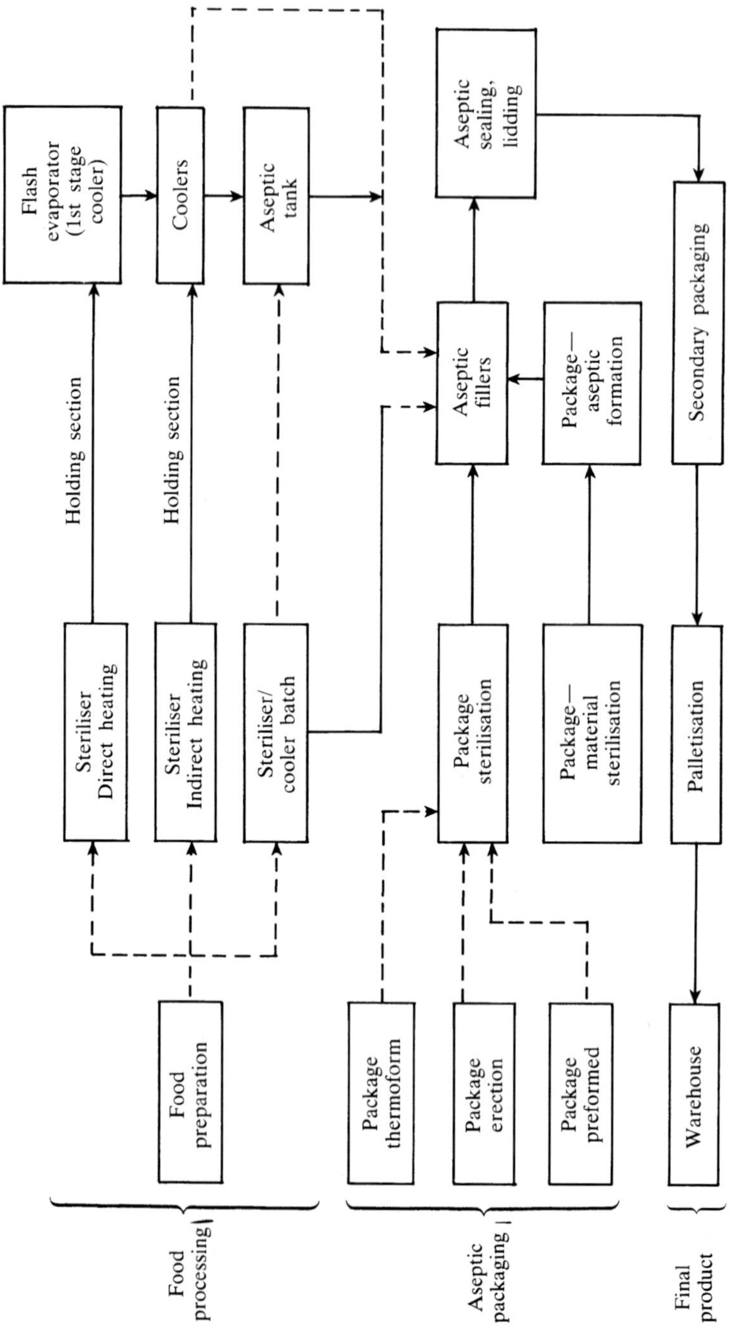

Figure 13.1 Aseptic flow diagram.

processes. These are set by a processor in respect of:

- Preproduction sterilisation of food contact surfaces.
- Heat processing of foods.
- Microbiological decontamination of non-food contact surfaces, e.g. in some designs of filling machines.
- Microbiological decontamination of packaging.
- Aseptic filling environments.

Also of major importance is the avoidance of reinfection of product at any stage within the production plant, or of packages in storage and distribution. The major potential hazard to aseptically packaged low-acid foods for control purposes is the food poisoning bacterium *Clostridium botulinum*. In this respect the hazard is not different to in-pack sterilised (canned) foods, although the risks are often different. For comparison, the hazard for acid products is spoilage, whether caused by acidophilic bacteria, moulds or yeasts. Cerf [13] said that non-sterility might be linked to the weakest part of an operation.

Commercial sterility is a condition achieved when food, equipment or packaging is rendered free from microorganisms capable of growing in the food at temperatures at which it is likely to be held during manufacture, distribution and storage. The heat process which is chosen by a processor for food is expressed as a time–temperature relationship. For a low-acid food, this should be sufficient to reduce the probability of survival of a spore of *Clostridium botulinum* to not more than 10^{-12}, although in practice this may have to be exceeded in order to control spoilage organisms of greater heat resistance than *C. botulinum*. The concern expressed by Bean [8] for the correctness or otherwise of assuming exponential heat resistance of *C. botulinum* was widely shared by the UK food industry, the Ministry of Agriculture, Fisheries and Food and the Department of Health [14]. In particular, worst case assumptions were thought to be necessary to ensure safety against *C. botulinum* by assuming the z value to be higher than the classic value of 10 Celsius degrees. This was mainly because of work by Neaves and Jarvis [15] which had indicated a continually increasing z with higher temperatures. Brown *et al.* [16] subsequently reported the construction of a computer controlled thermoresistometer for the determination of heat resistance of bacterial spores over the temperature range 100–150°C, and then Brown and Gaze [17] produced data which showed the z value of *C. botulinum* strain 213B spores to be constant within the range of 120–140°C. This provided support for the recommended classical process calculations being adequate if extrapolated to 140°C, when using a z value of 10 Celsius degrees.

During production, the safety of food products is established within a Quality Assurance programme that depends much more on instrument outputs used for on-line process control, rather than any retrospective

Table 13.1 Protection of product integrity

Area	Critical factors
1. Packaging	
Compatibility	Food must not cause breakdown of packaging materials
Package structure	Freedom from defects
Sealing method	Must form a microbiological barrier
2. Storage	
Identification	Product referable to records
Protection	Adequate for handling, stacking
3. Transportation	Package integrity to be maintained by correct handling

analysis of end product. It is most important that correct handling procedures should be established for packages, extending throughout the whole distribution chain, and which take into account any package susceptibility to abuse (Table 13.1). The guidance given in this area by a processor to those outside his direct control may be significant under the laws on public liability.

Apart, however, from the requirements of the UK Food Safety Act 1990, the only legislation relating to UHT processes are those of The Milk (Special Designation) Regulations 1989 and Statutory Instruments Nos. 1508 (1983), 1509 (1983), and 734 (1959), which refer to milk, cream milk-based drinks and ice cream respectively. Quite clearly, these minimal processes cannot be expected to be transferred safely to other low-acid foods.

There has been a rapid increase in the application of Hazard Analysis and Critical Control Point by the food industry in general over the past ten years. Food and Agriculture Organisation/World Health Organisation Codex Committee on Food Hygiene [18] confirmed the position of the HACCP technique as a major element in safe processing of foods and that its application within food processing establishments was essential. Baird-Parker and Mayes [19] criticised most current Codes of Practice for not providing users with the guidance on the relative importance of procedures and practices for control and hence the overall safety of the product. They concluded that HACCP throughout the food industry would bring food poisoning risk under better control. Rose [20] described the role of HACCP in aseptic Good Manufacturing Practice, concluding that there was an expectation of its increased usage in aseptic technology whilst cautioning that it was not a fool-proof system. Recommendations on the HACCP method have undergone subtle changes since their first introduction, but guidance on the most up to date nuances is available [21]. Software is available from this reference-source for use in documentation and training.

13.4 GMP guidelines

Campden Food and Drink Research Association has published Good Manufacturing Practice Guidelines for low acid foods produced by aseptic technology [11]. Clearly the approaches enunciated will also be useful to producers of acid products even though process temperatures may be lower. The basis of these documents is in the identification of those factors which determine the safety of food products, in order that the right questions should be asked and the necessary knowledge developed when preparing to use the technology. Test methods which originate from information that is generally available are included in a second part of the foregoing Campden Technical Manual. Their value was increased by another document for use at the stages of design and commissioning, and also throughout production [22]. It is fundamentally important that theoretical aspects of the technology should be reconciled by data acquired by practical tests (process validations), culminating in tests on the functional efficiency of the whole aseptic system before the commercial release of any product.

Other less comprehensive guideline documents that deal with aseptic technology include those of the US Food and Drug Administration [23], UK Department of Health and Social Security *et al.* [24], United States Department of Agriculture [25], Institute of Food Science and Technology of the UK [26] and Department of Health [27]. The requirements of all these confirm the responsibilities of food processors especially when they use the technology to make high-risk products. The standards to which aseptically packed low-acid foods are expected to perform are no lower than those which are applied to in-container heat sterilised foods in respect of microbiological hazards to public health and rates of pack spoilage. The risk factors are very different however, and these are specifically affected by the complexity of the technology.

The European Hygienic Equipment Design Group (EHEDG) [28] has provided guidance for microbiologically safe aseptic packaging of food products. It lays emphasis on the accumulated total of contaminant micro-organisms from numerous origins and their effects on the sum total of packs which spoil as a result. EHEDG [29] also defined the requirements of all components which comprise heat processing systems in terms of their suitability to perform safely, and identified the measures necessary to deal with deviations in key process parameters.

13.5 Design and development

A systematic approach to stages in the development of a new aseptic system is presented in Figure 13.2 [30]. This section identifies some critical factors associated with scheduled processes.

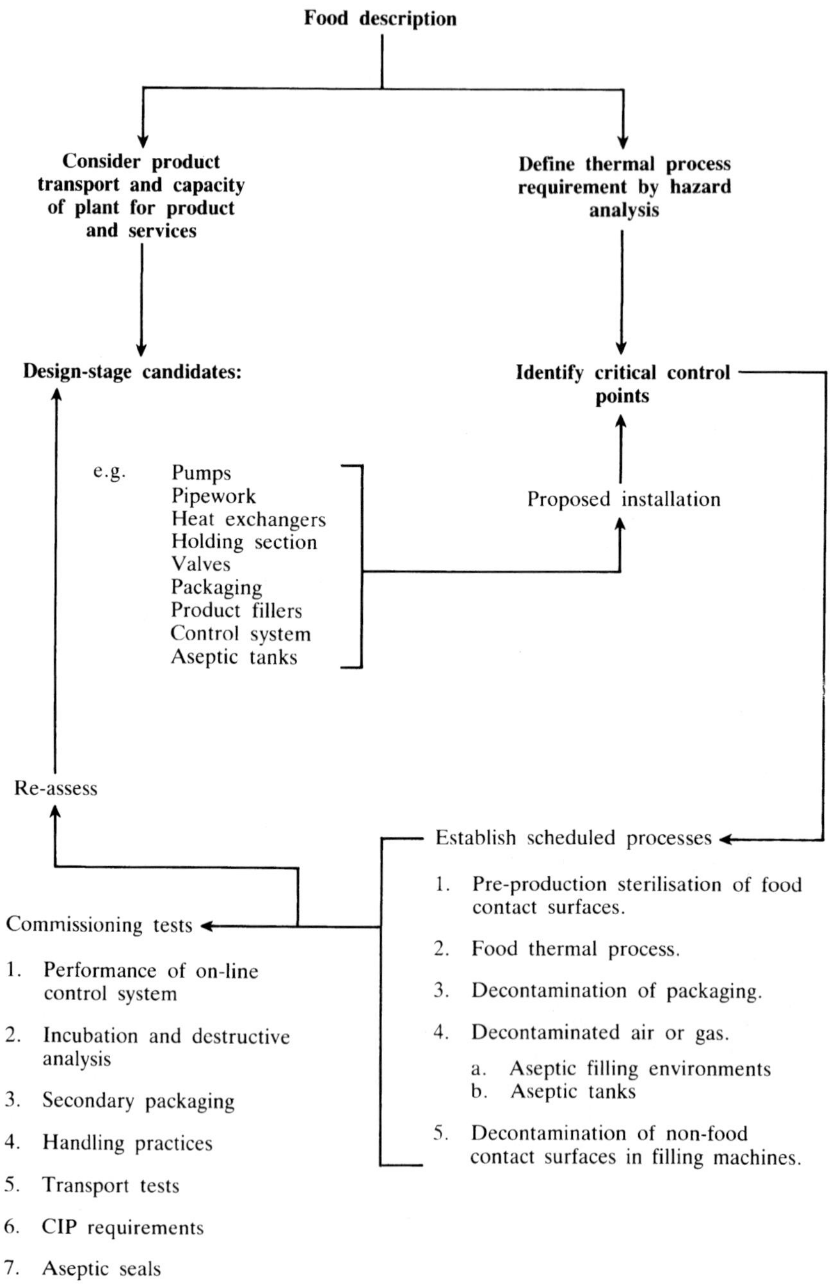

Figure 13.2 Stages in system development.

13.5.1 Food contact surfaces

Critical factors relating to preproduction sterilisation of food contact surfaces include correct assembly of equipment, and assurance that it is clean, and without crevices or dead-legs which are inaccessible for cleaning or heat sterilisation purposes. It should be possible to maintain sterility by means of sterile barriers using, for example, steam or sterile air as appropriate. It must be confirmed that all parts reach the required temperature for the required time, and achievement of this is established by a control system with reproducible accuracy. Process steam used in these applications must be generated within a boiler in which no hazardous chemicals have been used for water treatment.

13.5.2 Food process

Because the establishment of a thermal process resolves to a time–temperature requirement, any calculations must include flow properties and heat transfer coefficients which may be multiphase. This subject is well documented by Holdsworth [31] for whom it held special interest for three decades. Even with the electrical resistance heating method, some unevenness of temperature distribution may be anticipated to occur due to varying effects on different tissues, e.g. fat and lean meat.

Critical factors of the liquid phases which may wholly or partly comprise foods (including pastes) are density, rheological properties, viscosity; a_w, pH, thermal properties, specific heat, gas in fluid, and microbial loading, in addition to, thermal diffusivity, composition, shrinkage or swelling of discrete food particles, when present. Mixing techniques in product make-up should be sufficient to avoid aggregation of powders or gums in such a way that forms lumps because these carry the risks of internally dry environments which may protect microorganisms from the heat process as well as impairing sensory quality.

Much effort has been directed at methods to validate the temperature-history of particulates in situations of dynamic flow. Sastry [32] described ways to determine heat transfer coefficients from particles held stationary in a flowing fluid stream, and by using a moving thermocouple to monitor a particle moving with a flowing stream. Inevitably these methods created their own difficulties and limitations. He described other temperature indicators but in particular the biological method ascribed to pioneer work by Heppel such as given in [33], which was based on the use of test microorganisms immobilised in alginate beads. Earlier work by Dallyn, Falloon and Bean [34] had introduced this method under practical conditions for small alginate beads, and significant improvements in technique by Brown, Ayres, Gaze and Newman [35] included manufacture and testing of bigger particles (0.8–2.4 cm) which contained vegetable or meat puree. This conferred the properties of analogue food particles which after heat processing are suitable for analysis and which gave data close to the theoretical predictions. Results of such challenge tests on commercially available products were given by Skudder [36] and they demonstrated the high level of microbiological security of the ohmic heat process and the narrow range of lethality from the surface to the centre of the particulates.

In continuous-flow sterilisation, the additional factors which are critical to the scheduled process are control of feed rate, temperature sensing and control, pressure indication and control, avoidance of channelling by non-sterile products (especially through the holding section), and absence of microbiologically contaminating leaks particularly in heat exchangers used for product cooling. Plant design should be such that it prevents entrapment of air or gas, especially in the holding section. Residence time distribution of fluid elements of homogeneous liquid products in holding sections must be characterised for safe process design or alternatively, so that suitable safety margins are included, by assuming that laminar flow occurs and that the holding efficiency is 50% of the average residence time. When processing of Newtonian products is carried out at less than 100°C, it is normal to use a simple calculation for Reynolds number using rheology data from bench viscometers. Such data can cause uncertainty when attempting to measure batch-wise the viscosity at temperatures and pressure within the UHT band. Such very difficult problems can be avoided according to Tucker [37], who highlighted tube-viscometry as the means by which flow conditions may be monitored on-line by taking inlet and outlet pressure measurements and the product flow rate. When discrete particles form part of a food then the residence time of the fastest particle in the holding section is of critical importance according to Singh and Lee [38] who gave the measurement methods for this as visual observation, photography, playback video taping, radio-active tracer and magnetic response. Tucker [37] cited magnetic detection as the most promising technique for refining into a method for establishing residence time distribution of particles in holding sections. This would constitute a very important test for use in process development studies. Magnets are embedded into particles and incorporated with the food pumped through continuous flow heat exchange systems. Optional detection systems located at each end of the holding section to sense the magnetic fields are copper coils or Hall-effect sensors, which use semi-conductors. Tucker and Richardson [39] reported the successful use of such Hall effect sensors at room temperature in starch based fluids and the plans for testing in heated systems in conjunction with a tube viscometer, thus yielding options for modelling holding sections based on improved knowledge.

In batch sterilisation, the bulk mixing pattern and aseptic transfer of product may comprise additional critical factors to those of temperature sensing and control, pressure indication and control, and absence of microbiologically contaminating leaks. Hersom and Shore [40] described a process validation procedure involving insertion of capillary tubes containing 10^6 spores of *Bacillus stearothermophilus* within sealed cavities in carrot particles. Manvell [1] also reported on the heat process of the APV Jupiter batch sterilisation system, and the calculation of F_0 values by studies of heat penetration and spore reduction.

13.5.3 Non-food contact surfaces

The decontamination of some non-product contact surfaces is essential in avoiding product reinfection. This may relate to headspace areas of aseptic tanks and balance tanks or to the internal surfaces of aseptic filling machines. In the former case, tanks are usually sterilised in their entirety during the preproduction sterilisation and thereafter held in the aseptic condition by continuous maintenance of internal pressure by means of sterile air or gas.

For aseptic filling machines, individual solutions usually need to be found according to machine design. Decontamination methods such as chemical application with self-cleaning by sterile air should be evaluated for practical effect.

13.5.4 Decontamination of packaging

The design stage should establish the performance of the decontamination system, resulting in a proposed control system and specification relating to packaging material decontamination. During commissioning, some spore challenge tests may be needed to check conformance with specification.

It is always important to assess the compatibility between packaging materials and the method of decontamination employed.

Temperatures and humidities at which packaging is stored and handled must be controlled within acceptable limits, and safeguards made to prevent avoidance increases over base levels of microbial contamination, or, in the case of presterilised packaging, to prevent them from becoming contaminated in such a way that might result in product infection.

The critical factors which are involved depend on the decontaminating agencies employed, e.g.:

- *Heat* – moist or dry, temperature, pressure, time, air venting.
- *Chemical* – type, concentration, quantity and application, temperature, coverage, time.
- *UV* – waveband frequency, output, age of lamp, distance, time of exposure, temperature.
- *Ionising radiation and electron beam* – dose, protection from reinfection.

Hydrogen peroxide is generally considered to be an elegant chemical means for decontaminating packaging materials and containers, or for preproduction decontamination both of food-contact and also non food-contact surfaces of some aseptic packaging systems for food. The main reasons for this are its effectiveness as a biocide and its rapid decomposition into oxygen and water during activation and evaporative removal. This minimises any carryover of residual hydrogen peroxide into the product providing that systems in which it is used are correctly designed, installed and operated. The United States Food and Drug Administration does not permit residual concentrations of hydrogen peroxide in excess of 1.0 part per million when distilled water is packaged in the system. The analysis for this must be immediate, as decomposition occurs on storage.

When bacterial spores are used for testing decontamination efficiency, they should be characterised before use. Variations in resistance may be due to the conditions under which spores are produced and indeed variations between batches are sometimes found from otherwise standardised and controlled production procedures. *Bacillus subtilis* (NCA 72-52) has often been used as a challenge bacterium and work by Leaper [41] demonstrated not only the effects of temperature and concentration of hydrogen peroxide on the decimal reduction (*D*) value of *B. subtilis*, but also the powerful synergistic effect of low concentrations of peracetic acid.

13.5.5 Aseptic filling zone

Gaseous aseptic filling environments frequently depend on filtration for sterility, and in such cases it is usual for the machines to operate to the filter requirements that conform to Class 1 in the UK (Bs 5295). In the event of using gas incinerators or heated filling environments, then conditions also need to be established. In machines using such agencies, overpressure, air or gas velocity and direction should encourage removal of chance contaminants from the system, the design of which will prevent deposition of contaminants within an unclosed pack. Attention should be given to any recycling mechanisms which carry non-sterile materials into the system or aseptic packs away from it. The nature of high efficiency particulate air filters should be carefully considered in respect of leak-free installation and external factors such as oil or dust which may affect performance.

Filling mechanisms must be compatible with the product, and dispensing controlled in a way that yields satisfactory fill weights without soiling other surfaces if that introduces other risks. Special conditions may apply in the procedures for start-up, stand-by, filling or shutdown, e.g. in the application of heat, chemicals or supply of decontaminated air or gas.

Bell [42] criticised air standards which failed to put a bacterial dimension to a classification. He stated that the mechanism of bacterial contamination during aseptic fills is predominantly by sedimentation exacerbated by attachment of bacteria to larger particles which aggravate the net gravity effect, especially where turbulent conditions occur. Taking counts from settle plates he then gave a simple equation that would calculate the probability of contamination. Clearly not all types of aseptic filling machines have either the design or the means for inserting Petri dishes into the aseptic zone. Rose [43] showed that anemometry, smoke markers, thermometry, the use of airborne particle counters, bacteriological air samplers and tests on overall system performance are complementary to process design and the successful commissioning of plant. He considered the use of particle counters to be advantageous because the speed in analysis can provide greater assurance of probability of freedom from airborne microorganisms. Viable counting can provide data useful in risk analysis but raises problems in interpreting results that can arise through technical error. Nevertheless it is the only option in machines which depend on air sterilisation by incineration or in some designs where airflow from high pressure filters (as opposed to HEPA) is low and non-organic entrainment of particulates is high.

13.6 Commissioning tests

Reference to Table 13.1 shows tests typically carried out at the commissioning stage, often in connection with the establishment of scheduled processes and perhaps with reassessment of plant components. It is an important time for personnel training, establishment of plant records, detailing maintenance needs, and working out procedures to enact in the event of system failure. Initial checks, sometimes known as M & E (Mechanical and Electrical) or water trials, are the first stage in the commissioning. Because delays, from whatever cause, are likely to have occurred there will be considerable pressure if deadlines are to be met. This means that it is important to have both clear, well thought out plans for each of the stages of commissioning, and clear accountabilities for those who are involved, including the supplier of the equipment. In practice, it is convenient to consider the installation in two parts, the Process or Product Line and the Filler but not overlooking the requirement for correct interlinking of process control units that are necessary to synchronise the appropriate operational sequences.

Performance assessment investigates the microbiological status of:

- Basic machine function
- Production function
- Cleaning

This assessment requires the aseptic plant to be able to produce and maintain a microbiologically stable product [22].

13.7 Manufacturing directive

Production sequences should only be initiated following the issue of a comprehensive checklist. The all-day and every-day control functions result from on-line process monitoring, but auditing is important as a periodic assessment that safe and wholesome food is made. Instrument maintenance and calibration programmes are essential to ensure their accuracy and reproducibility. Procedures should be established to safeguard the commercial sterility of product during shutdown of production, whether this be scheduled or the risk of gradual or sudden failure within the plant. Routine inspection should be made of product and other manufacturing raw materials, and package seal strengths checked on a regular basis. All products must be referable to records which should be held indefinitely (Figure 13.2).

Zero product contamination is always the aim, but the manufacturer will have to choose an acceptable spoilage level at which action will be taken. It follows that decisions involving the release for sale of batches with more than zero infected containers will have two essential requirements. The first is that there is no chance that the contaminants detected can be of any public health significance. The second is that any level of contamination must be very low.

Any sampling plan trying to detect very low levels of contamination will inevitably require a high number of sample units. As an acceptable contamination level in this case is always below 1 in 1000 packs, it is certainly the case that such sampling plans are not going to be applied routinely because of the prohibitive high number of sample units needing to be examined to be confident of identifying an unacceptable batch. For example, with a critical defect level as high as 1 in 1000, at least 3000 packs must be examined to be 95% sure of identifying such a batch as unacceptable.

Sampling will therefore not be used as an acceptance criterion for routine production of aseptic packs; however, sample failure will be clear evidence of a problem [22].

13.8 Economics and market trends

Since the volumetric capacities of packages may range from as little as 6 ml to several tonnes, then clearly the markets they serve with commodities or recipe products are diverse in their business activities. For example they include retailers, caterers and reprocessing factories. Not surprisingly the interactions between technology-led manufacturing enterprises and their potential customers may not be expected to produce immediately clear or predictable outcomes in the marketplace. The complex factors involved often can be examined only in a speculative way, as very little data has been published that can give firm guidance to marketplace responses.

An examination of significance in market trends can only be realistic when some understanding of the underlying economics has been revealed. This can be done by cost modelling and comparing the results with other relevant unit operations if necessary.

Dumbleton [44] stated that functional and convenience factors play an important role in the acceptance of any package, but he emphasised that when the cost is sufficiently low then it will gain market acceptance. His analysis covered single service (6–8 oz, 170–225 g), moderate size (850 ml) and family size containers. In all cases plastic or plastic/paperboard/foil packages were cheaper than glass bottles or cans in respect of the costs that included packaging materials, royalties, filling and packaging waste.

Rose [45] produced elementary cost comparisons between a product canned or aseptically packed into laminate cartons. His form of analysis indicates how and where cost benefits or penalties accrue and revealed that aseptic technology is:

- capital intensive,
- automated with monitoring and control systems that minimise labour costs,
- occasionally energy efficient,
- offering competitively priced packaging materials.

The rationale explains why high volume commodities such as aseptically packaged fruit juices have developed strongly in the market place over the recent decade. It omits, however, the inclusion of any development costs which can be substantial especially when foods are of the non-dairy low-acid type which precludes any turn-key operations. For example up to £10m was reported [46] to be invested by one company for the launch of their range of soups.

Underlying this, however, is an inflexible technology dependent on purpose designed systems. The implications are that line-capacity will be maximised by processing the sorts of foods for which there are long, sustained demands. Where competition is heavy or new market sizes uncertain then utilisation of plant capacity may be maximised with a product range that, in food process engineering terms, amount to the same thing, even if flavours differ.

13.8.1 Bulk packaging

Aseptic technology has the inherent capability to pack products into almost any size of package without altering the thermal process of the product necessary to achieve commercial sterility.

Initiatives for bulk aseptic packaging can be ascribed to the following commercial pressures [47].

- Transforming bulk surpluses of food into stable form without refrigeration.
- The need for farmers to grow more profitable crops.
- A need to produce exotic fruit products or blend in order to win a greater market share.
- A means to develop exports.

The impact of bulk aseptic packaging developments is continually reflected in altered practices by their users. It would be inconceivable now that the large re-processing factories who buy tomato paste in consignments of hundreds or thousands of tonnes would revert to deliveries in 5 kg cans. The 'new' product in e.g. 220 kg bags offers cost advantages right the way through from storage, transport and utilisation to reduced disposal problems of used packaging. Increasing amounts of fruit pulp and juices, e.g. raspberry, pear and apple, are processed and stored for export or out of season reprocessing. In the latter case this cannot only be for reconstitution and reprocessing into retail size packs of juice or juice drinks but for commercial alcoholic fermentations to make cider.

Similarly, increasing amounts of wine from grape juice concentrate are unseasonally made away from the grape producing areas. The use of bulk bags for institutionalised catering has long been a serviced market, for example, UHT milk and flavour milk to chain-restaurants and hospitals, or cream supplied to bakeries.

13.8.2 Commodity, added value or niche product?

National trends may perhaps reflect different habits, tastes and preferences. They can also relate to factors such as climate, efficiency in food distribution chains, usage of refillable bottles, and ownership of refrigerators, freezers or microwave ovens. In aseptic cartons, the overall split in product groups were milk and dairy products (39%), juices (54%) and soup/sauces/tomato products (7%) according to Nielander [48]. Most of the new product launches in retail packs have been recorded by Campden Food Products Intelligence Centre and Bond [49] summarised the considerable extent of new product activity in liquid products.

(a) Milk and cream. In the UK around 2.5% of all milk is UHT proces-

sed. This is far less than in Germany or France where around 50–75% is so treated, but in sharp contrast to the position in the USA or Scandinavia where there is hardly any exploitation.

Skimmed milk retains good flavour after UHT processing and is a commodity showing some growth in the UK, perhaps assisted by dietary concerns. Long life dairy cream is a highly successful product with 35% of all production being aseptically packaged in the UK. An outstanding example of success in a packaging-led niche is aerosol dairy cream. Here, convenience and novelty appear to be dominant factors that enabled market growth of value added products which are distributed through the chill system to protect the properties of the emulsion. Such storage considerations are also important to other foods like non-dairy creams. Of recent interest in UK are the launches of baby milk food, heavily branded milk-based drinks and even flavoured milk drinks for dogs and cats.

Aseptic polyethylene bottles of UHT milk have a high market profile in France and Spain. Renewed efforts are being made to use these containers in order to bring French milk into UK supermarkets [50] with a brand aimed to compete directly against fresh pasteurised milk.

(b) Juices and nectars. Throughout the European Union the indications are that growth continues in the consumption of juices, juice-based drinks and nectars albeit at a reduced rate during periods of recession. In the UK consumption now amounts to 13 litres of juice each year per person whilst in Germany that figure is probably around 45 litres although a higher proportion of this will comprise 'nectars' whose composition must conform (in the UK) to the requirements of Statutory Instrument No. 1284, The Fruit Juices and Fruit Nectars (England, Wales and Scotland) (Amendment) Regulations 1991. In the UK market 68% of product is in cartons, 16% in glass bottles, 12% in plastic bottles, 3% bag-in-box and 1% in cans. The product mix is orange 70%, apple 12%, grapefruit 8%, pineapple 6% and others 4% [51].

Of these pack forms it can be regarded that most cartons and bag-in-box will be aseptic, and that glass, cans and most of the plastic will be hot-filled. There is, however, a steadily growing market for short shelf-life fresh squeezed orange juice which is regarded as a premium product with prices to match, albeit that to date it has not been subject to the burden of value added tax in the UK unlike other heat processed products that are generally reconstituted by the addition of water to concentrated juice imported in either bulk aseptic or frozen states.

(c) Recipe products. Aseptically packaged custard and creamed rice has been firmly established in the market place for many years. The packaging may comprise cans, cartons, plastic pots or bag in box for custard; or cartons, plastic pots or bag in box for rice. At least six brands of cook-in

and pour-over sauces enjoyed brief popularity in aseptic cartons but the market later became dominated by aseptic and in-pack sterilised sauces packed in glass containers. Soups account for 11% of the UK canned food market, but launches of four brands packed in aseptic cartons around 1986 failed to take satisfactory market share although the installed capacity could have produced up to 20% of the ambient shelf stable sector. Marketing efforts redirected at France, however, were entirely successful and a popularly held view was that substitution by consumers in that country was made more easy since there was not a well established market for soup in metal cans. Even more technologically spectacular achievements have led to launches of densely particulate low-acid foods such as ready meals in trays and dog food in cartons [52]. These were later reported as being product lines that were set up to test market conditions and for which further exploitation is not yet apparent. More recently a solids-containing baby food was launched in France within a foil lidded tray further protected by a thermoformed over-lid [53].

13.9 Conclusions

Although aseptic canning established the commercial feasibility of aseptic technology, the fuller potential began to be realised from the early 1960s. This was due to the use of newer packaging materials that utilised paperboard, aluminium foil, plastics and metallised films in various combinations. The development of markets with novel combinations of product and package took place through the persuasion of consumers that the perceived advantages of the products were enough to justify their substitution in place of those already existing.

Few aseptically packaged products can truly be described as tasting 'fresh' and this is not too surprising considering that either the raw materials may have been preprocessed or that some element of 'cooked' flavour is caused in a heat process which, although defined according to the minimum residence time and exit temperature from the holding section, can nevertheless incur several minutes residence in the heat exchangers. Nonetheless sensory quality is usually very good by comparison with canned foods and consumers may find that the more modern types of packaging have an appeal to their own life-styles. This was confirmed to some extent by the lower than expected popularity of aseptically canned foods. They could not be differentiated from conventionally canned products and offer no advantages of lightness in weight, convenience and ease of storing. It is after all the retail consumers who determine the commercial success of products. They do not ever have to rely on aseptic packages as a sole source since products have long been available alternatively in a canned, fresh, frozen, dried or chilled state. Continuing research and technical developments have led to increasing confidence in the capability of aseptic technology to deliver safe, wholesome foods, including those which are particulate and low-acid.

It is generally well understood that in the production of new combinations of food and package, processors especially must have sound technical responsibility. In the face of technical progress and machinery operations, it remains the more remarkable that the relevant regulatory authorities in the USA show continued resistance to accept for filing any process and aseptic packaging operations which include low-acid particulates larger than, for example, a rice grain. It is impossible to assess how much of a hindrance this creates elsewhere in the world by association, or when the situation will change.

However, the proliferation of ohmic heating systems in thirteen installations [54] may provide a key to the gaining of regulatory acceptance in the future, and could eventually demonstrate the level of technological circumvention of obstinate problems that becomes necessary to dispel any remaining doubts on safety.

The technology is one of interest worldwide which should sustain past initiatives whilst providing future opportunities even if the balance between commercial success and failure with both commodities and recipe products has often been both extraordinary and unpredictable. Maximisation of line productivity is critical to economic operation for processors. Retailers on the other hand may be increasingly drawn towards product sold through the chill chain even because of their need to pay for them only after the inflow of cash from sales. Also, not withstanding the evolution of aseptic technology to the point where complex ambient shelf stable ready meals can now be produced safely, consideration must be given to competitive pressure by the option of new in-pack sterilisation technologies. New designs of metal can are lighter, aesthetically designed, with easy-open ends and customary superb oxygen barrier performance. Heat resistant semi-rigid plastic packaging is available in a wide range of tray or pot designs. Modern retort systems can operate to temperatures of 140°C or more, with programmable controls that enable flexibility and safety in use. Overall, for the applications of aseptic technology to expand, it continually needs to offer the processor continued safety, high productivity, and economic demand on resources; and to the consumer safety, convenience, excellence in sensory quality and good value for money.

References

1. Manvell, C. *Food Science and Technology Today*, **1** (1987) 106.
2. Glen, D.H.G. *United States Patent* 4,255,459, 1981.
3. Manvell, C. *UK Patent* GB 2148 694B, 1985.

4. Bean, P.G. In *Food Microbiology: Advances and Prospects*, eds T.A. Roberts and F.A. Skinner, Academic Press, London and New York (1983) 97.
5. Anon., *Food Product Development*, **July** 1979.
6. Burton, H. *Dairy Science. Abstracts* **31**, 1969, 287.
7. Mulvaney, T.R., Schaffner, R.M., Miller, R.A. and Johnston, M.R. *Food Technology*, **June** 1978, 73.
8. Anon., *Information Letter* No. 2984, 1984. NPFA, 1401 New York Ave, N.W. Washington DC 20005.
9. Johnston, M.R. and Terry, R.C. In *Proceedings of the Conference Aseptic Processing and the Bulk Storage and Distribution of Food*, Food Sciences Institute, Purdue University, Indianapolis, 1978, March 15–16.
10. Davis, R.B. and Dignan, D.M. *Association of Food and Drug Officials Quarterly Bulletin*, **47** (1983) 12.
11. Johnson, F.M. Pack-Info Conference 1979, Washington DC.
12. Bernard, D.T. In *Recent Developments in Aseptic Technology Symposium* (1989), Campden Food and Drink Research Association, GL55 6LD.
13. Cerf, O. In *New Monograph on UHT Milk*, International Dairy Federation Document 133, Brussels (1981) 93.
14. Rose, D. *Technical Manual* No. 11, Part 1 (1986), Campden Food and Drink Research Association, GL55 6LD.
15. Neaves, P. and Jarvis, B. *Research Report 286* (1978), Leatherhead Food Research Association.
16. Brown, K.L., Gaze, J.E., McClement, R.H. and Withers, P. *International Journal of Food Science and Technology*, **23** (1988) 361.
17. Gaze, J.E. and Brown, K.L. *International Journal of Food Science and Technology*, **23** (1988) 373.
18. FAO/WHO Codex Alimentarius Commission, Eighteenth Session (1989) Geneva.
19. Baird-Parker, A. and Mayes, A. *Food Science and Technology Today*, **3** (1989) 23.
20. Rose, D. In *Aseptic Processing of Foods*, ed. H. Reuter, Springer Verlag (1993).
21. Leaper, S. *Technical Manual* No. 38 (1992), Campden Food and Drink Research Association.
22. Microbiology Panel Working Party, *Technical Manual* No. 24 (1989), Campden Food and Drink Research Association.
23. Food and Drug Administration, *Code of Federal Regulations Title 21* (1977) Washington.
24. Department of Health and Social Security, *Food Hygiene*, Code of Practice No. 10 (1981) HMSO, London.
25. United States Department of Agriculture, *Guidelines for Aseptic Processing and Packaging Systems in Meat and Poultry Plants*. FSIS (1984).
26. Institute of Food Science and Technology, *Food and Drink* GMP (1991) 3rd Edition.
27. Department of Health, *Guidelines for the Safe Production of Heat Preserved Foods*, HMSO, London (1994).
28. EHEDG, *Trends in Food Science and Technology*, **4** (1993) 21.
29. EHEDG, *Trends in Food Science and Technology*, **4** (1993) 115.
30. Rose, D. In *Aseptic Processing – A Rapidly Expanding Technology*, Europa-Pak 1988, Elsevier Seminars, Oxford.
31. Holdsworth, S.D. *Aseptic Processing and Packaging of Food Products*, Elsevier, London and New York (1992).
32. Sastry, S.K. In *Advances in Aseptic Processing Technologies*, eds R.K. Singh and P.E. Nelson, Elsevier, New York (1992) 63.
33. Heppel, N.J. In IUFOST Symposium, Tylosand (1985).
34. Dallyn, H., Falloon, W.C. and Bean, P.G. *Laboratory Practice*, **26** (1977) 773.
35. Brown, K.L., Ayres, C.A., Gaze, J.E. and Newman, M.E. *Food Microbiology*, **1** (1984) 187.
36. Skudder, P.J. In *Recent Developments in Aseptic Technology Symposium* (1989), Campden Food and Drink Research Association, GL55 6LD.
37. Tucker, G.S. *Trends in Food Science and Technology*, **4** (1993) 243.
38. Singh, K.S. and Lee, J.H. In *Advances in Aseptic Processing Technologies*, eds R.K. Singh and P.E. Nelson, Elsevier, New York (1992) 7.

39. Tucker, G.S. and Richardson, P.S. *Residence Time Distribution and Flow Behaviour of Food Containing Particles in Aseptic Processing*, Poster Paper, AI Chem E Conference of Food Engineering, Chicago (1993).
40. Hersom, A.C. and Shore, D.T. *Food Technology*, **May** (1981) 53.
41. Leaper, S. *Food Microbiology*, **1** (1984) 199.
42. Bell, N.D.S. In *Enough is Enough Conference* (1990), Scottish Society for Contamination Control.
43. Rose, D. In *Enough is Enough Conference* (1990), Scottish Society for Contamination Control.
44. Dumbleton, D.P. In *Proceedings of the 2nd International Conference in Aseptic Packaging* (1984), Schotland Business Research Inc., Princetown, New Jersey.
45. Rose, D. *Dairy Industries International*, **51**, (1986) 25.
46. Anon., *Packaging News*, **April** (1988), 6.
47. Selman, J.D. and Rose, D. *European Food and Drink Review*, **Autumn** (1989) 13.
48. Nielander, G. In *Aseptic Processing – A Rapidly Expanding Technology*, Europa-Pak 1988, Elsevier Seminars, Oxford.
49. Bond, S. In *UHT and Heat Processed Foods Symposium* (1991), Campden Food and Drink Research Association, GL55 6LD.
50. Anon., *Food Processing*, **August** (1993), 20.
51. Burton, C. In *Tetra Laval Fruit Juice Seminar* (1993), Uxbridge.
52. Hunt, J. *Packaging Week*, **9** (1993) 24.
53. Anon., *Packaging Week*, **9** (1993) 25.
54. APV Baker Ltd, *Eurofood*, **March** (1994).

14 Advances in modified-atmosphere packaging

A.R. DAVIES

14.1 Introduction

The ability of modified-atmosphere packaging (MAP) to extend the shelf-life of foods has been recognised for many years. In the 1920s, work at the Low Temperature Research Station, in Cambridge, UK, showed that the shelf-life of apples could be increased by storing them in atmospheres containing lowered levels of oxygen and increased levels of carbon dioxide, and in the 1930s beef carcases were transported in atmospheres containing carbon dioxide, which approximately doubled the storage life previously obtained. In the United Kingdom it was the retail chain of Marks & Spencer in 1979 that paved the way for Britain's pre-eminence today in the world marketplace for modified-atmosphere products, with its test launch of MAP meat (Parry, 1993). Since then there has been a marked expansion in the use and market share of MAP, partly as a result of the increasing consumer demand for fresh and chilled convenience foods containing fewer preservatives. This has led to a significant increase in the range of products packaged in modified atmospheres. Today, foods packaged in modified atmospheres include raw and cooked meats, poultry and fish, vegetables and fruit, fresh pasta, cheese, bakery products, potato crisps, coffee and tea.

The potential advantages and disadvantages of MAP have been presented by both Farber (1991) and Parry (1993) and these are summarised in Table 14.1.

There are several techniques by which the atmosphere surrounding a product can be modified, and often the terminology is confusing.

- *Modified-atmosphere packaging (MAP)*: the replacement of air in a pack by a different mixture of gases, where the proportion of each component is fixed when the mixture is introduced, but no further control is exercised during storage.
- *Controlled-atmosphere packaging (CAP)*: packaging in an atmosphere where the composition of gases is continuously controlled throughout storage. This technique is used primarily for the bulk storage of products and requires constant monitoring and control of the gas composition.
- *Equilibrium-modified atmosphere (EMA)*: used primarily for fruit and vegetables; either the pack is flushed with the required gas mix or the

Table 14.1 Potential advantages and disadvantages of MAP

Advantages
Shelf-life increase of 50–400% possible
Economic losses reduced (longer shelf-life to spoilage)
Products can be distributed over longer distances and with fewer deliveries, leading to decreased distribution costs
Provides a high quality product
Easier separation of sliced products
Centralised packaging and portion control
Improved presentation – clear view of product and all-round visibility
Little or no need for chemical preservatives
Disadvantages
Visible added cost
Temperature control necessary
Different gas formulations for each product type
Special equipment and training required
Product safety to be established
Increased pack volume – adversely affects transport costs and retail display space
Benefits are lost once the pack is opened or leaks

Adapted from Farber (1991) and Parry (1993).

produce is sealed within the pack with no modification to the atmosphere. Subsequent respiration of the produce and the gas permeability of the packaging allow an equilibrium-modified atmosphere to be reached.

- *Vacuum packaging (VP)*: the product is placed in a pack of low oxygen permeability, air is evacuated and the package sealed. The gaseous atmosphere of the vacuum package is likely to change during storage (from metabolism of the product or microorganisms) and therefore the atmosphere becomes modified indirectly.

14.1.1 Role of gases

The three major gases used commercially in MAP are oxygen, nitrogen and carbon dioxide, although several other gases have been investigated, e.g. carbon monoxide, sulphur dioxide, nitrous oxide, ozone and chlorine. However, the use of these has been limited by safety concerns, legislation, negative effects on organoleptic properties and cost.

Oxygen (O_2) will generally stimulate the growth of aerobic bacteria and can inhibit the growth of strictly anaerobic bacteria, although there is a very wide variation in the sensitivity of anaerobes to oxygen (Farber, 1991). Oxygen is very important in fresh meats to maintain myoglobin in its oxygenated form (oxymyoglobin), which gives fresh meat its bright red colour. Its presence may cause problems with oxidative rancidity or colour in some products (e.g. fatty fish and cured meats, respectively).

Nitrogen (N_2) is an inert tasteless gas with low solubility in both water and lipid. It is used to replace oxygen in packs so as to delay oxidative rancidity and inhibit the growth of aerobic microorganisms (Farber, 1991). Because of its low solubility, it is used as a filler gas to prevent the pack collapse which may occur in high CO_2-containing atmospheres.

Carbon dioxide (CO_2) is both water- and lipid-soluble and is mainly responsible for the bacteriostatic effects seen on microorganisms in modified atmospheres (Farber, 1991). The overall effect on microorganisms is an extension of the lag phase of growth and a decrease in the growth rate during the logarithmic phase (Farber, 1991). This bacteriostatic effect is influenced by the concentration of CO_2, the age and load of the initial bacterial population, storage temperature and the type of product to be packaged (Reddy *et al.*, 1992). Although the bacteriostatic effect of CO_2 has been known for many years, the precise mechanism of its action is still not clearly understood. Farber (1991) summarised the theories regarding the influence of CO_2 on the bacterial cell as:

- alteration of cell membrane function including effects on nutrient uptake and absorption;
- direct inhibition of enzymes or decreases in the rate of enzyme reactions;
- penetration of bacterial membranes, leading to intracellular pH changes;
- direct changes to the physicochemical properties of proteins.

With high-moisture/high-fat foods such as meat, poultry and seafood, excessive absorption of CO_2 can lead to the phenomenon known as ‘pack collapse’ (Parry, 1993). Increased in-pack drip is also caused by dissolution of the gas into the surface of fresh muscle foods which reduces their pH values sufficiently to weaken the water-holding capacity of the proteins (Parry, 1993).

14.2 Market status and potential

The market status of MAP is difficult to research because it is a relatively new sector of the packaging market and there is considerable confusion over definitions (Cakebread, 1994). However, Day (1990, 1993a), Brody (1993) and Cakebread (1994), have all reviewed the market, albeit largely from a European perspective. Day (1990) details the differences between the European and American markets that he suggests explain why MAP has been applied successfully for over 15 years in Europe while it is still in its infancy in America. The differences presented are:

1. Larger population but smaller land mass of the EEC.

2. European buyers shop more frequently than their American counterparts.
3. Higher refrigerated-foods awareness and acceptance by the European consumer.
4. Total system/quality approach for MAP technology in Europe.
5. Higher prices of food in the EEC market-place as compared to the USA has made MAP more economically feasible.
6. Different driving forces for MAP and refrigerated foods in Europe and the USA. The European market is retail driven while the USA market is packer/consumer driven.
7. The dominance and profitability of the giant retail food chains especially in the UK and France, who are dedicated to high-quality refrigerated foods.

As well as these differences between the American and European markets, there is also a highly uneven spatial development of MAP across Europe, with those countries with concentrated multiple retailer market shares (Cakebread, 1994), notably the UK and France, being dominant. The percentage of the market for MAP foods in Europe in 1990 as calculated from Brody (1993) are UK 37.5%, France 25%, Italy 15%, Netherlands 10%, Belgium 6.5% and (West) Germany 6%. Over the last 2–3 years there has been a considerable diversification in products packed in MAP both in France and in the UK (Cakebread, 1994). This diversification is reflected in Figure 14.1, which shows the segmentation of the UK MAP retail market by food product for 1992.

As mentioned previously, estimation of the market size is very difficult, particularly in view of the definitions used. Brody (1993) states that the market size for MAP ranges from one billion to ten billion package units, depending on the definition accepted. The larger figure includes vacuum-packaged primals, and the lower figure encompass only retail packages that have been flushed with mixtures of CO_2, N_2 and O_2 as required (Brody, 1993). From the lower figure, Brody estimates a 5–10% annual growth rate during the 1980s. In Europe it has been estimated that 2000 million MA packages were produced in 1990 in contrast to only 650 million units in 1986 (Day, 1993a). A further market research report estimated that 1930 million MA packages were sold in the UK in 1992 with significant growth in demand to 3400 million units predicted for 1996 (Day, 1993a). Cakebread (1994) has estimated (from received opinions of the industry) a total European market for MAP materials of about 4000 million units in 1992. The European market as a whole is thought to be growing at about 10–15% per annum – about 5% below 1991 levels (Cakebread, 1994).

Cakebread (1994) has reviewed the future prospects for the European MAP market and concludes that, despite a theoretically large potential, MAP will not continue to grow rapidly in Europe during the 1990s unless

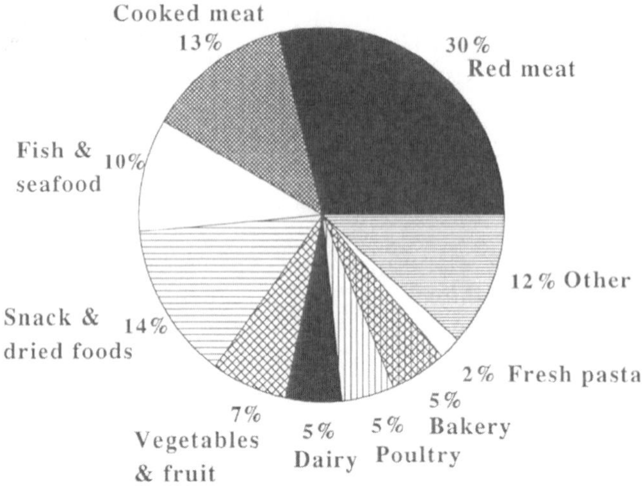

Figure 14.1 Segmentation of the UK MAP retail market by food product sector, 1992. Reproduced with permission from Day B.P.F. Recent Developments in MAP. *European Food and Drink Review*, Summer, pp. 87–95.

concrete solutions can be found to a number of factors that have restricted commercial development outside France and the UK. These factors include:

(i) the lack of a concentrated retail food and chill chain distribution structure in many European countries;
(ii) entrenched environmental prejudice against PVC, plastics and 'unnecessary' packaging generally;
(iii) competition from high quality unpacked fresh produce;
(iv) competition from vacuum-skin packaging in certain key sectors such as fresh and cooked meats and fish;
(v) concern over the safety of some chilled food products.

Several of these factors will obviously also have a major influence on the development of the MAP market outside Europe.

14.3 Microbiology of MAP

14.3.1 Microbial spoilage

In general, aerobic microorganisms are sensitive to CO_2 and it is this, along with their requirement for O_2, that is utilised in MAP to control the spoilage of foods. Gram-negative bacteria are generally more sensitive to

CO_2 than Gram-positive bacteria (Lambert *et al.*, 1991). In chill-stored proteinaceous foods such as meat and fish, this generally results in the inhibition of the Gram-negative *Pseudomonas*, Enterobacteriaceae and *Acinetobacter/Moraxella*, whilst the Gram-positive lactic acid bacteria and *Brochothrix thermosphacta* become the dominant organisms. A comparison of the developing microflora on rainbow trout (*Oncorhynus mykiss*) stored aerobically or in MAP (80% CO_2/20% N_2) at 5°C is presented in Figure 14.2. Packaging in 80% CO_2/20% N_2 inhibits growth of all the organisms examined, but particularly the Gram-negative Enterobacteriaceae and *Pseudomonas* sp.

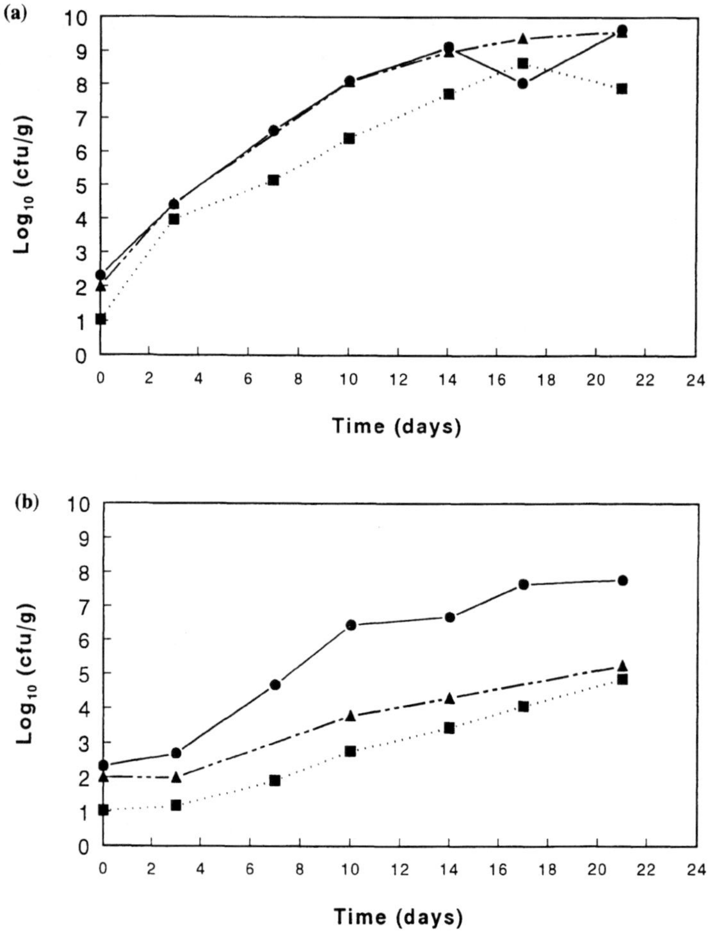

Figure 14.2 Microflora of (a) aerobic and (b) MAP (80% CO_2, 20% N_2) rainbow trout stored at 5°C. ■–■ = *Enterobacteriaceae*; ▲–▲ = *Pseudomonas* spp.; ●–● = lactic acid bacteria.

As moulds have an absolute requirement for O_2, in foods where mould spoilage is the major concern, e.g. bakery products or hard cheese, packaging in an anaerobic MAP can be extremely successful in delaying spoilage. If CO_2 is used to produce the MAP there is also the additional benefit of the anti-bacterial and mould activity of CO_2.

14.3.2 Microbial safety

Whilst the ability of MAP to extend the shelf-life of many products is well recognised, concern has been expressed by regulatory authorities, food industry groups and others that MAP may introduce undue safety hazards. The concerns are that suppression of the normal spoilage flora may result in an organoleptically acceptable product whilst either allowing or enhancing the growth of pathogenic organisms. Historically, the non-proteolytic, psychrotrophic strains of *Clostridium botulinum* have been the major safety concern. These strains can grow and may produce toxin without producing overt signs of spoilage, which may also be absent as a result of an inhibition of the normal spoilage flora. More recently, concerns have been expressed about the ability of the other psychrotrophic pathogens, e.g. *Aeromonas*, *Listeria* and *Yersinia*, to grow in MAP products.

14.3.3 Clostridium botulinum

Historically, MAP of fish and fish products has been the greatest cause for concern with respect to *Cl. botulinum*, and resulted in the US National Academy of Sciences recommending that, until the safety of the system was established, fish should not be packed in MAP (Davis, 1993). This resulted from the isolation of all types of *Cl. botulinum* from marine environments, which, although highly variable by geographic location and season, is frequent enough that processors must assume its presence.

Strains of type E and the non-proteolytic types B and F are the major concern in MAP as they are able to grow at temperatures as low as 3.3°C, albeit slowly, and, as they do not putrefy proteins, may not cause obvious signs of spoilage. Numerous studies have examined the relationship between time to toxin production and signs of organoleptic spoilage for MAP fish, and these have been reviewed (Reddy *et al.*, 1992; Stammen *et al.*, 1990). Unfortunately, because of the many variables between studies, e.g. fish type, size and site of inoculum, temperature, season, atmosphere, etc., direct comparisons between the studies cannot be made. However, Stammen *et al.* (1990) in their review, concluded that

> 'with few exceptions, at temperatures above 20°C, organoleptic spoilage coincided with toxin production in many fresh fishery products, regardless of the modified atmosphere used. However, at lower temperatures, organoleptic

spoilage preceded toxin development in all fresh fish products except cod and whiting fillet held in either an air, vacuum or CO_2 atmosphere. This trend was seen at storage temperatures from 4–12°C. The time interval between toxin development and organoleptic spoilage of MAP fish products generally decreased as storage temperatures increased. In contrast toxin development preceded organoleptic spoilage in cod and whiting fillets packaged in 100% CO_2 and held at refrigeration temperatures. These products were still acceptable for consumption even though botulinal toxin was found in them.'

Baker and Genigeorgis (1990) developed a predictive model from over 18 700 samples analysed over a 5-year period. The utility of the model was demonstrated by its ability to predict the time before toxigenesis in inoculated fish stored under different MAs as reported in the international literature. Temperature explained 74.6% of experimental variation in the final multiple linear regression model ($r_2 = 0.883$) but, surprisingly, the gaseous atmosphere was of little importance.

One approach that may provide the safety required for MAP of fish with respect to *Cl. botulinum* is the use of a pre-treatment in combination with MAP. For instance, potassium sorbate, sodium chloride and irradiation in combination with MAP have all been shown to be effective (Stammen *et al.*, 1990).

14.3.4 Other pathogens

Whilst there have been numerous studies on the effect of MAP on *Cl. botulinum* there have been few studies on the effect on the other food-borne pathogens, particularly the psychrotrophic strains.

At Leatherhead Food RA, we have recently completed two European Community projects that examined pathogen growth/survival on MAP fish and meat and their products. Studies with fish have concentrated on cod (*Gadhus morhua*) and rainbow trout (*Oncorhynus mykiss*) with storage at 0°, 5° and 12°C. In no instance was the growth/survival of any of the pathogens examined greater than that in the aerobically stored control and frequently growth was reduced in MAP. Studies on meat have concentrated on raw beef but have also included cooked and dry-cured hams. The control atmosphere used throughout for meats was vacuum pack.

As for fish, but with one exception, the growth/survival in MAP was never greater than that in the control, and occasionally growth was reduced in MAP. The one exception was verotoxigenic *E. coli* (VTEC) on beef stored at 12°C, in which growth in one of the modified atmospheres (80% O_2/20% CO_2) examined was greater than that in the vacuum-pack control. However, this atmosphere was included as an atmosphere recommended for the retail packaging of beef, and hence a more appropriate control for this atmosphere may be aerobic storage. A comparison of VTEC growth on aerobically stored and 80% O_2/20% CO_2 stored beef showed that growth in the MAP was reduced in comparison with that of the aerobically stored product.

As found by others (Luiten *et al.*, 1982; Eklund and Jarmund, 1983; Baker *et al.*, 1986), we observed a CO_2-dependent bacteriostasis of *Salmonella* at chill temperatures. We did not, however, see the marked difference reported (Wimpfheimer *et al.*, 1990) between aerobic and anaerobic MAPS on the growth of *Listeria monocytogenes*. Our results with *Aeromonas* and *Yersinia* on beef differ from those reported for high-pH beef (Gill and Reichel, 1989) and highlight the importance of assessing the safety of individual foods to be packaged in MAP.

Overall, our results and the majority of those reported in the literature indicate that the risks from the other foodborne pathogens in MAP are no greater and frequently are less than those from aerobically stored foods. These findings are substantiated by the excellent safety record, to date, of MAP (Church, 1993).

As is the case for almost all food production the Hazard Analysis Critical Control Point (HACCP) system is likely to play a major role in controlling the safety of MAP foods. This has been discussed by Rose (1990), Ooraikul (1991), Alli (1993) and many others.

14.4 Developments in MAP

A number of recent developments offer the potential to improve further the safety of MAP and extend the technology available to a wider range of products. These developments are discussed below.

14.4.1 Intelligent packaging

The development of 'smart' films (Sneller, 1986), more recently described as 'active' or 'intelligent' packaging (Labuza, 1989; Summers, 1992) probably account for the most significant area of advancement of MAP technology. These intelligent packaging techniques have the ability either to absorb or to emit gases and vapours, and have been categorised by Robertson (1991) under eight headings:

- oxygen scavenging
- carbon dioxide formation
- aroma removal
- off-flavour removal
- ethylene removal
- ethanol emitters
- water removal
- edible films

These techniques have been described by Labuza and Breene (1989), Labuza (1990), Robertson (1991) amongst others. Of these techniques, oxygen scavengers have probably received the widest attention and greatest commercial acceptance.

Two approaches to the development of oxygen-scavenging systems have been examined. The most successful commercially has been the use of sachets that are included in the pack; the second and most exciting area has been the development of oxygen-scavenging films. Rooney (1994) has reported on an evaluation of two types of oxygen-scavenging plastics. The first type uses visible light to excite oxygen diffusing from the headspace into the plastic packaging and scavenging it therein, whilst the second type is activatable in several ways regardless of illumination. The basic scavenger film composition was found to retain its activity for at least 100 days in the absence of oxygen.

The use of oxygen absorbers is not new; there have been reports on the removal of oxygen as far back as the 1920s (Robertson, 1991). Isherwood in 1943 issued a patent (BP 553,991) for removing oxygen from a container containing vacuum- or gas-packed food in which a metal (e.g. iron) absorbs oxygen to form an oxide. The most commercially successful system (Robertson, 1991) under the trade name Ageless was introduced in Japan in 1977 by the Mitsubishi Gas Chemical Company. Several companies now market such products and it is reported that the market size for oxygen absorbers in Japan was almost 7 billion packets a year (based on 100 cc absorption capacity) in 1989 in a market still growing at about 20% per year (Kondo, 1991). Abe (1990) discussed the attributes of oxygen absorbers that have led to their widespread use in Japan. These attributes include removal of oxygen from porous food products, simple operation at the packaging line, reduction in preservatives/additives, maintenance of low (100 ppm) oxygen concentrations, etc. An indicator system (Agelesseye) for the oxygen status of a pack is also marketed by the Mitsubishi Gas Chemical Company. The indicator can be included with the oxygen scavenger and is pink when there is no oxygen in the environment (0.1% or less) and blue if oxygen does exist (0.5% or more).

Abe (1990) also stated that oxygen absorbers enable new, innovative products to be introduced and can help the development of healthy products by reducing sugar, salt and some food preservatives, or sometimes by increasing the moisture content.

Idol (1991) critically reviewed the use of in-package oxygen scavengers, specifically addressing the resistance encountered to their adoption in the United States. He contends that the basic differences between the United States and Japanese food markets, such as demographics, pricing, and cultures, are at the root of the problem. Specifically, he stated that there are several common characteristics of oxygen absorbers that produce market place concern in the US. The enhanced fear in the USA of litigation arising from accidental ingestion of the scavenger, the relatively greater cost of packaged food in Japan, and the more receptive nature of the Japanese to a package of chemicals placed in their food have been the major factors. Of these concerns, the problems from both ingestion and the free-floating nature of the package would be resolved by firmly fixing the absorber to the packaging material or by incorporating the chemicals directly into the film. Research is continuing in both these areas and packages have been developed, for example, in which the oxygen sachet is presented in a separate compartment to the product so as to prevent its accidental consumption. Recently, oxygen-absorbing labels have been introduced into the UK market (Anon., 1994) for smoked and processed delicatessen meat products. The product is presented with a paper label with an absorbing label positioned on the lidstock under this paper label. Information on the outside label tells the consumer that the inside label helps retain freshness and is only seen by the user after peeling back the lid. It is claimed that this application opens the door for future usage in the United States because the fear of ingestion is eliminated and the accurate sizing of the label for each application allows the absolute minimum of reactive agent to be used (Anon., 1994).

New types of oxygen absorbers have recently been developed. These include some for use with frozen foods, which overcome the slow absorption speed of the standard types at low temperatures, some that are microwaveable, and some that emit the same volume of carbon dioxide as that of oxygen absorbed (Abe, 1990).

Oxygen scavengers can now be found in packages of many foods, including bakery items like bread, cakes, cookies, pizza crust and pastry, as well as packages of pre-cooked pasta, cured or smoked meats and fish, dry tea, dried fish, cheese, soft cookies, beans, potato chips, dried eggs, spices, nuts, coffee and chocolates (Labuza and Breene, 1989).

In parallel with the development of these oxygen-scavenging systems, several systems have been developed that are capable of either absorbing or generating carbon dioxide. Some of these systems are coupled with the ability to absorb or generate oxygen.

A commercially available sachet system capable of scavenging both oxygen and carbon dioxide has been marketed. This system has been used mainly for freshly roasted coffee beans in flexible bags and more than triples the high-quality shelf-life (Labuza, 1989).

Another commercial system (Verifrais; Codimer) uses a porous chemical pouch containing mineral salts affixed to the base of a tray and covered by a plastic grid. The pad is activated when water vapour from respiring fruits and vegetables or exudate from flesh foods drips through the grid. Depending on actual salts, which vary from one product to another, CO_2 and/or O_2 is released, or ethylene and/or CO_2 is absorbed (Robertson, 1991).

14.4.2 Predictive/mathematical modelling

Mathematical modelling affords the potential to make significant improvements in at least two areas of MAP. The first of these is in the optimisation (gas to product ratio, film permeability, etc.) of MAP for specific products, notably fruit and vegetables. The second aspect is in the prediction of the microbial safety and shelf-life of MAP foods.

At present, evaluation of MAP for fresh produce remains a largely empirical, trial-and-error exercise which is time consuming, subjective and often without unifying principles to guide research and development efforts (Kader *et al.*, 1989). Many variables interact within a modified-atmosphere produce pack and, because many of these variables feed back to alter related factors, it is difficult to manipulate all of them simultaneously in a quantitative manner (Day, 1993b). Zagory (1990) has reviewed what he considers are the most important variables (temperature, product respiration rate, plastic film permeabilities, film surface area, film thickness, product weight, product respiratory quotient and film CO_2/O_2 permeability ratio) that determine the package atmosphere. Computer models have been developed that aid in elucidating the salient variables active in a particular package design and in optimising the design while minimising development time and costly laboratory testing (Zagory, 1990). However, none of the mathematical models to date has been comprehensive enough to include all the salient intrinsic and extrinsic parameters (Day, 1993b).

Predictive microbiology uses mathematical equations to estimate the growth, survival or death of microorganisms as affected by extrinsic (processing and storage conditions) and intrinsic parameters of the food (e.g. salt concentration, pH or a_w; Willocx *et al.*, 1993). To date, few people have developed models that include gaseous atmospheres that would be relevant to most MAP products. Some models have, however, been developed and efforts are continuing in this area. For example Einarsson (1992) has developed a model that could be used to predict the shelf-life of MAP cod fillets, whilst Willocx *et al.* (1993) modelled the influence of temperature and carbon dioxide upon the growth of *Pseudomonas fluorescens*.

In the UK, the Ministry of Agriculture, Fisheries and Food (MAFF) initiated a nationally co-ordinated 5-year programme of research into the growth and survival of microorganisms in foods, with the aim of developing a computerised Predictive Microbiology Database in the UK (McClure *et al.*, 1994). The database developed was launched as a commercial service in 1992 (Food Micromodel) and a personal computer (PC) based version is currently being developed. For certain organisms the effect of gaseous atmosphere is being examined and may in the future be included within the commercial database.

The major difficulty with modelling the effect of the gaseous environment is the very dynamic nature of the atmosphere; particularly with proteinaceous foods and oxygen-containing atmospheres. It is, therefore, necessary to have a good understanding of the changes in the gaseous composition with time (which in itself is dependent on numerous other variables), and a sufficiently sophisticated model to be able to predict the effect of this on microbial growth/survival.

14.4.3 Combination treatments

As mentioned previously, one interesting area, with the potential to improve both the safety and shelf-life of MAP products, is the use of a pre-treatment in combination with MAP. Such an approach would be extremely attractive if it could aid in controlling pathogen growth, as reliance on temperature control alone is not desirable particularly as this cannot be effectively controlled beyond the point of sale. Several workers (Fey and Regenstein, 1982; Regenstein, 1982; Stathem *et al.*, 1985; Sharp *et al.*, 1986; Barnett *et al.*, 1987) have examined the effect of chemical pre-treatments in combination with MAP, particularly for fish, whilst others (e.g. Eklund, 1982; Licciardello *et al.*, 1984; Przybiski *et al.*, 1989; Patterson, 1990; Lambert *et al.*, 1992; Grandison and Jennings, 1993) have examined the potential of irradiation or irradiation and chemical pre-treatment in combination with MAP.

Chemical pre-treatments that have been examined include the addition of potassium sorbate, polyphosphates, sodium chloride, linoleic acid and nisin. All these combination treatments (chemical and irradiation) have been shown, on occasions, to provide a benefit in shelf-life, safety, quality or a combination of these factors for various products. Occasionally a synergistic effect of the treatments and MAP has been observed. Because of the paucity of information on the effect of these combination treatments, however, it is currently necessary to evaluate any proposed treatments for individual foods. Further studies are required to gain a fuller understanding of the true potential of these combined treatments for a wide range of products.

14.4.4 Packaging films/equipment

In the last few years there have been several developments in the packaging film and equipment area and these have been reviewed by Roberts (1990), Day (1993a), Hastings (1993), Greengrass (1993), and Cakebread (1994) amongst others. Many of these developments have been in the design of high-permeability films for the packaging of fruit and vegetables. These include films in which the moisture and gas permeability change under specific temperature conditions that are designed to match or exceed the respiratory rate of the product (Cakebread, 1994). Much of the development work has centred on retail packs and includes the development of microwaveable (Roberts, 1990) and resealable (Cakebread, 1994) MAP packs and the development of triple-web packs (Roberts, 1990; Cakebread, 1994) which combine the advantages of both MAP and vacuum-skin packaging (VSP). Recently, several companies have also developed new bulk MAP systems and the New Zealand Captech system has entered the European market (Cakebread, 1994). One of the most significant advances in equipment is the development of in-line non-destructive leak detectors. The detection of leakers is one of the major challenges facing the MAP producer.

14.4.5 Indicators

Whilst not a new development, time/temperature indicators (TTIs) still appear to have potential for use with chill-stored MAP products. To date, market penetration of TTIs has been low (Selman, 1990; Church, 1993). Church (1993) has attributed this mainly to cost, uncertainty over performance, poor correlation between product and air temperatures, economics, potential for litigation, wisdom of use and consumer confusion. He concludes that, in the short term, indicators are likely to be used for bulk package monitoring and indication of safety at point of sale, including indication by colour that the best-before/use-by date has not been exceeded.

Carbon-dioxide-sensitive labels that change colour if modified-atmosphere packs are opened or damaged and labels capable of detecting increases in CO_2 concentration are also being developed (Church, 1993).

14.5 The future

All the evidence indicates that there remains considerable potential for MAP worldwide. The increased diversification in the product range packaged in MAP is likely to continue and this should expand the market potential. Current developments in films and equipment, particularly the development of intelligent packaging, should also favour an expansion of the potential of MAP. However, concerns about the safety of chill-stored MAP products still remain. Indeed, in the UK, the Advisory Committee on the Microbiological Safety of Food has recently published a report on vacuum packaging and associated processes. This report states that, for products with a shelf-life of greater than 10 days, that are not subjected to a heat treatment that will sufficiently inactivate the spores of psychrotrophic *Clostridium botulinum*, combinations of preservative factors such as pH, water activity, and salt should be established to prevent the growth of psychrotrophic *Cl. botulinum*. The development of predictive microbiology and research on the effects of combination treatments should go some way to answering these concerns. Likewise, the increased use of HACCP should help in understanding and controlling the safety of such products. Two other factors are also likely to play a crucial role in the future of MAP. The first of these is the requirement for a concentrated retail and chill chain distribution structure. In Europe, many multiples are now trying to develop pan-European networks; however, this is a very slow process (Cakebread, 1994). The second factor is environmental concern about packaging materials (White, 1990; Greengrass, 1993) both from a legislative and consumer perception viewpoint.

Acknowledgement

The author is grateful to *Culture* for kind permission to reproduce material from an article presented by the author and Dr Paul A. Gibbs. The financial support of the EC, MAFF, MLC and individual members of the Leatherhead Food Research Association is gratefully acknowledged.

References

Abe, Y. (1990) *Ageless – oxygen absorber*. Food Packaging Technology International. Interpack 90. The World of Packaging, Dusseldorf, 7–13.6.90, pp. 183–184.

Alli, I. (1993) Quality control of MAP products. In *Principles and Applications of Modified Atmosphere Packaging of Food*, Parry, R.T. (Ed.). Blackie Academic & Professional, Glasgow, pp. 101–113.

Anon. (1994) Britons report successful use of an oxygen-removing label. *Food Engineering*, **March**, 68, 70.

Baker, D.A. and Genigeorgis, C. (1990) Predicting the safe storage of fresh fish under modified atmospheres with respect to *Clostridium botulinum* toxigenesis by modelling length of the lag phase of growth. *J. Food Prot.*, **53**, 131–140.

Baker, R.C., Quereshi, R.A. and Hotchkiss, J.H. (1986) Effect of an elevated level of carbon dioxide containing atmosphere on the growth of spoilage and pathogenic bacteria at 2°, 7° and 13°C. *Poult. Sci.*, **65**, 729–737.

Barnett, H.J., Conrad, J.W. and Nelson, R.W. (1987) Use of laminated high- and low-density polyethylene flexible packaging to store trout (*Salmo gairdneri*) in modified atmosphere. *J. Food Prot.*, **50(8)**, 645–651.

Brody, A.L. (1993) The Market. In *Principles and Applications of Modified Atmosphere Packaging of Food*, Parry, R.T. (Ed.). Blackie Academic & Professional, Glasgow, pp. 19–40.

Cakebread, D. (1994) *MAP Market Developments and Opportunities*. Leatherhead Food RA, Modified-atmosphere Packaging Course, T072.

Church, P.N. (1993) Meat Products. In *Principles and Applications of Modified Atmosphere Packaging of Food*, Parry, R.T. (Ed.). Blackie Academic & Professional, Glasgow, pp. 229–268.

Davis, H.K. (1993) Fish. In *Principles and Applications of Modified Atmosphere Packaging of Food*, Parry, R.T. (Ed.). Blackie Academic & Professional, Glasgow, pp. 189–228.

Day, B.P.F. (1990) A perspective of modified atmospheric packaging of fresh produce in Western Europe. *Food Science and Technology Today (November)*, **4(4)**, 215–221.

Day, B.P.F. (1993a) Recent Developments in MAP. *European Food & Drink Review*, **Summer**, 87–95.
Day, B.P.F. (1993b) Fruit and Vegetables. In *Principles and Applications of Modified Atmosphere Packaging of Food*, Parry, R.T. (Ed.). Blackie Academic & Professional, Glasgow, pp. 114–133.
Einarsson, H. (1992) Predicting the shelf life of cod (*Gadhus morhua*) fillets stored in air and modified atmosphere at temperatures between −4°C and +16°C. In *Quality Assurance in the Fish Industry*, Eds H.H. Huss *et al.* Elsevier Science Publishers B.V.
Eklund, M.W. (1982) Significance of *Clostridium botulinum* in fishery products preserved short of sterilization. *Food Technol. (Aust)*, **36**, 12.
Eklund, T. and Jarmund, T. (1983) Microculture model studies on the effect of various gas atmospheres on microbial growth at different temperatures. *J. Appl. Bacteriol.*, **55**, 119–125.
Farber, J.M. (1991) Microbiological aspects of modified-atmosphere packaging technology – a review. *J. Food Prot.*, **54**, 58–70.
Fey, M.S. and Regenstein, J.M. (1982) Extending shelf-life of fresh wet red hake and salmon using CO_2–O_2 modified atmosphere and potassium sorbate ice at 1°C. *J. Food Sci.*, **47**, 1048–1054.
Gill, C.O. and Reichel, M.P. (1989) Growth of the cold-tolerant pathogens *Yersinia enterocolitica*, *Aeromonas hydrophila* and *Listeria monocytogenes* on high-pH beef packaged under vacuum or carbon dioxide. *Food Microbiol.*, **6**, 223–230.
Grandison, A.S. and Jennings, A. (1993) Extension of the shelf life of fresh minced chicken meat by electron beam irradiation combined with modified atmosphere packaging. *Food Control*, **4(2)**, 83–88.
Greengrass, J. (1993) Films for MAP of foods. In *Principles and Applications of Modified Atmosphere Packaging of Food*, Parry, R.T. (Ed.). Blackie Academic & Professional, Glasgow, pp. 63–100.
Hastings, M.J. (1993) Packaging machinery. In *Principles and Applications of Modified Atmosphere Packaging of Food*, Parry, R.T. (Ed.). Blackie Academic & Professional, Glasgow, pp. 41–60.
Idol, R.C. (1991) A critical review of in-package oxygen scavengers and moisture absorbers. In *Proceedings of CAP 91*, Sixth International Conference on Controlled/Modified Atmosphere/Vacuum Packaging, San Diego, California, Schotland Business Research Inc., pp. 181–190.
Kader, A.A., Zagory, D. and Kerbel, E.L. (1989) Modified atmosphere packaging of fruits and vegetables. *CRC Crit. Rev. Food Sci. Nut.*, **28(1)**, 1–30.
Kondo, Y. (1991) A Japanese perspective on oxygen scavengers. In *Proceedings of CAP 91*, Sixth International Conference on Controlled/Modified Atmosphere/Vacuum Packaging, San Diego, California, Schotland Business Research Inc., pp. 191–196.
Labuza, T.P. (1989) Active Food Packaging Technologies. In *Engineering and Food Vol. 2. Preservation Processes and Related Techniques*. Eds. W.E.L. Spiess and H. Schubert. Elsevier Applied Science, London, pp. 304–311.
Labuza, T.P. (1990) *Action Packs for Longer Life*. Food Packaging Technology International. Interpack 90. The World of Packaging, Dusseldorf, 7–13.6.90, pp. 190–194.
Labuza, T.P. and Breene, W.M. (1989) Applications of 'active packaging' for improvement of shelf-life and nutritional quality of fresh and extended shelf-life foods. *Journal of Food Processing and Preservation*, **13**, 1–69.
Lambert, A.D., Smith, J.P. and Dodds, K.L. (1991) Shelf-life extension and microbiological safety of fresh meat – a review. *Food Microbiol.*, **8**, 267–297.
Lambert, A.D., Smith, J.P., Dodds, K.L. and Charbonneau, R. (1992) Microbiological changes and shelf life of MAP, irradiated fresh pork. *Food Microbiology*, **9**, 231–244.
Licciardello, J.J., Revesi, E.M., Tuhkunen, B.E. and Racicot, L.D. (1984) Effect of some potentially synergistic treatments in combination with 100 krad irradiation on the iced shelf life of cod fillets. *J. Food Sci.*, **49**, 1341–1346, 1375.
Luiten, L.S., Marchello, J.A. and Dryden, F.D. (1982) Growth of *Salmonella typhimurium* and mesophilic organisms on beef steaks as influenced by type of packaging. *J. Food Prot.*, **45**, 263–267.
McClure, P.J., Blackburn, C. de W., Cole, M.B., *et al.* (1994) The UK approach to modelling the growth, survival and death of microorganisms in foods. *Int. J. Food Micro.* (in press).

Ooraikul, B. (1991) Further research in modified atmosphere packaging. In *Modified Atmosphere Packaging of Food*. Ooraikul, B. and Stiles, M.E. (Eds). Ellis Horwood, New York.

Parry, R.T. (1993) Introduction. In *Principles and Applications of Modified Atmosphere Packaging of Food*, Parry, R.T. (Ed.). Blackie Academic & Professional, Glasgow, pp. 1–18.

Patterson, M. (1990) Combination of irradiation with MAP. In *Proceedings of International Conference on Modified Atmosphere Packaging*, Campden Food and Drink Research Association, Chipping Campden, Gloucester, UK.

Przybiski, L.A., Finerty, M.W., Grodner, R.M. and Gerdes, D.L. (1989) Extension of shelf-life of iced fresh channel catfish fillets using modified atmospheric packaging and low dose irradiation. *J. Food Sci.*, **54**, 269–273.

Reddy, N.R., Armstrong, D.J., Rhodehamel, E.J. and Kauter, D.A. (1992) Shelf-life extension and safety concerns about fresh fishery products packaged under modified atmospheres: a review. *J. Food Safety*, **12**, 87–118.

Regenstein, J.M. (1982) The shelf-life extension of haddock in carbon dioxide–oxygen atmospheres with and without potassium sorbate. *J. Food Quality*, **5**, 285–300.

Roberts, R. (1990) An overview of packaging materials for MAP. In *Proceedings of International Conference on Modified Atmosphere Packaging*, Campden Food and Drink Research Association, Chipping Campden, Gloucester, UK.

Robertson, G.L. (1991) The really new techniques for extending food shelf life. In *Proceedings of CAP 91*, Sixth International Conference on Controlled/Modified Atmosphere/Vacuum Packaging, San Diego, California, Schotland Business Research Inc.

Rooney, M.L. (1994) *Oxygen-Scavenging Plastics Activated for Fresh and Processed Foods*. IFT, Annual Meeting, June 26–29, Atlanta, Georgia, USA.

Rose, S.A. (1990) Hazard analysis and safety control of MAP chilled perishable foods. In *Proceedings of International Conference on Modified Atmosphere Packaging*, Campden Food and Drink Research Association, Chipping Campden, Gloucester, UK.

Selman, J.D. (1990) An overview of temperature and time–temperature colour indicators. In *Proceedings of International Conference on Modified Atmosphere Packaging*, Campden Food and Drink Research Association, Chipping Campden, Gloucester, UK.

Sharp, W.F. Jr., Norback, J.P. and Stuiber, D.A. (1986) Using a new measure to define shelf life of fresh whitefish. *J. Food Sci.*, **51**, 936–939, 959.

Sneller, J.A. (1986) Smart films give a big lift to controlled atmosphere packaging. *Paper Film Converter*, **12**, 58–59.

Stammen, K., Gerdes, D. and Caporaso, F. (1990) Modified atmosphere packaging of seafood. *Crit. Rev. Food. Sci. Ntr.*, **29**, 301–331.

Statham, J.A., Bremner, H.A. and Quarmby, A.R. (1985) Storage of Morwong (*Nedadactylus macropterus* Bloch and Schneider) in combinations of polyphosphate, potassium sorbate and carbon dioxide at 4°C. *J. Food Sci.*, **50**, 1580–1584, 1587.

Summers, L. (1992) *Intelligent Packaging*. Centre for Exploitation of Science and Technology, London.

Willcox, F., Mercier, M., Hendrickx, M. and Tobback, P. (1993) Modelling the influence of temperature and carbon dioxide upon the growth of *Pseudomonas fluorescens*. *Food Microbiology*, **10**, 159–173.

Wimpfheimer, L., Altman, N.S. and Hotchkiss, J.H. (1990) Growth of *Listeria monocytogenes* Scott A, serotype 4 and competitive spoilage organisms in raw chicken packaged under modified atmospheres and in air. *Int. J. Food Microbiol.*, **11**, 205–214.

White, R. (1990) Modified atmosphere packaging and the environment. In *Proceedings of International Conference on Modified Atmosphere Packaging*, Campden Food and Drink Research Association, Chipping Campden, Gloucester, UK.

Zagory, D. (1990) Application of computers in the design of modified atmosphere packages for fresh produce. In *Proceedings of International Conference on Modified Atmosphere Packaging*, Campden Food and Drink Research Association, Chipping Campden, Gloucester, UK.

Index